US National Security

SIXTH EDITION

US National Security

Policymakers, Processes, and Politics

John Allen Williams
Stephen J. Cimbala
Sam C. Sarkesian

BOULDER
LONDON

Published in the United States of America in 2022 by
Lynne Rienner Publishers, Inc.
1800 30th Street, Suite 314, Boulder, Colorado 80301
www.rienner.com

and in the United Kingdom by
Lynne Rienner Publishers, Inc.
Gray's Inn House, 127 Clerkenwell Road, London EC1 5DB
www.eurospanbookstore.com/rienner

Library of Congress Cataloging-in-Publication Data
Names: Williams, John Allen, 1945– author. | Cimbala, Stephen J., author. | Sarkesian, Sam C. (Sam Charles), 1927–[2011] author.
Title: US national security : policymakers, processes, and politics / John Allen Williams, Stephen J. Cimbala, Sam C. Sarkesian.
Description: 6th edition. | Boulder, Colorado : Lynne Rienner Publishers, Inc, 2022. | Includes bibliographical references and index. | Summary: "Introduces and explores the full range of actors, processes, and politics involved in the US national security process"— Provided by publisher.
Identifiers: LCCN 2021062999 | ISBN 9781955055369 (paperback)
Subjects: LCSH: National security—United States—Decision making. | United States—Military policy—Decision making.
Classification: LCC UA23 .S275 2022 | DDC 355/.033073—dc23/eng/20220103
LC record available at https://lccn.loc.gov/2021062999

British Cataloguing in Publication Data
A Cataloguing in Publication record for this book is available from the British Library.

Printed and bound in the United States of America

The paper used in this publication meets the requirements of the American National Standard for Permanence of Paper for Printed Library Materials Z39.48-1992.

5 4 3

To Sam C. Sarkesian:
soldier-scholar, mentor, and friend

Contents

Part 4 Long-Range Perspectives

Tables and Figures

Tables

Figures

Preface

Our late colleague, mentor, and friend, Professor Sam C. Sarke- sian, remains very much a coauthor of this book. This latest revision evolved from our earlier efforts together, and we continue to rely on the solid theoretical framework he constructed. We miss him greatly, both professionally and personally, and hope that this book will contribute to his legacy. He will always be our example of a soldier-scholar.

* * *

US national security concerns have changed greatly since the previous edition of the book was published. The August 2021 withdrawal of US military involvement in Afghanistan and the February 2022 Russian invasion of Ukraine serve as stark reminders of the limits of US power. These developments reinforce our basic narrative: Concerns with homeland security, antiterrorism, and nation building that preoccupied the United States since the attacks of September 11, 2001, have given way to preparing for traditional state-on-state conflict. In addition, US public opinion is sharply divided along partisan and ideological lines, areas of common agreement are harder to find, and the nature of truth itself is questioned. This makes it an especially challenging time to reach consensus on controversial issues of public policy.

The Russian invasion of Ukraine—the largest European war since World War II—marks a violent departure from the postwar norm against changing state borders by force. Russian hopes for a quick decapitation of the Ukrainian government were soon dashed by determined Ukrainian military resistance and political resolve, coupled with initial Russian ineptitude. As the war shifts to eastern and southern Ukraine, the final military outcome and eventual political settlement remain uncertain.

Some wider political and military changes have become apparent. These include remarkable cohesion among the members of the North Atlantic Treaty Organization (NATO), harsh economic sanctions against Russia, the possibility of formerly neutral Sweden and Finland joining NATO, increased defense spending, and a willingness to send vast amounts of sophisticated defense equipment to Ukraine over strenuous Russian objections. For the foreseeable future, the United States will regard Russia as an adversary to be contained rather than as a potential partner. In the event of a Russian victory on the battlefield, there will likely be widespread resistance from Ukrainians, who are more supportive of their nationhood than ever. From the Russian perspective, a war to bolster their western flank may have an effect opposite of what was intended.

Although we have many strong opinions, neither of us has been active in partisan politics, and we consider ourselves to be political moderates. We hope that our analysis will stimulate further thinking on the important and controversial issues dealt with here. Our opinions and conclusions are based on our academic work and experience in various governmental and nongovernmental security-related organizations. They do not necessarily reflect the views of any organization with which we are or have been associated.

Several colleagues, friends, and past students were particularly helpful in the preparation of this edition. We wish to thank Joseph J. Collins, Ryan Evans, Lawrence J. Korb, Adam Lowther, James E. McPherson, Michael P. Noonan, Keith Payne, David P. Redlawsk, David R. Segal, Mady Wechsler Segal, Michael St-Pierre, and Brian Wright. Grateful acknowledgment is made to Marilyn Wells, chancellor, and Wiebke Strehl, director of academic affairs, at Penn State Brandywine, and to Lee Ann Banaszak, head of the Department of Political Science, at Penn State University Park, for research support. We deeply appreciate everyone's assistance. Their participation does not imply their or their organizations' approval of our analysis. Of course, we are responsible for the use that we made of their contributions.

We are indebted most of all to our families for their love and continuing support. John Williams thanks his wife, Karen Williams; their daughters, Anissa and Emily; and their sons-in-law, Will Jordan and David Jordan. Stephen Cimbala thanks his wife, Betsy Cimbala; their sons, Chris and David; and their daughter-in-law, Kelly.

—John Allen Williams
—Stephen J. Cimbala

1

Defining and Defending the National Interest

National security challenges in the United States have evolved considerably in the last few years, and policies to meet them have changed in response. Earlier priorities of nation building and fighting a global "war on terror" have been supplanted by dealing with a resurgent Russia and the rise of China as a peer competitor. Strategies, force structure, and training are shifting to reflect the emphasis on state-on-state conflict as a primary concern. At the same time, the challenges of nonstate actors remain and must be considered. Competing state actors and adversary nonstate actors are also learning from one another, posing strategic challenges for US military planners and policymakers.

The international strategic landscape continues to be shaped by complex and contradictory forces. The world is characterized by unrest and changing patterns of interstate relationships, as well as conflicts within states caused by ethnic, religious, and nationalistic differences. International terrorism, drug cartels, population flows, and vulnerabilities to threats made possible by information-age technology add to the turmoil. It is difficult to devise coherent national security policies in this environment, let alone muster the political will and resources to implement them. It is hardly surprising that US national security policies are often complicated, ambiguous, and inconsistent.

US attention has shifted toward nuclear weapons programs in North Korea and Iran, a resurgent Russia seeking hegemony in its "near abroad," and the increasing power and assertiveness of a rising China that no longer hides its geopolitical ambitions. The United States has undergone numerous cyber attacks attributed to one or more of those countries, and tensions over Russian intervention in Ukraine, Chinese territorial claims in the South China Sea, and the status of Taiwan could eventually result in a military confrontation. Relations with all of those countries have diplomatic as well as military dimensions, and these need to be coordinated wisely.

National Security

The challenging international security landscape of the twenty-first century has clouded the concept and meaning of US national security and the protection of national interests by meaningful national security policies has become more difficult.[1] Recognizing the problems of conceptualizing national security, we offer a preliminary definition that includes both objective capability and perception: US national security is, first, the ability of national institutions to prevent adversaries from using force to harm Americans or their national interests, and, second, the confidence of Americans in this capability.

There are two dimensions of this definition: physical and psychological. The first is an objective measure based on the military capacity of the nation to challenge adversaries successfully, including going to war if necessary. It also includes a prominent role for intelligence, economics, and other nonmilitary measures and the ability to use them as political-military levers in dealings with other states. The psychological dimension is subjective, reflecting the confidence of Americans in the nation's ability to remain secure relative to the external world. The wisdom and political will of national leaders is critical in the development and implementation of effective national security policies, as well as the willingness of the American people to support such policies.

National security must be analyzed in the context of foreign policy, defined as the policies of a nation that encompass all official relations with other countries. The goal is to enhance conditions favorable to US national interests and to reduce those conditions detrimental to them. The instruments of foreign policy are primarily diplomatic and political, but include a variety of psychological and economic measures.

In the past, national security policy was more distinct from foreign policy. National security purposes were narrower and focused on security and safety, and national security policy was primarily concerned with potential adversaries and their use of force to threaten national interests. There was a clear military emphasis, which is not usually the case in foreign policy. National security policy increasingly overlaps with foreign policy, however, sometimes blurring any distinction. But much of foreign policy requires compromise and negotiations—the dynamics of give-and-take—associated with traditional diplomacy. This kind of work is primarily a matter for the US Department of State, with long-range implications for national security policy. These relationships are shown in Figure 1.1.

Historically, most Americans felt that imposing US values on other states was a low priority unless survival was at stake. For reasons discussed here, national security goals were increasingly seen to include the projection of US values abroad. Given the cost of such efforts in Vietnam, Iraq, Syria, and Afghanistan, this perspective is increasingly called into question and highlights the interrelationship among foreign, domestic, and national security policies. An observation made more than two decades ago remains relevant: "America's concept of national security today is infinitely more complex than at any time in its history. The same is true for the relationship between the foreign and domestic components of national security."[2]

The difficulties of determining US national interests and establishing national security priorities are compounded by the increasing number of linkages between national security and domestic policies. The domestic economic impact of certain national security policies links US domestic interests and policies to the international security arena. This is seen in economic sanctions, embargos on agriculture exports to adversaries or potential adversaries, diminished foreign oil sources, border security, and the export of technologically advanced industrial products.

Owing to the special characteristics of our democratic system and political culture, it is increasingly difficult to isolate national security issues from domestic policy. Besides the relationship between foreign and national security policies, domestic interests are important in establishing national security priorities and interests. The primary distinction between foreign and domestic policy and national security policy rests in the likelihood of the use of the military as the primary instrument for implementing national security policy. Although many other matters are relevant to US national interests, they are best incorporated into foreign policy and the overlap between such policy and national security policy.[3]

These observations are the basis for defining national security policy, expanding on the concept of national security: National security policy is primarily concerned with formulating and implementing national strategy involving the threat or use of force to create a favorable environment for US national interests. An integral part of this is to

Figure 1.1 National Security and Foreign Policy

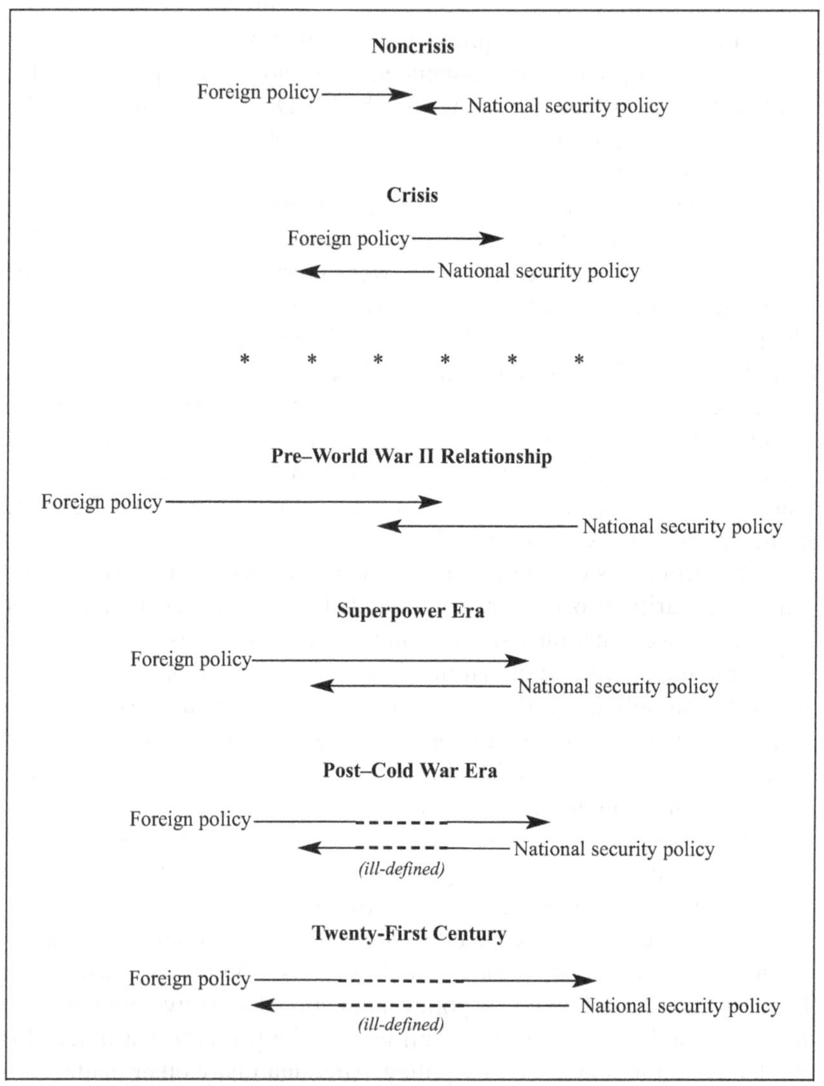

Source: Adapted from Col. William J. Taylor Jr., "Interdependence, Specialization, and National Security: Problems for Diplomats, Soldiers, and Scholars," *Air University Review* 30, no. 5 (July–August 1979), pp. 17–26.

Note: The gap between foreign policy and national policy indicates the relative degree of "closeness" between foreign and national security policy. The arrows indicate the relative degree of overlap. As shown, during times of crisis, the gap between foreign and national security policy is minimal and virtually nonexistent. In the twenty-first century it is often difficult to clearly separate foreign policy and national security because the use of force has become closely connected with a variety of peacekeeping missions, humanitarian crises, operations of war, and operations other than war; many such missions are extensions of foreign policy or a combination of national security and foreign policy, particularly in combating international terrorism.

prevent the effective use of military force and/or covert operations by adversaries or potential adversaries to obstruct the ability of the United States to pursue its national interests.

National security means more than the capacity to conduct international wars. Given the characteristics of the international arena and contemporary conflicts, challenges to US national security might take any number of nontraditional forms. Therefore, cyber warfare, international terrorism, weapons of mass destruction (WMD)—including chemical and biological warfare—and information warfare are important dimensions of national security policy. Still, the capacity to deter nuclear war and wage conventional conflicts remains essential for the conduct of US national security policy. Indeed, given the rise of China and difficulties with a resurgent Russia and nuclear programs in Iran and North Korea, these latter capabilities have moved to the forefront of US national security thinking.

National security policy must be carefully developed and implemented according to priorities distinguishing survival (that is, vital) interests from others. Too often, national security is used synonymously with any interest, suggesting that all interests are survival priorities. Taking a page from Sun-tzu, if almost everything is a matter of national security, then the concept of national security becomes virtually meaningless.[4] If national security policy and strategy followed such a pattern, the United States would have to defend everything everywhere; as a result, it would be unable to defend anything. Resources and personnel would be scattered across the globe and rarely be in a position to bring sufficient force to bear, even if survival were at stake. Additionally, a perception that blood and treasure are being expended for nonvital interests would greatly reduce the public's political will to support national security policies generally.

Short of clear threats to US territory, Americans often disagree over priorities. Even when there is agreement on priorities, there may be disagreement on resource commitment and strategy. Yet a system of priorities provides a way to identify levels of threats and helps in the design of strategies. The relationship between national interests and national security is particularly important. As former US national security advisor and secretary of state Henry A. Kissinger wrote, "What is it in our interest to prevent? What should we seek to accomplish?"[5] The same questions continue to challenge policymakers, scholars, and elected officials. The answers were elusive at the start of the post–Cold War period and became even more complicated after September 2001. The US war against terrorism became the dominant theme then, but this was complicated by the US involvement in Iraq and Afghanistan, troubling issues with Iran, North Korea, and Syria, and a variety of issues linked to homeland security. As noted earlier, conflicts with peer competitors have now become the most important metric for force planning.

Just what is the US national interest? At first glance the answer seems relatively simple: it is to promote US values and objectives. Promoting these includes implementing effective national security policies. Upon closer examination, these answers raise additional questions. What are US values? How are they reflected in national interests? What is the relationship between national security and national interests? How should US national security policy be implemented? Not surprisingly, there is no agreement on the answers to these questions.

Each generation of Americans interprets national values, national interests, and national security in terms of its own perspectives and mindset. Although there is agreement about core elements, such as protection of the US homeland, opinions differ about the meaning of national security, the nature of external threats, and the best course of conduct for security policy. The answers to Kissinger's questions are even more elusive today. National interests encompass a wide range of elements in a complex society such as the United States.

It is to be expected that in a country with a diverse population and multiple power centers there will be different opinions about security issues. Recognizing that these matters are ambiguous and rarely resolved by onetime solutions, we explore the concepts of national security in the context of national interests and national values. In the process, we design a framework for analyzing national security policy.

Regardless of the policies of any administration, the United States has political, economic, cultural, and even psychological links to most parts of the world. What the United States does or does not do has a significant impact on international politics. Americans can neither withdraw from external responsibilities nor retreat into isolation without damage to the national interest. As Henry Kissinger wrote:

> No country has played such a decisive role in shaping contemporary world order as the United States, nor professed such ambivalence about participation in it. Imbued with the conviction that its course would shape the destiny of mankind, America has, over its history, played a paradoxical role in world order: it expanded across a continent in the name of Manifest Destiny while abjuring any imperial designs; exerted a decisive influence on momentous events while disclaiming any motivation of national interest; and became a superpower while disavowing any intention to conduct power politics.[6]

National Interests

US national interests are expressions of US values projected into the domestic and international arenas. The notion of interests includes the creation and perpetuation of an international environment that is most favorable to the peaceful pursuit of US values. Americans generally believe that their own democracy is safer in an international system that expands democracy and open systems. (An important controversial issue is under what conditions open systems can be successful and what costs are justifiable to support them.) Similarly, the United States wishes to prevent the expansion of closed systems by their use of force or indirect aggression. The domestic arena has become an important consideration in pursuing national interests affected by asymmetrical threats, information-age challenges, international terrorism, and rising national competitors.[7]

Three statements serve as reference points. First, US values as they apply to the external world are at the core of national interests. The US public expects American policies to be consistent with their view of morality. Policy critics at home and abroad often regard this as simply a justification for policies decided on other bases. Sometimes this is true, but policies as diverse as the Vietnam War and the treatment of asylum-seekers at the US southern border are criticized on moral rather than simply prudential grounds. Second, pursuing national interests does not mean that US national security strategy is limited to the homeland. It may require power projection into various parts of the world. Third, the president is the focal point in defining and articulating US national interests. He or she will justify policies as much as possible on whether they are consistent with US values.

National interests can be categorized in order of priorities as follows. The first order is vital interests. Vital interests include protection of the homeland and areas and issues directly affecting this interest. This may require total military mobilization and resource commitment. In homeland defense, this also may require a coordinated effort of all agencies of government, especially in defense against terrorist attacks and cyber and information warfare. The homeland focus was highlighted by the creation of a new cabinet-level Department of Homeland Security (DHS) by President George W. Bush following September 11 to coordinate the efforts of a number of agencies in countering terrorism in the United States. Interagency and interservice information sharing and coordination greatly improved, but this was not all done under the auspices of DHS.

The second order is critical interests. Critical interests do not directly affect the survival of the United States or pose a threat to the homeland, but in the long run can become first-order priorities. Critical interests are measured by the degree to which they affect the systems of the US and its allies. Some examples include US economic competitiveness, energy availability, and the emergence of regional hegemons.

The third order is serious interests. These are issues that do not critically affect first- and second-order interests yet cast some shadow over them. US efforts are focused on creating favorable conditions to preclude third-order interests from developing into higher-order ones.

All other interests are peripheral in that they have no immediate impact on any order of interests but must be watched in case events transform these interests. In the meantime, peripheral interests require few US resources.

Categories of priorities such as these can be used as a framework for a systematic assessment of national interests and national security and also as a way to distinguish immediate from long-range security issues. They can provide a basis for rational and systematic debate regarding the US national security posture and are useful in studying national security policy. There is rarely a clear line between categories of interests or complete agreement on what interests should be included in each category, however. Many changes have expanded the concept of national interests to include several moral and humanitarian dimensions, among others. From this perspective US support for authoritarian regimes that benefit the US in some way (economically, militarily, or politically) should be minimized.

A realistic assignment of priorities can be better understood by looking at geopolitical boundaries of core, contiguous, and outer areas (see Figure 1.2). In specific terms, at the core of US national interests is the survival of the homeland and political order. But survival cannot be limited to the "final" defense of the homeland. In light of international terrorism and today's weapons technology, weapons proliferation, and chemical/biological and cyber warfare, homeland survival precludes simply retreating to the borders and threatening anyone who might attack with total destruction. By then it is too late for national security policy to do much good, and in some cases the attacker can be difficult to identify.

If national interest is invoked only when the homeland is directly threatened and survival is at stake, then the concept may be of little use and a US response too late to overcome the peril. If the concept is to have any meaning for policy and strategy, then it must be something more. Developing this broader view can spark a great deal of disagree-

Figure 1.2 US National Security Priorities

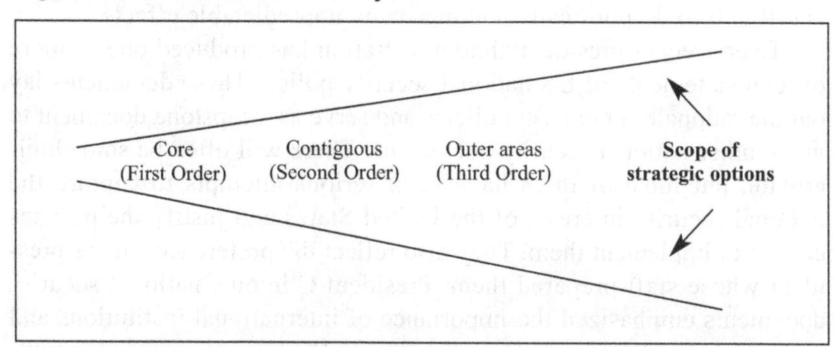

ment between the executive and legislative branches of government and in the US political arena. The media also frequently become involved. Policymakers rarely have the luxury of endless debate, nor do they have unlimited time or all necessary facts in a given situation. Yet policy must be made and strategy options examined, chosen, and implemented, even in the face of uncertainty and while disagreements remain intense.

Policy must be determined and implemented at some point. Before that, national interests for the particular situation must be identified and articulated. At the same time, national interests over the long range must be considered. Custom, usage, and constitutional powers give the president a basis for articulating their meaning. Initiatives in foreign and national security policy usually rest with the president as the commander in chief of US armed forces, chief diplomat, and chief of state. Importantly, the president can take action while the other branches of government deliberate. This illustrates the key role of the president in national security decisionmaking and the importance of the checks-and-balances system as a limitation on presidential power.

To be sure, Congress has an important role, but the president must take the lead and is the country's only legal representative with respect to foreign relations. For better or for worse, the president articulates the national interests, and Congress responds. The same holds true with respect to the president and the variety of interest groups in the government bureaucracy and public arena. Members of Congress find it very difficult to force a president committed to a course of action to change direction in national security policies, even in the case of a long war that has become unpopular. In particular, Congress finds it difficult to stop a war that the president feels should be continued.[8] The "nuclear option" of shutting off funding, as was done to stop the war in Vietnam, is difficult to do politically and has many unpredictable effects.

Every recent presidential administration has produced one or more official statements of US national security policy. These documents lay out the rationale for current policies and serve as a capstone document to guide other national security documents. There will often be some boilerplate, but most of them have been serious attempts to capture the national security interests of the United States and justify the policies chosen to implement them. They also reflect the preferences of the president whose staff prepared them. President Clinton's national security documents emphasized the importance of international institutions and the need to work through them. George W. Bush's emphasized the willingness of the United States to launch preventive war to deal with national security threats and to go it alone if necessary. President Trump's 2017 statement was notable for a sharper political tone and its emphasis on "defending America's sovereignty without apology" and putting America first in national security strategy.[9] President Biden put out an interim strategic guidance soon after his inauguration that reflected his priority of reengaging with allies to pursue common goals.[10]

US Values and National Interests

US values underlie the philosophical, legal, and moral bases of the US system. These attributes are deeply engrained in the US political system and domestic environment; they also apply to the way in which the public perceives justice in the international system and "just cause" for the use of force. In other words, values are principles that give the US political system and social order their innate character and are the basis of further principles upon which to base national interests.

The Value System

The growing heterogeneity of US society has affected all aspects of US culture and enriched the country considerably. All cultures have contributed to the US value system, but it has been most affected by the dominant social groups as it evolved. Accordingly, modern US values derive primarily from the Judeo-Christian heritage, the Anglo-Saxon legacy (including the Reformation, the Renaissance, and the philosophies of John Locke, Jean-Jacques Rousseau, and others), and the principles rooted in the American Revolution, the Declaration of Independence, and the US Constitution. We identify at least six fundamental values that define the United States and its role in the international world.[11]

First is the right of self-determination, a dual concept in this context: it applies not only to the nation-state but also to people within that state. Each nation-state is presumed to have the right to determine its own policy and to govern as it chooses as long as it does not threaten neighbors or oppress its own people. At the same time, people within that nation-state also have the right of self-determination. From the US perspective, this means that through free and fair elections, people in a nation-state have the right to determine how and by whom they will be ruled, with the option to replace rulers as they see fit.

There is another dimension as well: an emerging right claimed by minority groups to demand autonomy as a matter of self-determination. This duality of self-determination and state sovereignty creates serious problems in determining appropriate and legitimate action on the part of the United Nations, regional organizations, and the United States. This duality also has important implications for US military strategy. Moreover, this duality can lead to a dangerous confrontation between minority groups within a state demanding self-determination and the state itself, as occurred in the former Yugoslavia (between Albanians and Serbians in Kosovo, then a province in Serbia) and in Iraq, among other states.[12] Ideally, self-determination is accomplished within a system of laws and peaceful change. The peaceful partition of the former Czechoslovakia into the Czech Republic and Slovakia is an example of this. There is also the possibility of Scotland, Wales, and Northern Ireland spinning off from the United Kingdom. Not all aspirations for national determination by nonstate nations will be resolved without great struggle, however. The attempt of Catalonia to achieve independence from Spain continues, generally peaceably so far. Attempts by the Kurds to carve a homeland out of territory in Iraq, Iran, Turkey, and Syria will be resisted militarily by the countries involved. The possibility of a peaceful "two state" solution for Palestinians and Israelis grows ever dimmer, and a peaceful resolution of that conflict seems unlikely.

Second, it follows that there is an inherent worth of each individual in his or her relationship to others, to the political system, and to the social order. What does this mean? Put simply, every person is intrinsically a moral, legal, and political entity to which the system must respond and whose rights must be respected. Each individual has the right to achieve all that he or she can, without encumbrances other than protection of fellow citizens, homeland protection, and survival. Individual worth must therefore be reflected in economic, political, and legal systems. This notion has been challenged in connection with the treatment of persons entering the country without authorization, in particular with respect to the separation of families at the border.

Third, rulers owe their power and accountability to the people, which is the essence of democratic political legitimacy. The people are the final authority. There is a continuing responsibility by elected and appointed officials to act according to moral and legal principles, and the people have the right to change their leaders by regular constitutional processes. Furthermore, individual worth necessitates limited government with no absolute and permanent focal point of power. To ensure this, governance must be open. That means, with only a few justifiable exceptions, decisions and policies must be undertaken in full public view, with input from a variety of formal and informal groups. The system of rule must be accessible to the people and their representatives. This is the essence of what are called "open systems."

Fourth, policies and changes in the international environment must be based on the first three values just outlined. Therefore, peaceful change brought about by rational discourse among nation-states is a fundamental value. The resort to war is acceptable only if it is clearly based on homeland protection and survival or other core values, and only if all other means have failed. In this respect, diplomacy and state-to-state relationships must be based on mutually acceptable rules of the game.

Fifth, systems professing such values and trying to function according to them should be protected and nurtured. Nation-states whose values are compatible with US values are thought to be best served by an international order based on those same values. The United States continues to discover the limits of the possible in trying to spread democracy to areas of the world unprepared to nurture it, however. Hopeful notions of "nation building" have been severely challenged by their very disappointing results in Iraq and Afghanistan.[13]

Sixth, while US values are grounded in the Judeo-Christian heritage that predated the founding of the republic in the late eighteenth century, these characteristics are consistent with the precepts of other religious traditions, including Islam. For many Americans, they instill a sense of humanity, a sensitivity to the plight and status of individuals, and a search for divine guidance.

We do not suggest that these values are perfectly embodied in the US system, although they remain aspirational. There are many historical examples of value distortions and their misuse to disguise other purposes. But these values are esteemed in their own right by most Americans and are embodied in the political-social system. Furthermore, the system of rule and the character of the political system have institutionalized these values, albeit imperfectly. The expectations of most Americans and their assessments of other states are, in no small measure, based on these values.

American Values Today

The early years of the republic saw little need to translate values into the external world, as the interest of the United States rarely extended beyond its own shores. This changed as the United States became a great power, partly as the result of acquiring territory in the Spanish-American War in 1898. Within two decades, US involvement in World War I was seen as a way to make the world safe for democracy and subdue a tyrannical Old World power.

The collapse of the old order in Europe following World War I set the stage for the continental evolution of both democratic regimes and authoritarian Marxist-Leninist and Fascist systems. Until that time, Pax Britannica had provided a sense of stability and order to European affairs as well as a security umbrella for the United States in its relationships with Europe. But for many Americans, involvement in a conflict to save Europe seemed in retrospect to be a mistake. The United States withdrew into isolationism with the "Back to Normalcy" policy of President Warren G. Harding in 1921 and sank into an economic depression in 1929, which was itself exacerbated by economic nationalism and high tariffs. Although the United States participated in disarmament negotiations during the interwar period, many view the failure to join the League of Nations and participate actively in it as a contributing factor to World War II—a serious step back from President Woodrow Wilson's Fourteen Points as the basis for a new world order.[14]

Even in the aftermath of World War I, Americans were accustomed to a world dominated by a European order compatible with US values and interests. Although an imperfect order, it did not offend the US value system until the rise of Fascist Italy and Germany.[15] At the beginning of the twentieth century, US values were expressed by progressivism in Theodore Roosevelt's presidency and later by Franklin Roosevelt's New Deal. Franklin Roosevelt's "Four Freedoms" from his 1941 State of the Union address—freedom of speech, freedom of worship, freedom from want, and freedom from fear—remain excellent reflections of US values.

Between the two world wars, Americans presumed that US interests were also world interests. US values were viewed as morally unassailable and therefore to be sought after by the rest of the world. In this context, US national security was primarily a narrow focus on the protection of the homeland. Given US geographic isolation from Europe, this required few armed forces and a simple military strategy. Furthermore, there was little need to struggle with issues over US values and how to protect them in the external world, except occasionally for the sake of international economics. The US passed responsibility to others, primarily Britain and France, for keeping the democratic peace. Opinions began to change with the gathering clouds of World War II and President Franklin D. Roosevelt's policies on rearmament and support for the British and French. Many Americans wanted no part of the "European War" that started in 1939, but the Japanese bombing of the US Pacific Fleet in Pearl Harbor on December 7, 1941, caused a rapid shift in US opinion about the war.

Regardless of the US desire to return to isolation following the successful conclusion of World War II, US interests were increasingly threatened. Parts of Europe and Asia were smoldering from the war, and it soon became clear that US interests and responsibilities extended beyond the nation's borders. In addition, it was perceived that democracy and US values could not be nurtured and expanded by disengagement; if democracy needed to be defended, then it seemed to require a US presence in all parts of the world. Beyond protection of the US homeland, then, what did the United States stand for? And how did it intend to achieve its goals, whatever they were?

The United States was against Marxist-Leninist and other authoritarian political systems determined to subvert or overthrow the international order. The policy of containment reflected a US policy consensus to prevent the expansion of the Soviet Union and its Communist system. The United States played a vital role in rebuilding Europe, especially with the economic recovery program known as the Marshall Plan. All of this placed the United States in the leadership role of the West and was consistent with the earlier Puritan view of Americans as a chosen people.[16] For many, the second half of the twentieth century was the "American Century," and the containment policy provided the rationale for involvement in the Korean and Vietnam Wars. Challenges to that policy involved where and how it was implemented, but there was strong support for the policy in principle. Even today there are echoes of the containment policy in US concerns about the possibility that the government of China would attempt to force the island of Taiwan—which it considers a renegade province—to unite with mainland China.

But with the end of the Cold War, in roughly 1989, the emergence of a new security landscape and domestic economic and social challenges caused many Americans to refocus on domestic issues. There was a turning inward, reinforced by the conviction that the United States had won the Cold War and the danger of a major war had diminished considerably. But this new landscape was ambiguous and difficult to comprehend. US political scientist John Mersheimer argued presciently that the United States would miss the Cold War, with its moral certainties and predictable (if difficult) responsibilities.[17] The reemergence of peer competitors to the United States and a refocusing on large-scale war makes the strategic landscape even more complicated and difficult to deal with, since lower-level challenges still remained.

Turning inward, Americans faced issues of diversity: gender and gender identity, race, sexual orientation, and the integration of various groups with non-Western heritages. Some argued that the United States might never have been a true melting pot of culture, yet it had benefited greatly from the waves of immigrants who brought along their rich heritage. Others argued there was the risk of cultural erosion from the increasing prominence of non-Western cultures.[18] Arguments about "multiculturalism" and the degree to which that is consistent with "Americanism" continue to flourish. Unfortunately, these issues are often dealt with based on dogmas, political soundbites, and platitudes rather than rational analysis. The US motto *E pluribus unum* ("one out of many") is accepted in principle, but Americans continue to disagree on the extent to which they should focus on the *pluribus* or the *unum.*

The New Era

In the new era it is difficult to agree on the principles of US values as they apply to the international order. Issues of multiculturalism and diversity have become more salient and controversial in the wake of publicized examples of excessive police force used against persons of color. While there is a wide consensus on the need to address racial issues, there are great differences in opinions as to how they should be addressed. For example, in viewing the US domestic system, former chairman of the US Joint Chiefs of Staff and later secretary of state the late Colin Powell wrote:

> And Lord help anyone who strays from accepted ideas of political correctness. The slightest suggestion of offense toward any group . . . will be met with cries that the offender be fired or forced to undergo sensitivity training, or threats of legal action. Ironically for all the present

> sensitivity over correctness, we seem to have lost our shame as a society. Nothing seems to embarrass us; nothing shocks us anymore.[19]

As some critics point out, spokespersons for various groups in the United States often use terms such as "our people" or "my people" in referring to their particular racial, ethnic, or religious group to the exclusion of others. This tends to distinguish and separate one group from Americans in general. But as President Franklin D. Roosevelt remarked in 1943 when activating the predominantly Japanese-American 442nd Combat Team, "Americanism is a matter of the mind and heart. Americanism is not, and never was, a matter of race or ancestry."[20]

Clearly, demographics and cultural issues have an impact on US national security policy and strategy. When the national interest is clear and the political objectives are closely aligned with that interest, however, there is likely to be strong support by Americans for US action. But US involvement in cultures and religions abroad can have domestic repercussions, such as involvement in the conflict in the Middle East between Israel and the Palestinians and US involvement in Iraq and Afghanistan. This makes it more difficult to project US values into the international arena. In sum, the commitment of the US military in foreign areas will not draw support from the public unless it is convinced that such actions will support the vital interests of the United States.

The Study of National Security

We consider three approaches to the study of national security: the concentric-circle approach, the elite-versus-participatory policymaking approach, and the systems-analysis approach. All concentrate on the way in which policy is made. They should be distinguished from studies that examine national security issues, such as US nuclear strategy or US policy in the Middle East. The three approaches should be further distinguished from studies of government institutions.

The concentric-circle approach places the president at the center of the national security policy process (see Figure 1.3). The president's staff and the national security establishment provide advice and implement national security policy. This approach shows the degree of importance of various groups as the primary objects of national security policy. For example, a major objective is to influence the behavior and policies of allies as well as adversaries. At the same time, Congress, the public, and the media have important roles in the national security policy process. The more distant circles represent government structures and agencies, constituencies, and the media. The farther the institutions are from the center, the less their direct influence on national security policy. The problems with this approach are its oversimplification of the national security policy process and its presumption of rationality in decisionmaking.

The elite-versus-participatory policymaking approach is based on the view that the policy process is dominated by elites (see Figure 1.4). National security policy is undertaken by elites within the national security establishment, but that elite group must in turn develop support in the broader public. On the one hand, the elites have the skill and access to information to formulate national security policy, in contrast to an

Figure 1.3 Concentric Circle Approach

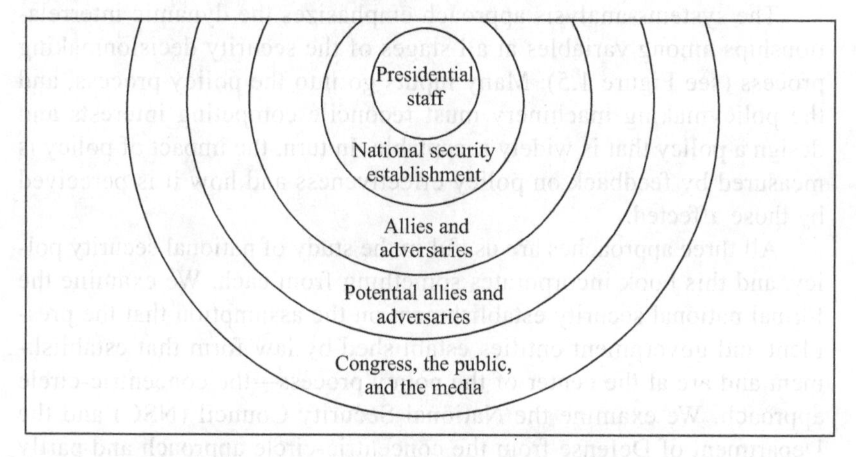

Figure 1.4 Elite and Participatory Models

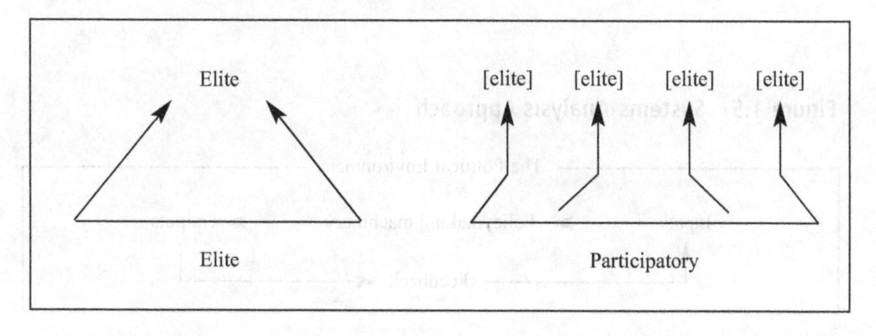

uninformed and unorganized public. On the other hand, for national security policy to be successful in the long run, there must be some degree of participation by the public and political will within the body politic. The elite model sees national security policy as being made by a small circle that includes the president, his or her staff, key members of Congress, high-ranking military officers, and influential members of the business community. These elites—who may or may not be cohesive—are assumed to operate on the basis of their own self-interest, often overriding other considerations. The participatory model assumes the existence of a variety of elites who represent various segments of the public, interest groups, and officials. In this model, the same elites rarely control all aspects of national security policy. Coalitions are formed for particular issues, then reformed for other issues. This approach attempts to reconcile the skill and power of the elite with the demands of participatory democracy.

The systems-analysis approach emphasizes the dynamic interrelationships among variables at all stages of the security decisionmaking process (see Figure 1.5). Many inputs go into the policy process, and the policymaking machinery must reconcile competing interests and design a policy that is widely acceptable. In turn, the impact of policy is measured by feedback on policy effectiveness and how it is perceived by those affected.

All three approaches are useful in the study of national security policy, and this book incorporates something from each. We examine the formal national security establishment on the assumption that the president and government entities established by law form that establishment and are at the center of the policy process—the concentric-circle approach. We examine the National Security Council (NSC) and the Department of Defense from the concentric-circle approach and partly from the elite-versus-participatory approach. Finally, in analyzing the formal policy process, we give the most attention to the national secu-

Figure 1.5 Systems-Analysis Approach

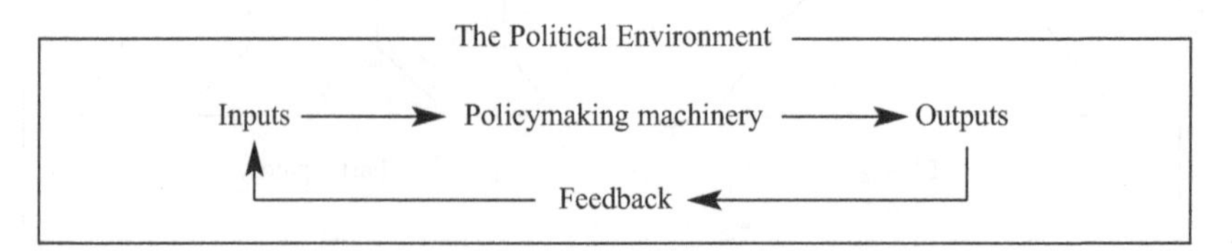

rity network—a systems-analysis approach that considers many power clusters within the governmental structure, the political system, and the international environment that have an impact on the national security establishment and the policymaking process.

The term *national security establishment* refers to those responsible for national security decisionmaking as well as a descriptive term that identifies a set of actors and processes that actually produce security policy outcomes. The character and personality of the president can also lead to the creation of informal and parallel structures and processes for developing national security policy. This sets up a series of policy power clusters that form a national security network that drives the national security establishment and the formal policymaking process. The relationships among and within these power clusters and their actual powers are dependent upon the way the president exercises his or her leadership and his or her views on how the national security establishment should function.

We consider four major power clusters within the US command structure, whose powers vary according to presidential leadership and preferences: (1) the policy triad, consisting of the secretary of state, the secretary of defense, and the national security advisor; (2) the director of national intelligence and the chairman of the Joint Chiefs of Staff; (3) the president's closest White House advisors, such as the White House chief of staff and the counselor to the president; and (4) the secretary of Homeland Security.

These four power clusters are important in shaping national security policy (see Figure 1.6). They represent critical parts of the national security establishment but operate in ways that reflect presidential leadership style and the mind-sets of those within the three power clusters. As such, they may or may not be compatible with the formal national security establishment.

Put another way, the national security establishment is fluid and dynamic, and the policymaking process is not as rational and systematic as one is led to believe or as one might hope. For example, President Franklin D. Roosevelt had numerous advisors reporting directly to him, some of whom may have had the same assignment (unbeknownst to the others); President Dwight D. Eisenhower preferred a highly structured staff organization similar to those he was accustomed to from his Army experience; President John F. Kennedy was comfortable with a much looser organization; and President Donald Trump resisted attempts to structure his decisionmaking process, relying instead on his own instincts and informal advisors outside of government. President Biden

Figure 1.6 Policy Power Clusters and the National Security System

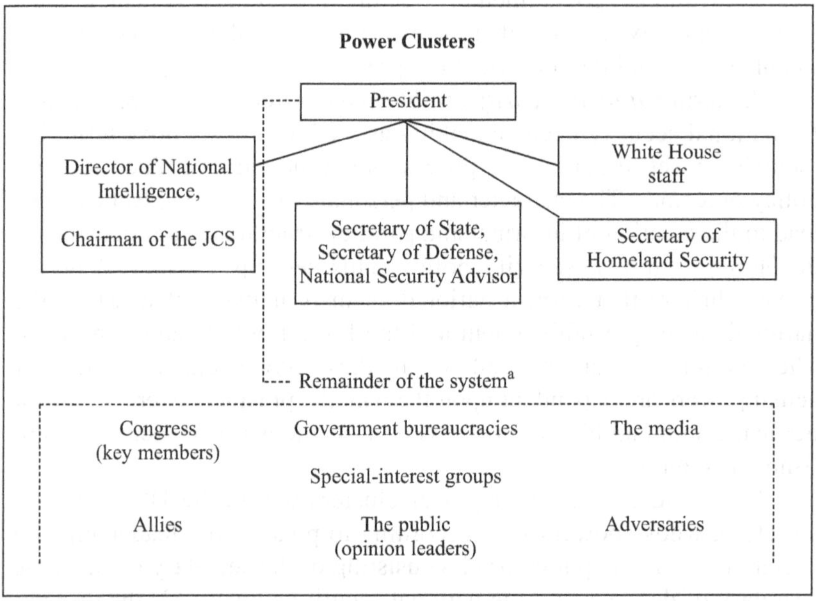

Power Clusters

President

Director of National Intelligence,

Chairman of the JCS

White House staff

Secretary of State, Secretary of Defense, National Security Advisor

Secretary of Homeland Security

Remainder of the system[a]

Congress (key members)

Government bureaucracies

The media

Special-interest groups

Allies

The public (opinion leaders)

Adversaries

Note: a. Objects of national security policy and inputs into national security policy.

relies on his own experience, but does so in the context of a more structured and analytical process than his predecessor.

The observation of Frederick Hartmann and Robert Wendzel on the defense planning process remains relevant today: "The defense planning process . . . is beset with multiple dilemmas. Assessing the threat and acquiring the force structure to meet that threat require an efficient crystal ball—not only in the sense of defining the future in the here and now in terms of events and dangers; the process also requires accurately estimating the national mood years before the critical event."[21]

Conclusion

There is a set of limitations that cannot be separated from the operations of the US national security establishment. The policy process cannot be understood apart from these considerations. As a result, there is likely to be internal disagreement within the national security establishment, between the establishment and other branches and agencies of government, and between all of these and the public. When we add the differing views of allies and adversaries, it is clear that simply examining the actors or the policy process is not sufficient to explain the complexities inherent in making and implementing US national security policy.

All of this is exacerbated by the diffusion of power within the US political system, within and among the branches of government, and within the general population. Participatory politics and single-issue politics, changing domestic demographics, the policy role of the media, the rise of social media and the twenty-four-hour news cycle, and internal power problems within government have made it very difficult for the president to undertake foreign policy or national security initiatives that are perceived as outside the mainstream or as requiring a new kind of military posture or preparedness. To place a stamp on national security policy, the president must build a political base within the government and activate the general public as well as convince the media of the appropriateness of new policies and strategies. This usually means that they must be seen as major national security issues, with the US position clearly proper and morally correct, and must involve acceptable risk and a high expectation of success. Very few policies meet all of these tests—a frequent source of presidential frustration.

The US fear of concentrated power is ingrained in the constitutional principles of separation of powers and checks and balances, which have provided clear limits to the exercise of power of any one branch of government. Yet these restraints can also prevent effective responses to challenges that require a concentration of power to succeed. Thus, the problem is self-contradictory. The legal niceties of US constitutional practice can be problematic in the international security setting, where power and politics are often inextricable. It is in this context that the US national security establishment and the process by which security policy is formulated and implemented meet their greatest test. Such a test is evident in the continuing struggles between the president and Congress over the war power and how to meet our national security objectives.

In this book our primary concern is the US national security establishment and the security policy process. In addition, we examine the international security setting, the factors that affect the substance of US national security policy, and the presidential mandate. All of these matters have become complicated by the disagreements within the United States on military operations abroad, how to deal with a rising and ambitious China, proper responses to international terrorism, and other troublesome issues. The chapters on the national security establishment and the national security process are focused on these issues.

Notes

1. See John Allen Williams, "The Military and Society Beyond the Postmodern Era," *Orbis: A Journal of World Affairs* 52, no. 2 (Spring 2008), pp. 199–216.

2. David Jablonsky, "The State of the National Security State," in David Jablonsky, Ronald Steel, Lawrence Korb, Morton H. Halperin, and Robert Ellsworth, *US National Security: Beyond the Cold War* (Carlisle, Pa.: Strategic Studies Institute, US Army War College, July 26, 1997), pp. 39–40.

3. See Charles W. Kegley Jr. and Shannon L. Blanton, *World Politics: Trend and Transformation,* 14th ed. (Stamford, Conn.: Cengage Learning, 2013).

4. *Sun Tzu: The Art of War,* translated and with an introduction by Samuel B. Griffith (New York: Oxford University Press, 1971).

5. Henry A. Kissinger, *American Foreign Policy: Three Essays* (New York: Norton, 1969), p. 92. This was written before Kissinger became assistant to President Richard Nixon for national security affairs (a position that is known widely as national security advisor) and then secretary of state.

6. Henry Kissinger, *World Order* (New York: Penguin, 2014), p. 234. See also John Lewis Gaddis, *On Grand Strategy* (New York: Penguin, 2018), chap. 9, pp. 255–293.

7. See David E. Sanger, *The Perfect Weapon: War, Sabotage, and Fear in the Cyber Age* (New York: Crown, 2018); Fred Kaplan, *Dark Territory: The Secret History of Cyber War* (New York: Simon and Schuster, 2016); Thomas E. Copeland, ed., *The Information Revolution and National Security* (Carlisle, Pa.: Strategic Studies Institute, US Army War College, August 2000); John Allen Williams, "Understanding Asymmetric Warfare: Threats and Responses," *National Strategy Forum Review* (Summer 2007), pp. 23–26; and John Allen Williams, "Terrorism: The New Threats," *National Strategy Forum Review* (Winter 2004), pp. 7–9.

8. See Richard E. Grimmett, *Congressional Use of Funding Cutoffs Since 1970 Involving U.S. Military Forces and Overseas Deployments* (Washington, DC: Congressional Research Service, January 16, 2007).

9. *National Security Strategy of the United States of America* (Washington, DC: December 2017), https://www.whitehouse.gov/wp-content/uploads/2017/12/NSS-Final-12-18-2017-0905-2.pdf.

10. Joseph R. Biden Jr., *Renewing Americas's Advantages: Interim National Security Guidance* (Washington, DC, March 2021), https://www.whitehouse.gov/wp-content/uploads/2021/03/NSC-1v2.pdf.

11. See Eugene R. Wittkopf and James M. McCormick, eds., *The Domestic Sources of American Foreign Policy: Insights and Evidence,* 5th ed. (Lanham, Md.: Rowman and Littlefield, 2007).

12. David Scheffer concluded, "I propose that we are witnessing the end of sovereignty as it has been traditionally understood in international law and in state practice. In its place we are seeing a new form of national integrity emerging." David Scheffer, "Humanitarian Intervention Versus State Sovereignty," in United States Institute of Peace, *Peacemaking and Peacekeeping Implications for the United States Military* (Washington, DC, May 1993), p. 9.

13. For a succinct explanation of the US failure in Afghanistan, see Joseph J. Collins, "What Went Wrong in Afghanistan," September 2, 2021. https://www.defenseone.com/ideas/2021/09/what-went-wrong-afghanistan/185061.

14. See James M. McCormick, *American Foreign Policy and Process,* 4th ed. (Itasca, Ill.: F. E. Peacock, 2004), pp. 28–30.

15. Certainly many actions of the Soviet Union during this period were inconsistent with US values, but they were not yet an influential part of the European order.

16. See, for example, Kenneth D. Wald and Allison Calhoun-Brown, *Religion and Politics in the United States,* 5th ed. (Washington, DC: Congressional Quarterly, 2006). Also see Douglas Johnston and Cynthia Sampson, eds., *Religion: The Missing Dimension of Statecraft* (Oxford: Oxford University Press, 1994).

17. See John Mearsheimer, "Why We Will Soon Miss the Cold War," *The Atlantic,* August 1990, pp. 35–50.

18. Samuel P. Huntington, *The Clash of Civilizations and the Remaking of World Order* (New York: Simon and Schuster, 1996). See also Samuel P. Huntington, "The Clash of Civilizations?" *Foreign Affairs* 72, no. 3 (Summer 1993), pp. 22–49. For a critique of the Huntington thesis, see Shireen T. Hunter, *The Future of Islam and the West: Clash of Civilizations or Peaceful Coexistence?* (Westport, Conn.: Praeger, 1998).

19. Colin Powell with Joseph E. Persico, *My American Journey* (New York: Random, 1995), p. 610.

20. See https://content.libraries.wsu.edu/digital/collection/propaganda/id/221. Roosevelt's comment was consistent with American values, but it is sadly ironic that at the time these words were spoken some 120,000 persons of Japanese ancestry from the West Coast, citizen and noncitizen alike, were incarcerated in internment camps by a 1942 presidential order for the duration of the war. This policy was approved in *Korematsu v. United States* (December 18, 1944). See http://www.law.cornell.edu/supct/html/historics/USSC_CR_0323_0214_ZO.html, especially the blistering dissents.

21. Frederick H. Hartmann and Robert L. Wendzel, *Defending America's Security* (Washington, DC: Pergamon-Brassey's, 1988), p. 146.

PART 1

The National Security Context

2

International Actors

Despite the emergence of peer competitors, the United States remains in a class by itself with respect to its economic strength and its military power.[1] Yet potential power does not always translate into influence over policy outcomes, and the United States is often powerless to affect many international issues, especially if acting alone. This disparity between potential and actual power hit home on September 11, 2001, and again in the face of the Covid-19 pandemic beginning in 2020. The limit on the actual power of the United States comes from it coexisting with hundreds of other state and nonstate actors within the international system, all of whom are pursuing their own agendas. The term *system* simply means that states and other actors interact with one another in more or less regular and predictable ways.

The international system, which includes some 200 states and many nonstate actors, is a complex web that places the United States in relationships that can advance or constrain US goals and the means to attain them. Even though the United States is a superpower with vast resources to implement national security policy, its power is limited in the international system. First, several major powers (e.g., Russia, China, Great Britain, Germany, France, India, and Japan) have their own agendas and national interests that they project into the international arena. Second, not all states share the view of the world held by the United States. US culture, for better or worse, is unique: a blend of pragmatism, economic and technological assertiveness, and political inclusiveness, preferring future-oriented instead of traditional perspectives on social development.

Third, even the United States has limited resources and must be judicious about how and when they are used to achieve national security objectives. Fourth, regional powers pursue interests that may be contradictory to those of the United States. Fifth, democratic proprieties require that the United States use diplomacy, negotiations, and consensus-building as the primary means to achieve its national security goals—which means compromise and recognition of the national interests of other states. Sixth, the United States shares power and responsibility with a variety of other states with common purposes and some cultural affinity—the most visible and successful example of which is the North Atlantic Treaty Organization (NATO) alliance that binds the United States and democratic Europe. Seventh, if vital interests dictate and war is unavoidable, US policymakers and commanders must understand the nature of war. As H. R. McMaster has noted, war is political, human, uncertain, and involves a test of wills between combatants.[2] It is easier to get into wars than it is to bring them to a satisfactory policy-relevant termination at an acceptable cost.

The United States must operate in a world environment that is disorderly. Many governments do not share US concepts of the proper world order. Within the context of US national security, sovereign states can be grouped into broad categories: allies, adversaries, potential adversaries, and others. *Allies* are best illustrated by fellow members of NATO and others such as Japan, South Korea, Australia, New Zealand, and Israel. With the demise of the Soviet Union in 1991, NATO seemed to have no clear adversary. However, a more assertive Russia under the leadership of Vladimir Putin has revived NATO's primary mission of deterrence and defense in Europe, especially after the Russian invasion of Ukraine in 2022.[3] *Potential adversaries* include states that have adopted explicit anti-US policies and have the military means to pursue their goals. Some are motivated by strategic cultures profoundly suspicious of the Western cultural heritage, or even hostile to it.[4] Some might evolve into US adversaries. *Others* include virtually the entire third world, with the exception of some states in the Middle East as well as other states that have the potential to become adversaries (Iran) or allies (Egypt), depending upon the specific security issue.

In this chapter, we discuss the effect of the international environment on US security policymaking. First, we review major US alliance commitments, especially those related to European security and NATO. Because NATO is exceptional in its durability and success, we also look at other US commitments with the flavor, although not the texture, of an alliance such as NATO. Second, we consider past and possible future

US adversaries and some of the problems they might pose for US defense and foreign policymaking. Third, we discuss the other actors who fall outside the category of allies or adversaries as used here but whose relationships with the United States on security matters might be important in the future.

The Current State of Alliances

Historically, the United States has been suspicious of what George Washington called "entangling alliances." Even with respect to Europe, whose cultural affinity with the United States has deep historical roots, the United States has tried to keep its distance. Until World War II, Old World Europe retained its tarnished image of monarchy and extreme ideologies. US isolationism in the 1920s and 1930s was intended to keep the United States at a distance from Europe. After World War I the United States did not ratify the Versailles Treaty (which included the League of Nations), but simply maintained an observer status during debates at the League. US military forces were primarily concerned with defending US boundaries rather than engaging in offensive contingencies and global power projections. There were exceptions, of course, especially at the turn of the twentieth century with respect to the Southern Hemisphere and parts of the Pacific, and Theodore Roosevelt's dispatch of the Great White Fleet to the Far East from 1907 to 1909 was a classic demonstration of US military might.

In the aftermath of World War II, an era of permanent US international involvement during peacetime began. The United States became one of the prime movers in the creation of an international system with the objective of spreading and supporting market democracies and liberal values. US provincialism gave way in the face of international threats and US involvement in a variety of Cold War confrontations, including several Berlin crises and the Cuban missile crisis. George F. Kennan's grand strategy of containment was the lodestar of US policy planning for the duration of the Cold War: the Soviets were to be fenced within their spheres of interest in Europe and Asia and not permitted to advance their security perimeter against vital US and allied interests. The US containment strategy, with wide support in subsequent presidential administrations, expedited the demise of the Soviet Union and its empire without a World War III.[5]

The end of the Cold War, the dawn of the twenty-first century, and the terrorist attacks on September 11 made it obvious that much of US security strategy and the assumptions on which it was based needed rethinking. Various concepts of grand strategy competed for priority among scholars and policymakers. During the 1990s the Bill Clinton administration favored a grand strategy of "engagement and enlargement" that required a robust peacekeeping presence or assertive humanitarian intervention in various countries outside previously defined US zones of vital interest. The George W. Bush administration preferred a higher-profile grand strategy after September 11, characterized by its critics as imperialism and by its advocates as US liberal hegemony as a United States destined to play the role of "sheriff of world order."[6] On the other hand, some evaluated both the Clinton and second Bush presidencies as adopting grand strategies that were inconsistent with the best interests of the United States and a realistic appraisal of US capabilities. For example, Harvard University professor Stephen M. Walt argued that both the Clinton grand strategy of selective engagement and the George W. Bush grand strategy of global military preeminence overextended US commitments and risked alienating allies and potential adversaries. Instead, Walt favored "offshore balancing" as a US grand strategy consistent with US traditions.[7]

Barack Obama's strategy of "leading from behind" in the multinational effort to depose Libyan dictator Muammar Gaddafi, or Donald J. Trump's "America first" approach to international politics, might be examples of a more restrained US profile in foreign military intervention. On the other hand, the intervention in Libya was flawed because it was sufficient to depose the regime, but insufficient to provide for post-conflict stability. And Trump's nationalist and unilateralist approach to foreign policy had traditional US allies scratching their heads and seeking alternatives to impulsive US presidential demarches. A lighter military footprint does not necessarily translate into an improved political outcome, regardless of preferred grand strategy. President Biden began undoing this nationalist approach upon taking office, but allies and opponents alike wonder whether a new administration will revert back to previous policies.

A strategy of offshore balancing would limit US military deployments and interventions to those situations in which vital US interests were at risk. Offshore balancing assumes that "only a few areas of the globe are of strategic importance to the United States (i.e., worth fighting and dying for)."[8] These areas include those regions having major concentrations of power and wealth or critical natural resources: Europe, industrialized Asia, and the Persian Gulf. Offshore balancing does not require that the United States control these areas directly: the United

States need only ensure that no hostile great power, and especially no adversary "peer competitor," controls these vital regions.[9] A similar argument for a more selective US policy of military engagement, compared to the post–September 11 assertive interventionism, is made by Barry Posen. His preferred foreign policy of "restraint" emphasizes the need for the United States to control the sea and air "commons" enabling more careful picking and choosing among possible military uses of ground forces.[10]

Offshore balancing is not isolationism: the United States would still be a major player in world politics and a leader in multilateral institutions such as NATO and the United Nations (UN); it would also be able to act unilaterally when necessary in response to threat or attack. One important component of a US balancing grand strategy is control of strategic waterways, including important trade routes and littoral as well as "blue water" capabilities for naval warfare or deterrence. China's forceful assertion of sovereignty claims in the South China Sea and elsewhere poses a serious challenge to this capability.

North Atlantic Treaty Organization

The most important alliance for the United States is NATO, formed in 1949. Designed primarily as a defensive military alliance of sixteen countries against the Soviet Union, NATO moved toward a more expansive definition of its purpose in the 1990s following the dissolution of the Soviet Union. In 1994, NATO initiated its Partnership for Peace program, intended to provide a closer relationship with several Eastern European countries and the former republics of the Soviet Union.[11] Partnership for Peace turned out to be among NATO's most important and successful innovations of the decade. Partner agreements allowed countries not yet eligible for NATO membership to establish routine military cooperation with NATO and, in some cases, helped pave the way for later admission. This was the case, for example, with Poland, the Czech Republic, and Hungary, which joined NATO in 1999. Those on the waiting list, and even those with no aspirations for permanent membership, regarded the joint military exercises and other experiences under Partnership for Peace as favorable to military stability and arms control transparency in Europe. Partner agreements helped NATO to expand its membership to thirty countries by 2020, with additional candidate members waiting offstage.[12]

In addition, NATO also moved to establish post–Cold War consultative mechanisms with Russia. The NATO-Russia Council (NRC) was established in 2002 as a forum for consultation, consensus-building, cooperation, and joint decisionmaking among the various NATO members and Russia. The council promotes continuing dialogue on security issues of common interest, including terrorism, peacekeeping, theater missile defenses, and arms proliferation. Following Russia's war with Georgia in 2008, NATO temporarily suspended formal meetings of the NRC and cooperation with Russia in some areas. The "reset" of US policy toward Russia by Barack Obama's administration beginning in 2009 temporarily established a more favorable climate for the renewal of NRC activities, but Russian actions deemed aggressive have caused the organization to reduce the number of members to its mission at NATO.

Russia's military annexation of Crimea and destabilization of eastern Ukraine in 2014 and thereafter, together with charges of Russian interference in the 2016 US presidential elections, severely damaged US-Russian relations toward the end of the Barack Obama and the beginning of the Donald J. Trump administration. Despite candidate Trump's campaign reassurances that, if elected president, he would be able to get along well with Russia, post-election disappointments awaited. Congress imposed economic sanctions against Russia over election meddling and other issues, and Russia's military intervention in Syria and its continuing aggression in eastern Ukraine derailed further cooperation between Washington and Moscow. The 2022 Russia-Ukraine crisis marked a low point in US-Russian relations and had the effect of strengthening support for NATO in the West and causing Germany, in particular, to increase its defense budget. It may also cause the accession of other states to the alliance, such as nominally neutral Sweden and Finland.

In 2019 both states announced their intention to abrogate the Intermediate Nuclear Forces Treaty of 1987, a cornerstone of Cold War international security and arms control. The New Strategic Arms Reduction Talks (START) agreement on strategic nuclear arms limitation, signed by the United States and Russia in 2010 and set to expire in 2021, was also an uncertain survivor of the deteriorated relationship between the United States and Russia as of 2020. The death of New START would mean the demise of the last nuclear arms control agreement between the United States and Russia, opening the door to a renewed nuclear arms race and encouraging nuclear proliferation in other states.

NATO Enlargement

NATO's expanded membership in the post–Cold War era reflected its success in redefining its Cold War mission for a different Europe and a changing world. In addition to its Cold War "Article 5" mission (collective self-defense), NATO announced a willingness to take the lead in conflict prevention, conflict management and resolution, and preventing or assuaging humanitarian disasters. These broader missions were tested in Bosnia, where NATO deployed the 60,000-person Implementation Force (IFOR) beginning in December 1995 pursuant to the Dayton Peace Accords. IFOR was changed to Stabilization Force (SFOR) one year later. Russia and other nonmembers of NATO also participated in the SFOR deployments in Bosnia.

NATO's reputation as the most successful and enduring of America's international alliances was tested by its commitments after the attacks of September 11 to the wars in Afghanistan and Iraq. These extended and expensive "out of the area" operations involved NATO allies and others in support of counterinsurgency, counterterror, and postconflict stability operations. US and NATO allied combat operations have ceased in Iraq, where the US maintains a training and advisory presence as of this writing. The withdrawal of US forces from Afghanistan in August 2021 by the Biden administration was consistent with a Trump administration agreement with the Taliban, but its abrupt and chaotic nature was widely criticized in the United States and abroad.

Throughout the 1990s the subject of Europe's post–Cold War security relationship with North America became an important agenda item. NATO was supportive of the establishment of a "European pillar" within the Atlantic alliance under which Europeans might become more self-reliant in regional security issues that did not call for direct US participation. European members and nonmembers pursued the related but distinct European Security and Defense Identity (ESDI), which has ties to the European Union (EU).[13] NATO's European-pillar concept, or a more assertive version of European security and defense identity under the aegis of the Western European Union (WEU), would permit various coalitions among EU members to undertake conflict prevention and conflict resolution missions with purpose-built forces (perhaps combined joint task forces). Arranging for NATO-related or NATO-congruent, but not NATO-controlled, security architectures was complicated by the problem of partially overlapping memberships among NATO, the WEU, and the EU.

An additional problem in integrating European defenses without breaking European ties to NATO is the diversity of competencies among European NATO defense forces. The most obvious disparity is the technology gap between the United States and the other NATO members relative to high-end conventional warfare. This disparity in long-range precision strike and advanced command, control, communications, computers/intelligence, surveillance, and reconnaissance (C4/ISR) systems was apparent to observers during Operation Allied Force, NATO's air campaign against the former Yugoslavia in 1999, and to some extent remains the case. The gap was also apparent in US and allied NATO operations in Afghanistan, and against Libyan dictator Muammar Gaddafi in the wake of a civil war in Libya in the spring and summer of 2011.

In addition to differences among NATO members in their respective military competencies, there is also the contentious matter of "burden sharing" among NATO members with respect to their financial contributions to national and alliance defenses. The Trump administration raised the profile of this burden sharing issue into a matter of considerable contention as between Washington and its NATO allies. NATO members had previously committed themselves to spending at least 2 percent of their gross domestic products on national and allied NATO defense, but most had still fallen short of that target by 2020 (the deadline was in the early 2020s). On the other hand, NATO did accomplish a number of defense and deterrence initiatives with respect to Russia and Europe during the Obama and early Trump years, including the stationing of additional ready response forces in frontline East Europe and Baltic member states and beefing up infrastructure and exercises. NATO's capabilities in C4/ISR have also improved and deterrence-defense missions have been supported by additional deployments of sea and shore based missile defenses under the European phased adaptive approach supported by the United States.

Notwithstanding these accomplishments, NATO in the Trump era remained as uncertain of future US foreign and defense priorities as was the rest of the international community. It was not so much that Trump was an isolationist or an ultranationalist, but the more subtle but potentially debilitating issue of his unpredictability. Alliance commitments require assurance among the relevant members that in the event of threats to collective security, the various member states will act with a united will and purpose. The Trump administration's uncertain trumpet across various issues left allies, potential adversaries, and others grasping to understand what, if any, central organizing concept existed in Washington. The issue is important because "adhocracy" in foreign policy invites confusion about US ends and means, allowing events to manage US leaders instead of the reverse. The Biden administration moved quickly to shore up US relations with its NATO allies. This was welcomed, but lingering doubts remain about the solidity of US commitments. Biden's moves to strengthen NATO allies nearest to Russia as the Russian military threat to Ukraine unfolded in 2022 were specifically designed to reassure them of the US commitment to their defense.

NATO has been the most enduring, but by no means the only, US security alliance commitment since the end of World War II. But given the changing international landscape, permanent alliances may be the exception. Instead, the emerging norm may be alliances and coalitions for a specific mission or purpose. During the 1991 Gulf War, for example, President George H. W. Bush was forced to improvise a "coalition of the willing" that cut across traditional Cold War alliances. President Clinton supported peace operations using US, NATO, and other forces in coalitions tailored for particular missions, even when NATO accepted formal responsibility for peacekeeping and peace enforcement. Such was the case with Operation Joint Endeavor in Bosnia after 1995 and in Kosovo in 1999: in both cases a significant number of participants came from outside NATO, including Russia. Subsequently, President George W. Bush used a coalition of the willing in Operation Iraqi Freedom to topple the regime of Saddam Hussein in 2003.

Despite NATO's growth spurt in membership in the two decades following the end of the Cold War, harmony among alliance members on security issues cannot be taken for granted. The US-European relationship is fraught with concerns about "burden sharing" with respect to the distribution of military tasks and economic costs among members. In addition, adding new members sometimes increased the difficulty of obtaining consensus on particular issues. For example, NATO deliberations over both the George W. Bush administration and Obama administration plans for European missile defense were encumbered by intra-alliance disagreements about Russian goals and about Russia's trustworthiness as a negotiating partner. The US decision for war against Iraq in 2003, despite lack of UN Security Council or NATO support, also alienated several long-standing US European allies, including France and Germany.

As noted earlier, several rounds of NATO enlargement pushed its membership to thirty countries by 2020. This demonstrated that NATO had become, despite Russia's wishes and some doubts among NATO traditionalists, the all-purpose security guarantor for post–Cold War Europe—and perhaps beyond. NATO's "out of the area" profile of activities was raised higher when the alliance assumed formal military responsibility in providing security for the Afghan regime of President Hamid Karzai and undertook a variety of support missions in postwar Iraq, including the training of 1,500 middle- and upper-level officers for the new Iraqi security forces.[14] NATO's responsibility for postconflict stabilization in Libya after the fall of Gaddafi in 2011 was unclear. NATO had expedited Gaddafi's departure by providing air assets and intelligence support to indigenous Libyan anti-Gaddafi rebels, but the regime change left an immediate power vacuum with contending militias and political factions struggling for power that remain unresolved as of this writing.

In the world order that evolved in the early 1990s, the Cold War alliance system unraveled and NATO underwent changes. A new system of alliances and relationships is emerging, with the United States and Russia shifting from almost-partners in the Yeltsin years to adversaries under Putin. Additionally, the United States–China relationship is becoming increasingly important and is changing as China's economy now ranks as the world's second largest and its military power increases in East Asia. Furthermore, regional groupings and regional alliances have emerged, often with little reference to the United States (e.g., in East Asia and West Africa). Economic associations and alliances have become increasingly important in light of global interdependence.

The Gulf War of 1991 demonstrated the effectiveness of temporary coalitions in responding to overt aggression. This "coalitions of the willing" approach may be a model for future responses to international crises. In such cases, it is likely that prevailing alignments will encourage the formation of temporary coalitions to achieve mutually acceptable objectives. The victorious Gulf War coalition and other coalitions of the willing are dependent, however, on immediate circumstances, personal leadership, and influence among heads of state. The chances for a repeat in the twenty-first century are dim, as US experience in Iraq in 2003 and beyond has proved.

The September 11 attacks caused the George W. Bush administration to mobilize a coalition that included NATO, Russia, and China, with support from some Arab and Islamic states, against the Taliban of Afghanistan for sheltering and supporting the Osama bin Laden terrorist network. However, the level of international support for US policy in Afghanistan did not set a precedent for future conflicts. The US military operation to depose Saddam Hussein in 2003 was opposed by major NATO allies, by Russia, and by important US allies and friends in the Arab and Islamic worlds. Nevertheless the George W. Bush administration assembled its own minicoalition of the willing, including Britain as a major military contributor, for Operation Iraqi Freedom and its follow-on effort to pacify and rebuild Iraq. As time passed and history moved beyond the George W. Bush and Barack Obama administrations, US public and elite opinion doubts about the prolonged conflicts in Afghanistan became more visible and contentious. President Trump campaigned explicitly against further military involvement in Middle Eastern wars, although his administration authorized limited strikes against Syrian military targets on two occasions for alleged use of chemical weapons against civilians. He also authorized limited strikes in Iraq to protect the few US combat forces remaining. In Afghanistan, he struck a deal with the Taliban to release a large number of prisoners and remove all US troops by March 2021. President Biden did not change Afghanistan policy other than to move the date to remove all US troops from Afghanistan to the end of August 2021. Even with this delay, there was not time to evacuate all US citizens and Afghans who helped in the war effort, especially as the Taliban quickly captured the most important cities in the country and surrounded the capital, Kabul. The crush at the Kabul airport proved a tempting target, and a suicide bomber killed thirteen US soldiers and some 200 Afghani civilians.

Allies, Friends, and Temporary Alignments

National security interests demand that the United States do more than develop and maintain close relationships with its allies who are members of formal alliances. Other states may be reluctant to enter formal arrangements (such as treaties and executive agreements) with the United States; nevertheless, some short-term issues may require US involvement. Furthermore, circumstances may dictate indirect relationships to pursue independent but mutually supportive policies and strategies.

National interests can evolve and emerge in unanticipated ways—and sometimes they rise unexpectedly to crisis proportions. The lengthy and complicated treaty process is too cumbersome for timely and effective responses. Additionally, national security issues can arise when formal US arrangements with foreign states are not feasible, leaving informal arrangements as the only solution. Since the end of World War II, the United States has signed many treaties, which require a two-thirds vote in the Senate for ratification. Treaties that involve long-range US commitments or other controversial matters are likely to provoke opposition. Because there is no assurance that treaties will be ratified, presidents have turned to executive agreements, which allow them to commit the country to a course of action through formal agreement with foreign nations without Senate approval; this offers considerable flexibility, but is subject to change by a new administration with different policy priorities.

In cases of controversy or failure, however, the president is vulnerable to criticism from the legislative branch and the body politic. In addition, many senators abhor the overuse of executive agreements, arguing that they violate the spirit of the Constitution and bypass elected representatives. Even though executive agreements give the president a degree of freedom, he or she will eventually have to explain any actions to the public.

There are many examples of nontreaty agreements, including the 1991 Gulf War coalition and other "coalitions of the willing" during the presidencies of Bill Clinton and George W. Bush. Likewise, US support of the mujahidin (Afghan resistance forces) in the 1980s during the Soviet occupation was consistent with the policies of China and Iran. The point is that national security policy can be pursued in many ways and, when necessary or expedient, through unexpected coalitions. Although Congress plays an important role, the moving force is the president, operating through the national security establishment.

Some states are not adversaries, although they certainly are not allies or even friendly. This group includes third world states that are neither democratic nor established authoritarian regimes, as well as nationalist states that fear any alliance, especially with the West—a vestige of the colonial experience. It is in these relationships that unexpected alliances can evolve, causing concern and disagreement within US political circles. Similarly, indirect relationships with states that are nominally friendly but otherwise unfriendly may require the president to explain the US role to the public. Such circumstances can create contradictory US national security policy and strategy—as in the case of Pakistan and the support for the Afghan Taliban provided by its military and intelligence organs.

For example, what was especially difficult for many Americans to understand during the 1980s was the US policy of supporting freedom fighters in Africa and Central America against established regimes, on the one hand, and the support of established regimes, as in El Salvador, against Marxist-Leninist revolutionaries, on the other hand. These complex relationships show that the United States was involved in treaty obligations and commitments through such structures as NATO while it also pursued national security interests through other relationships—friends, potential adversaries, and others. As for friendly Arab states, formal treaty arrangements might not be as feasible in light of Arab nationalism and the close relationship between the United States and Israel. Nonetheless, an understanding was reached by the first President Bush in 1990 for positioning US forces in the Gulf and for overflying

Arab airspace. His son, President George W. Bush, faced similar issues after September 11, 2001, in obtaining official Saudi support for US military retaliation against transnational terrorists. Such friendly relationships can become well established and form an integral part of the US national security effort. On the other hand, unofficial Saudi government and charities' support for the teaching of militant versions of Islam and allegations of complicity between influential Saudis and terrorists complicated US-Saudi relationships after September 11. More recent Saudi actions, such as the murder and dismemberment in their Embassy in Turkey of a Saudi writer for the *Washington Post,* have complicated the US-Saudi relationship further.

As for potential adversaries, the United States cannot adopt a simply military posture: it must establish formal relationships and use the range of diplomatic, political, economic, and psychological tools available to push for change in their policies and even their government, if necessary. Some regimes resist even the most persistent diplomatic overtures, especially if hard-pressed by revolution or civil war. Regime change is also a policy with uncertain chances of success. The United States and al-Qaeda both scrambled to comprehend the Arab Spring outbreaks of dissidence against previously durable regimes in 2011, including governments in Tunisia, Egypt, Syria, and Yemen. Turbulence in Yemen was exacerbated by Houthi rebels supported by Iran threatening to overthrow the Yemeni government, provoking a large military intervention by Saudi Arabia to suppress the rebels. In turn, Saudi military strikes created a major humanitarian disaster embarrassing to the United States on account of US-supplied military assistance to Riyadh. The ousting of President Hosni Mubarak in Egypt in 2011 left an apparent power vacuum filled at least temporarily by the armed forces, creating uncertainty about US future relations with one of its most durable allies in the Arab world. Mubarak's fall also created anxiety in Tel Aviv, since Egypt had been Israel's most reliable negotiating partner among Arab Middle Eastern powers, and it temporarily regenerated the political fortunes of the Muslim Brotherhood, as the formerly outlawed or marginalized Islamic organization competed effectively for power in the post-Mubarak Egyptian parliament and presidency. Eventually the explicitly Islamist government of Muhammed Morsi was overthrown by the Egyptian military in a coup led by General Abdel Fattah al-Sisi, restoring the status quo ante insofar as US-Egyptian relations were concerned.

An important dimension of both permanent and temporary alliances is that each imposes an obligation and commitment on each party, which constrains the ability to shift policies and strategies. Put simply, the more political actors that are involved, the more likely it is that the United States will have its options limited. Thus, for any relationship established with foreign states in pursuit of US national security interests, the advantages must be weighed carefully against the need for flexibility and maintaining options.

These characteristics of national interests and policy goals illustrate that national security policy and strategy can appear contradictory and confusing. The general public, predisposed to straightforward distinctions between good and evil, can find national security policy to be incoherent, thereby generating domestic opposition, especially in Congress. This is a consequence of the disparate character of national interests and national security policy and the public's lack of understanding of these complex issues, as well as an ingrained partisanship among certain groups.

Finally, US commitments, whether they derive from formal or informal arrangements, are customarily honored by each succeeding president, although this is not guaranteed. Each new president takes office with the network of alliances and geostrategic commitments that are already in place. Usually, it is only through major changes in the security environment that such commitments are altered. When President Donald Trump withdrew the United States from the Iran nuclear deal negotiated in 2015 by the Obama administration as part of the "P-5 plus Germany" coalition, and supported by the European Union and the UN, some US European allies scrambled to keep the treaty alive despite the threat of US economic sanctions against Iran and secondary sanctions against other states trading with Iran. The US abrogation of the Iran nuclear deal thus created intra-alliance friction that included both nonproliferation and economic issues. President Trump's equivocation on other matters related to NATO solidarity left the door open to divisions among NATO allies that facilitated the Russian goal of creating political chaos within and among NATO countries. Thus, by the time the Biden administration assumed office, the Nord Stream 2 gas pipeline deal between Russia and Germany was an all but completed fait accompli, despite US objections. The Biden administration is attempting to rejoin the nuclear agreement with Iran, but the outcome of this attempt is unclear.

There is some general agreement among policymakers and scholars as to the more important dimensions of future US national security policy. The following might be considered the critical relationships for US national security interests:

- United States–Russia; United States–China. US national security strategy in the Trump and Biden administrations identified these rising peer competitors as the principal US national security challenges.
- United States–Western Europe, which carries consequences for the United States–Russia relationship.
- The United States–China–Japan triangle, which is important in shaping the security environment in the Pacific.
- United States–South Korea. South Korea is a strong US ally that is in the crosshairs of the North Korean military.
- US relationships on the Indian subcontinent with both India and Pakistan.
- United States–India. India is a rising counterweight to China.
- The Western Hemisphere, which cannot be taken for granted politically due to the increased significance of border security since September 11, including the security environment in Mexico and cross-border cartel attacks.
- United States–Middle East, evolving from the US-Israeli alliance and the on-again, off-again Israeli-Palestinian peace process that seems to be a perennial aspiration falling short of achievement.
- Environmental despoliation and demographic change, including possible long-term catastrophic climate change, creating resource scarcities overlapping with ethnic and religious conflicts in third world countries and elsewhere.

These relationships are not listed in order of priority—indeed, the importance of each will vary over time. Balancing US efforts and dealing with each relationship, individually and collectively, make designing national security policy a complex and often frustrating task. Considerations range from trade to foreign policy to technology, yet each relationship will impact US national security. Even nonmilitary issues, as noted in the last bullet point, can create national security–related problems for US policymakers.

The end of the Cold War and the demise of the Soviet Union changed the character of the international system's relationships in complex patterns that have yet to reveal themselves fully. These include expanded NATO membership and reevaluation, Middle East peace initiatives, and the rush to armaments by several states, including nuclear proliferation in North Korea and potential nuclear weaponization in Iran. North Korea's acknowledged status as a nuclear power has raised Asian uncertainties about future containment of nuclear weapons spread.

Aggressive nuclear and missile testing by North Korean leader Kim Jong-un raised US and allied concerns about the possibility of a North Korean nuclear first use or increased nuclear coercion of US allies in Asia. President Trump held three summits with the impetuous North Korean leader, allegedly to expedite the process of North Korean denuclearization insisted upon by the United States. However, despite memorable "photo ops" between the two heads of state, North Korea continued to resist any meaningful steps toward denuclearization or even a full accounting of its nuclear weapons, materials, and infrastructure. In addition to the challenge from North Korea, in the aftermath of India's and Pakistan's nuclear testing and declared nuclear status in 1998, their relationships with one another, as well as with China, create a security arc of major importance to US policy and threaten regional security and stability. These developments, and others in Russia and Eastern Europe, raise questions about the utility and continued relevance of formal treaties and alliances entered into after World War II. Indeed, the very notion of friend, ally, and adversary has become muddled.

In sum, the new world order has created a strategic landscape that does not lend itself to Cold War–style alliances. In addition, there is a prevailing theme in US politics that resists alliances that enlarge the role of the United States in uncertain political-military contingencies, such as Somalia in 1993, or involve a combat role. Alliances during the early twenty-first century may very well have an entirely new character: flexible in composition, adaptable to short-term missions, and reversible if conditions dictate.

Adversaries and Potential Adversaries

The United States at the turn of the twenty-first century was in the unusual position of having no official enemy or adversary state, or hostile coalition of states, against which to benchmark its security threat assessments. The singularity of its military power and economic potential also induced complacency or even hubris on the topic of international security among members of Congress, the media, and the public at large. Relations with Russia, the core of former Soviet military power and influence, were friendly during the 1990s as Presidents Bill Clinton and Boris Yeltsin largely supported one another, at least until NATO's war against Russian ally Serbia in 1999. By the time of the Obama administration, if not earlier, this honeymoon between the United States and Russia had ended, and Russia continued to assert its broader claims to recognition as a regional and global power under Vladimir Putin. As well, China's bursting economy enabled an expansion in its military assertiveness and regional ambitions under President Xi, who made no secret of Chinese ambitions.

The nature of warfare throughout history warns us that potential opponents will resist US dominance in one aspect of the art of war by resorting to "asymmetrical" strategies that offset US strengths and exploit US weaknesses. These asymmetrical strategies include traditional kinds of insurgencies and terrorist attacks, as well as the use of cyber attacks on information systems and the manipulation of social media for political and military purposes. Therefore, US military strategists and leaders have gone back to the drawing board in order to plan for the impact of information warfare or cyber attacks on military and civilian targets, including attempts to influence public opinion through use of cyber deception (trolls, bots, and other components of information strategy already widely employed by state and nonstate actors).[15]

As Carl von Clausewitz wrote, politics causes hostile intentions between states; military forces are merely the instruments that express that hostility.[16] And the paradox of US power at this time is that it motivates envy and resentment on the part of dissatisfied state and nonstate actors. Security is highly context dependent, and part of that context is the uncertainty of future US relations with post–Cold War Russia and China.

The world breathed a sigh of relief when the nuclear weapons temporarily dispersed among Belarus, Kazakhstan, and Ukraine by the demise of the Soviet Union were relocated under Russian control, with specific security guarantees by the United States, the United Kingdom, and Russia for Ukraine—since violated by Russia. After the election of Vladimir Putin to succeed Boris Yeltsin as president in 2001, the Russian Duma (parliament) ratified the START II agreement, reducing both sides' strategic nuclear forces (forces based on launchers of intercontinental or transoceanic range) to 3,000–3,500 warheads each. Although START II was never completed, many expected that the two states could proceed immediately toward a START III agreement, reducing their respective arsenals even more. In May 2002, US president George W. Bush and Russian president Vladimir Putin signed the Strategic Offensive Reductions Treaty (SORT), which required each party to reduce its operationally deployed nuclear warheads to between 1,700 and 2,200 by the end of the year 2012. SORT was superseded in 2010 by the New START agreement signed by US president Barack Obama and Russian president Dmitri Medvedev, lowering the numbers of strategic nuclear weapons deployed by each state to a maximum of 1,550 on 700 nuclear capable launchers.[17] New START was extended by Presidents Biden and Putin in 2021 for another five years, until 2026. Keeping New START alive was even more significant due to the American and Russian exodus from the Intermediate Nuclear Forces Treaty in 2019, a cornerstone of the Cold War signed in 1987 by Presidents Ronald Reagan and Mikhail Gorbachev. The treaty banned the deployment of ground-launched missiles with ranges of 500–5,500 kilometers worldwide. The Obama and Trump administrations charged Moscow with cheating on the agreement by deploying cruise missiles of prohibited ranges; Russia, in turn, said the United States was cheating by deploying Aegis-Ashore batteries that could be repurposed as offensive weapons.[18]

The United States had an equal interest in the safety and security of Russia's nuclear weapons inventory and in the accurate accounting of Russia's fissile materials (enriched uranium and plutonium). The former Soviet system of accounting for warheads, and especially for fissile materials, left much to be desired. After the Soviet Union was dissolved in December 1991, US experts feared that former Soviet weapons or their constituent elements would trickle outside Russia's borders and into the hands of state and nonstate purchasers. These buyers of ill-gotten former Soviet nukes might include terrorists with agendas hostile to the United States or frustrated state actors with equally malign intentions. Therefore, through a variety of programs under the so-called Nunn-Lugar legislation passed by Congress during President Clinton's first term, the United States provided Russia with military aid and technical expertise to improve materials and weapons accountability, to increase the safe and secure transport and storage of fissile materials, and to retrain Russian nuclear scientists for work on environmental or other nondefense projects. The George W. Bush administration continued this program of cooperative threat reduction as a win-win for both sides but at a level of funding that many considered inadequate.

President Clinton in 1999 signed legislation that urged prompt US deployment of a nationwide ballistic missile defense system. Clinton was less than enthusiastic about actually deploying missile defenses, but he was willing to let research and development proceed. President George W. Bush felt otherwise. Bush termed the Anti–Ballistic Missile Treaty a relic of the Cold War and declared his intent to begin deploying missile defenses in 2004. The Bush program initially deployed ground-based defenses in Alaska and California against light attacks or accidental launches. Meanwhile, the Pentagon's Missile Defense Agency pursued research and development on ground-based, airborne, and sea-based components of a more comprehensive system. US interest in theater missile defenses, against weapons of shorter range than intercontinental systems, also continued, with refinements to the Patriot PAC-3 system. In 2009 the Obama administration announced its intention to shift US plans for European missile defenses to a phased adaptive approach based on progressive development and deployment of sea- and land-based SM-3 missile interceptors.[19]

Russia's military doctrine and national security concepts under the presidency of Vladimir Putin indicate that the United States looms large in Russia's security concerns. In the immediate aftermath of the September 11 attacks, it appeared that the United States and Russia would find common ground as security partners in the war against terrorism. But after 2007, Russian President Putin began to take a harder line against NATO expansion, against US missile defenses deployed in Europe, and against what he saw as America's expectation of global hegemony or unipolarity. US-Russian relations soured during the Obama and early Trump administrations in response to Russian aggression against Ukraine, continuing Russian opposition to US missile defenses, and US allegations of Russian interference in US domestic politics—especially the presidential election of 2016. US nostalgia for the comparatively weaker Russia of the 1990s crashed into Russia's larger twenty-first-century ambitions and rebuilt military potential, as well as Russia's uses of "new generation" or hybrid combinations of conventional and unconventional warfare in Europe.[20] US economic sanctions against Russia, especially those targeted against cronies of President Putin and their financial empires, were a special irritant in US-Russian relations, and they continue under President Biden. Indeed, they may become more severe in the wake of continuing Russian interference in US elections and failure to curtail "private" cyber hackers attacking US and allied computer networks.

Russia's war against Georgia in 2008 broke the logjam against military reform that the defense bureaucracy had resisted even during Putin's early years as president. Since then Russia has taken major steps to modernize its conventional and nuclear forces and to streamline its training, recruitment, and assumptions about force structure. The old Soviet-style force based on a cadre organization as the core of a mass mobilization army has been superseded by a more modern concept of a lighter, more rapidly deployable force with upgraded equipment and command-control-communications (C3) systems. Russian military exercises in recent years have included new strike and control systems and scenario-based planning for offensive and defensive maneuvers on Russia's western, eastern, and southern fronts. Russia's military capabilities were on display during their 2022 invasion, but they had more difficulty than expected against the outnumbered and out armed Ukrainian defenders. Russia has also undertaken a major modernization of its intercontinental and shorter range nuclear forces. Russia has also conducted military exercises with China as one signal of increasing Sino-Russian collaboration and resistance against perceived US hegemony, even as Russia remains wary of China's economic competition via China's Belt and Road Initiative of transnational infrastructure projects intended to spread Beijing's influence across Central Eurasia and beyond. Russia's oil and gas revenues in the new century have made additional resources available for all state purposes, including military modernization, training, and force restructuring.

Russia still faces threats from the East and South, the latter having the potential to exacerbate turbulence in Chechnya and elsewhere in the Caucasus or in former Soviet Central Asia. Russia wants to ensure that Central Asia remains mainly under its shadow as an extended security space and not a forward outpost for permanent US or Chinese military ambitions. Russia's geopolitics are required to balance among regional actors of significant military power or potential (China, India); a Japanese economic powerhouse with growing military capability; an acknowledged nuclear state in North Korea, whose example might be imitated by others in Asia, including South Korea and Japan; and the United States, with its strong commitments to the defense of allies in Asia (Japan, South Korea, and possibly Taiwan).[21]

Russia's special sensitivity toward foreign intervention in its "near abroad" of contiguous former Soviet states was on display during its war with Georgia in August 2008 and against Ukraine in 2022. Russia regards the rimlands of the former Soviet Union as a zone of privileged interests and is especially concerned about further expansion of NATO to include Georgia and Ukraine. After the Russo-Georgian clash of 2008, Russia recognized the breakaway Georgian republics of South Ossetia and Abkhazia as independent states, but few others followed this diplomatic precedent, and the final status of South Ossetia and Abkhazia remains unsettled. The alliance has not abandoned its long-term policy objective to offer membership to both countries. The degree to which this is a good idea remains contentious.

Others

Most of the states in the "other" category are non-European or fall outside the US-Western cultural system. The tendency to place all states outside Europe and North America into the third world prevails in much of the literature. The categorization is useful in broad terms but is of little use in close examination of US national security interests and in assessment of US policy. To place India, Chad, and Taiwan, for example, in the same third world category ignores the cultural, political, and economic differences, as well as geographic distinctions, among those three states. The same reasoning applies to many other developing and partly developed states in Africa, Asia, the Middle East, and Latin America.

China, with its vast human resources and geostrategic position in Asia, is in a special category. China's burgeoning economy and growing military power have raised US concerns that it could grow into a regional hegemon or even a peer military competitor by the middle of the twenty-first century and perhaps much sooner. Indeed, for many Western observers, little can be done in Asia and Southeast Asia without considering the role of China. China has an ambivalent relationship with the United States. On the positive side, both states benefit from trade and investment flows that are essential for the growth and development of both economies. On the negative side, the two disagree over Taiwan, with the United States insistent that Taiwan must not be forcibly annexed to the People's Republic of China (PRC) and China increasingly determined to force this issue and investing in military capabilities that would be useful in a military assault on the island. In addition, US military planners worry that a stronger China might engage in "access denial" of US power projection into the Pacific Rim and are wary of increased Chinese assertiveness in support of their disputed territorial claims in the South China Sea.

US policy planners will have to deal with a twenty-first-century China that has escaped the confines of purely Maoist strategic military thinking built around a people's war and national liberation.[22] China has modernized its armed forces and continues to improve its nuclear capabilities and ballistic missile forces.[23] China's theater ballistic missiles could serve as deterrents to US conventional military deployments in the Pacific supported by the implicit threat of US nuclear arms (i.e., as anti-access and area denial, or A2AD, forces). The range of Chinese, Indian, and other Asian ballistic missiles and weapons of mass destruction (WMD) has the potential to redefine the entire concept of geostrategic space in Asia. China (and Russia) may even be ahead of the United States in the development of "hypersonic" nuclear missiles that can be launched from space and elude current antiballistic missile systems. Asian countries armed with WMD and medium- or long-range ballistic missiles might be transformed from "map takers" (states largely acted upon by others) to "map makers" (states that determine the geostrategic policy agenda in a region).[24]

Dealing with China requires nuanced diplomacy and sensitivity to Chinese culture. The impatience of Western negotiators and their short-term objectives are often misplaced in China, whose diplomatic-strategic behavior befits the world's oldest civilization and focuses on long-term purposes.[25] The Clinton administration made favorable US relations with China on trade and other issues a high priority, and to some extent it succeeded. But serious differences also surfaced, including China's resentment over the inadvertent US bombing of the Chinese embassy in Belgrade during NATO's air war against Serbia in 1999. Also, a scandal over alleged Chinese espionage against US nuclear weapons laboratories clouded US relations with Beijing during Clinton's second term. The uproar led to congressional Republican demands for investigation of alleged security lapses and resulted in some tweaking of Chinese diplomatic sensitivities. The US and Chinese governments also disagreed on other issues, including Chinese export of ballistic missile technology to rogue states with anti-US agendas and US claims of Chinese human rights abuses, especially with respect to the treatment of the Muslim Uyghur population in the northwestern region of Sinkiang province.[26] The latter issue is of increasing importance as more information is received.

The George W. Bush administration had relatively untroubled relations with China, enlisting China's participation in six-party talks to contain North Korean nuclear weaponization and enlarging US-Chinese political and cultural ties. On the other hand, some US observers remained concerned by the high proportion of US federal debt held by Chinese creditors and by China's military modernization and doctrine foreshadowing rising military and political aspirations regionally in Asia and worldwide. During the Obama and Trump administrations, US-Chinese relations were complicated by trade issues and by US allegations of Chinese cyber mischief: including espionage, theft of intellectual property, and penetration of classified databases and exfiltration of confidential information from United States government agencies.

The Biden administration is continuing to exert pressure on China with respect to human rights, their expansive claims to sovereignty over a vast swath of the South China Sea, and the 2020–2021 crackdown on democracy in Hong Kong. China, in turn, faced the problem of maintaining perestroika without glasnost: continuing to restructure and modernize its economy along capitalist, free-market lines without releasing the control over political life held by the Chinese Communist Party and central government. Capitalism without democracy might work in the short run, but the long-term compatibility of free markets with totalitarian politics is highly questionable. Once having been freed of the myths that propped up Soviet power, and fed up with the lack of consumer goods easily available to other Europeans, Russians and other nationalities revolted against the entire concept of the Soviet Union. The Chinese leadership wants to avoid Mikhail Gorbachev's fate at all costs, but it also wants to be a great power.

India is not a US adversary, but it will play a larger role in this century in US security calculations. India is now an acknowledged nuclear weapons state (it publicly tested a nuclear device in 1998, soon followed by Pakistan). India sees China and Pakistan as potential threats to its security, and nuclear weapons are now a part of India's defense posture. Pakistan, in turn, sees India as the major military threat. China during the Cold War was mostly allied with Pakistan on security issues; the Soviet Union allied with India for the most part. Whether those patterns will persist in the twenty-first century is not clear, but China and Pakistan are assumed to have cooperated on the transfer of nuclear and missile technology to Iran and other states with potentially anti-US or anti-Western agendas. The US war on terror aligned the Pervez Musharraf government in Pakistan with the United States against al-Qaeda and other groups motivated by Islamic extremism. However, relations between the two governments deteriorated under successor governments in Islamabad, including considerable friction between Pakistan and the Obama administration over US drone missile attacks against suspected terrorists within Pakistan and other issues. The US special operation to kill Osama bin Laden in May 2011 by a direct-action raid into Pakistan exacerbated ill will between Pakistan's military and intelligence services (suspected of concealing and abetting bin Laden's hideaway) and US officials.

The growing ballistic missile arsenals in the Middle East, South Asia, and North Asia, and the possibility of nuclear proliferation in those regions, change US geopolitical calculations. US regional conflict strategy is no longer as dependent on permanent forward basing, such as Subic Bay and Clark Air Base in the Philippines. Instead, it now assumes that rapid deployment of forces from the homeland, or other bases such as Guam, can meet a crisis, getting forces into the theater of operations in sufficient time to make a difference.

The potential spread of WMD and ballistic missiles in Asia, however, challenges the assumption of unopposed US power projection along the Pacific Rim or into the Middle East and Southwest Asia. If, for example, Saddam Hussein had been able to fire nuclear-capable ballistic missiles at

Saudi Arabia in 1990, the massive US forces deployed in that country to expel Iraq from Kuwait would have been at risk.

The Middle East has its own set of political and geostrategic characteristics, many of which impact US national security interests. The US connection to Israel, combined with the US need to maintain reasonably friendly relationships with some Arab states, ensures that the Middle East is a fragile security area. Islamic links among the Arab states and sensitivity to non-Arab involvement in the Middle East give the area a degree of cultural homogeneity and offer a bastion against Israeli and Western power projections and influence. Included in the Middle East imbroglio are the Persian Gulf and its oil reserves, the increasing political influence and military capability of Iran, and the prospect of further weapons proliferation (including possible Iranian nuclear weapons). The importance of this region to US national security has been well documented, although with increased production of oil and natural gas and a shift away from fossil fuels Western dependence on Middle East oil is diminishing. Still, the United States is a major actor in the area, whether it likes it or not.

The US military interventions in Afghanistan in 2001 and in Iraq in 2003, and related US deployments in Central Asia, have redistributed power in the geopolitical "great game" that once was dominated by colonial European powers. Motivated by antiterrorism or desires for regime change in Iraq, the United States found itself depicted in much Arab and Islamic public discourse as an "imperial" or occupying power. The George W. Bush administration had significant allied and international support in building a post-Taliban Afghanistan, including the willingness of NATO to undertake a major role in providing security, although critics charged Bush with shifting US diplomatic and strategic emphasis to the war in Iraq after 2003. President Barack Obama sought to reboot US regional strategy after 2009 by increasing the US military commitment to Afghanistan, although his "surge" of 2009 was tied to a plan for beginning a phased withdrawal of US and allied NATO forces by 2012. Few imagined at the time that the United States and its allies would remain in Afghanistan until the summer of 2021, trying to manage a transition that would enable the Afghan government to deploy loyal, credible, and effective military and police forces against insurgents and terrorists and, at the same time, taking part in multi-sided negotiations including the Afghan Taliban that would if successful restore a security environment permissive of US departure. In 2021 President Biden decided to cut US losses in Afghanistan and remove all combat forces from the country by the end of August 2021.

A complete Taliban takeover of the country quickly ensued, causing great difficulties in extracting US citizens and Afghans who had assisted the United States and showed the difficulty of extracting political assets once military assets had been removed.

In Iraq after the fall of Saddam Hussein, the United States, Britain, and a coalition of smaller powers faced a protracted insurgency that had not been foreseen by Pentagon planners. Many critiques of US policy in Iraq pointed to devastating insufficiencies in plans and operations for postconflict reconstruction.[27] The situation was only redeemed by the high quality of US military performance in the field, regardless of gaps in political and economic planning for the postwar period. US forces adapted to the demanding tasks of urban warfare or "close quarter battle," performed a wide variety of social and humanitarian tasks, rebuilt infrastructure, and tried to fight a counterinsurgency while distinguishing between genuine Iraqi nationalists and foreign terrorists sent into Iraq by al-Qaeda or other Islamic radicals for the purpose of destabilizing the state and destroying the new regime. The surge by the Islamic State (ISIS) in Iraq and Syria beginning in 2014–2015 caused some anxious moments for the Iraqi armed forces and government, but US and other assistance to the Iraqis helped Baghdad hold firm during several years of fighting before the ISIS "caliphate" in Syria and Iraq was finally toppled in 2019. As of mid 2021, some 2,500 US troops remained in Iraq to assist in stability operations in support of the Iraqi government, but that is not likely to continue indefinitely.

A key frustration to US policy in the Middle East remained the inability of Israelis and Palestinians to agree on a peace process leading to the creation of a democratically based Palestinian state coexisting with Israel. In the summer of 2000, President Clinton nearly brokered a peace agreement between Yasir Arafat's Palestinian Authority and the Israeli government to resolve the issues that divided the two sides. The status of Jerusalem was among the difficult questions that kept the negotiators short of success. Nevertheless, the intensive involvement of the United States created a favorable climate for peace. But the second Palestinian uprising—the intifada—began in September 2000 and was met by Israeli military escalation. The Ehud Barak government fell in 2001, replaced by the more hard-line administration of Prime Minister Ariel Sharon. The crisis came to a head in early 2001 as hundreds of casualties from Palestinian suicide bombs and Israeli military attacks were added to the decades-long death toll in troubled Palestine. The battles continue on and off, and are unlikely to end short of an unexpected peace agreement.

The George W. Bush administration tried to restart the peace process, but with mixed success. The Sharon government conceded Palestinian self-rule over Gaza in 2005 and promised to be more accommodating in negotiations on the final status of the West Bank. Health problems forced Sharon to withdraw from the political stage in 2006, however, and elections for the Palestinian legislature in January 2006 complicated the picture by returning a majority of candidates from Hamas, the Islamic social movement and terrorist organization. The Obama administration indicated in 2010 that it would seek renewed international support for brokering Israeli-Palestinian peace agreements, but a hard-line Israeli government led by Prime Minister Benjamin Netanyahu until 2021 and a divided house among Palestinians (as between Hamas and al-Fatah factional leaderships) promised a challenging tapestry of frustration for peace negotiators.

Since then, Israeli and Palestinian positions have only hardened. The Trump administration tried the novel approach of transcending normal diplomatic channels, using Trump son-in-law Jared Kushner as plenipotentiary point man for brokering Middle Eastern peace agreements. Trump's favorable relationship with Israeli prime minister Benjamin Netanyahu was helpful in dealing with Tel Aviv, but Palestinian representatives were divided into factions with different visions of an acceptable peace process or outcome. No Rosetta Stone solution to this impasse has yet appeared and President Biden put this issue on the back burner.

The worst-ever terrorist attack on US soil, on September 11, 2001, added fuel to the combustible mixture of politics, religion, and nationalism that kept the Middle East and Southwest Asia in turmoil. Fighting this new nonstate enemy without seeming to launch a holy war of Western Christendom against Islamic and Arab communities was the challenge facing former president George W. Bush and later presidents. Military action in the war on terrorism, including special operations, airpower, and ground fighting, had to be contained within the larger policy framework of political de-escalation to avoid entrapping the United States in an open-ended series of unwinnable conflicts. The US Army and Marine Corps rethinking of US experience in Iraq and elsewhere produced a revised and ambitious counterinsurgency doctrine in 2007, and the favorable turn of events in Iraq from 2007 to 2009 was regarded by some as vindication of this new approach.[28]

On the other hand, the new counterinsurgency doctrine was less obviously a blueprint for success in Afghanistan. In that conflict, the Obama administration found itself divided between the emphatic embrace of robust counterinsurgency with high-aspirational social, economic, and political obliques, on the one hand, and a more restrictive counterterror strategy that emphasized smaller military "footprints" of permanently deployed US troops, advisory and training instead of combat missions, more reliance on drone attacks and special forces, and prompt transfer of responsibility for closure to host government forces, on the other.[29]

The Donald Trump administration was not so much concerned to update counterinsurgency doctrine as it was to wind down America's troop commitment in Afghanistan. Trump took the position that US involvement in large-scale ground wars in the Middle East under his two immediate predecessors had been mistaken overreach. His search for an exit strategy in Afghanistan was a mixed success. The US was able to establish ongoing negotiations with the Taliban through intermediaries, but the more difficult step of getting the Afghan government and Taliban into serious bilateral discussions remained elusive. There was also uncertainty among US military experts as to whether the Afghan government could withstand the opposition of Taliban and other groups operating in Afghanistan without a continuing US and allied commitment of substantial combat troops. When the Biden Administration started removing the last of its troops in August 2021, that question was answered: the Afghan government soon fell to the Taliban and the evacuation of US and allied forces and Afghans who had assisted the US became a chaotic nightmare.

Conclusion

The United States brings an impressive resume to the international table: a hyperpower of unprecedented military and economic strength, and, at least in the eyes of some, an example of democratic leadership. But not all situations are amenable to the use of US power, especially military power, whether in support of diplomacy or actual war. Military power is an effective persuader under certain, and very selective, conditions. These conditions can include the willingness of allies to help carry the burden. NATO has been among the most successful military alliances in modern history. The peaceful dissolution of the Soviet Union in 1991 opened the door for NATO to transform itself from a defense guarantor against Soviet attack into a promoter of a pan-European security community. Toward that end, NATO has worked with the European Union to develop options for contingency operations that might not necessarily involve the United States or formally commit NATO forces. These plans continue, but have not yet reached fruition.

History shows that the decision to label other states as US allies, friends, enemies, or others is a matter of national interest, but it is also influenced by the cultural, social, and political characteristics of other regimes. Currently, the United States has no official enemy comparable to the former Soviet Union, but the complexity of international politics in the twenty-first century argues against complacency. China and Russia are emerging "peer competitors" with the United States for regional and global influence and their security objectives are inconsistent with those of Washington on many issues. "Enemies" (plural) who are opposed to aspects of US policy will certainly appear, and some may be willing to risk everything—including annihilation—for their cause. The September 11 suicide hijackers are testimony to that.

But the United States can meet such challenges more effectively if it can call upon existing alliances such as NATO, or if it can assemble purpose-built coalitions, as during the wars against Iraq in 1991 and 2003. Relations with allies can also lead to strain (e.g., over burden-sharing in peacekeeping and postconflict reconstruction in Iraq and Afghanistan). And the "war" against terrorism meant that new frameworks for cooperation had to be constructed. The future structure of US allies, adversaries, and others will also be determined by additional clarity on the part of US policymakers as to the nature and extent of America's role in the world—as the "indispensable nation" accepting a unique burden for the maintenance of world order; as a strategic balancer against aspiring hegemons who threaten the peace and stability of core regions and states; or as a "systems integrator" that creates and expedites diplomatic and strategic syntheses across the old fault lines of past power blocs and regional antagonisms. Or perhaps the United States will need to combine these approaches depending on the strategic situation.

Notes

1. Eliot A. Cohen, *The Big Stick: The Limits of Soft Power and the Necessity of Military Force* (New York: Basic, 2016), esp. pp. 63–97.
2. H. R. McMaster, *Battlegrounds: The Fight to Defend the Free World* (New York: HarperCollins, 2020).
3. Angela A. Stent, *Putin's World: Russia Against the West and With the Rest* (New York: Twelve-Hachette, 2019), p. 6.
4. See Samuel P. Huntington, "The Clash of Civilizations?" *Foreign Affairs* 72, no. 3 (Summer 1993), pp. 22–49.
5. John Lewis Gaddis, *The Cold War: A New History* (New York: Penguin, 2005).

6. The US role as sheriff of world order is explained in Colin S. Gray, *The Sheriff: America's Defense of the New World Order* (Lexington: University Press of Kentucky, 2004).

7. Stephen M. Walt, *Taming American Power: The Global Response to US Primacy* (New York: Norton, 2005), chap. 5. See also Walt's *The Hell of Good Intentions: America's Foreign Policy Elite and the Decline of U.S. Primacy* (New York: Farrar, Straus, and Giroux, 2018).

8. Walt, *Taming American Power,* p. 222.

9. Ibid.

10. Barry R. Posen, *Restraint: A New Foundation for U.S. Grand Strategy* (Ithaca, N.Y.: Cornell University Press, 2014). For counterpoint, see McMaster, *Battlegrounds,* passim.

11. See, for example, Daniel Burroughs, "Joining the Club: NATO Debates Terms for Welcoming Former Foes," *Armed Forces Journal International* (December 1993), p. 25. See also John G. Roos, "Partnership for Peace: Cautious Movement Toward the 'Best Possible Future,'" *Armed Forces Journal International* (March 1994), pp. 17–20.

12. See https://www.nato.int/cps/en/natohq/topics_52044.htm.

13. Peter Schmidt, "ESDI: Separable but Not Separate?" *NATO Review* (Spring/Summer 2000), pp. 12–15.

14. "NATO Trains over 1,000 Iraqi Officers," *NATO Update,* January 18, 2006, http://www.nato.int/docu/update/2006/01-january/e0118a.htm.

15. P. W. Singer and Emerson T. Brooking, *LikeWar: The Weaponization of Social Media* (Boston: Houghton Mifflin Harcourt, 2018).

16. Carl von Clausewitz, *On War,* edited and translated by Michael Howard and Peter Paret (Princeton, N.J.: Princeton University Press, 1976).

17. *Treaty Between the United States of America and the Russian Federation on Measures for the Further Reduction and Limitation of Strategic Offensive Arms* (Washington, DC: US Department of State, April 8, 2010), http://www.state.gov/documents/organization/140035.pdf.

18. For perspective on this, see Stephen Blank, "Arms Control and Russia's Global Strategy After the INF Treaty," June 19, 2019, https://www.realcleardefense.com/articles/2019/06/19/arms_control_and_russias_global_strategy_after_the_inf_treaty_114513.html.

19. Robert M. Gates, "A Better Missile Defense for a Safer Europe," *New York Times,* September 19, 2009, http://www.nytimes.com/2009/09/20/opinion/20gates.html.

20. US Defense Intelligence Agency, *Russia: Military Power—Building a Military to Support Great Power Aspirations* (Washington, DC, 2017), https://www.dia.mil/Portals/110/Images/News/Military_Powers_Publications/Russia_Military_Power_Report_2017.pdf. For a parallel DIA analysis on China, see *China Military Power—Modernizing a Force to Fight and Win* (Washington, DC, 2017), https://www.dia.mil/Portals/110/Images/News/Military_Powers_Publications/China_Military_Power_FINAL_5MB_20190103.pdf.

21. With respect to Taiwan, the traditional policy of the United States is "strategic ambiguity." That means the United States will not declare in advance what actions it would take in response to an attack from China. A miscalculation in this area poses a most serious danger of escalation.

22. Michael Pillsbury, "PLA Capabilities in the 21st Century: How Does China Assess Its Future Security Needs?" in Larry M. Wortzel, ed., *The Chinese Armed Forces in the 21st Century* (Carlisle, Pa.: Strategic Studies Institute, US Army War College, December 1999), pp. 89–158.

23. Lt. Gen. Robert P. Ashley Jr., Director Defense Intelligence Agency, "Russian and Chinese Nuclear Modernization Trends," remarks as prepared for delivery, Hudson Institute, May 29, 2019, https://www.dia.mil/News/Speeches-and-Testimonies/Article-View/Article/1859890/russian-and-chinese-nuclear-modernization-trends/.

24. Paul Bracken, *Fire in the East: The Rise of Asian Military Power and the Second Nuclear Age* (New York: HarperCollins, 1999), passim.

25. Graham Allison, "What Xi Jinping Wants," *The Atlantic,* May 2017, https://www.theatlantic.com/international/archive/2017/05/what-china-wants/528561.

26. "Who Are the Uyghurs and Why Is China Being Accused of Genocide?" *BBC News,* June 21, 2021, https://www.bbc.com/news/world-asia-china-22278037.

27. See, for example, Thomas E. Ricks, *Fiasco: The American Military Adventure in Iraq* (New York: Penguin, 2006); and David C. Hendrickson and Robert W. Tucker, *Revisions in Need of Revising: What Went Wrong in the Iraq War* (Carlisle, Pa.: Strategic Studies Institute, US Army War College, December 2005).

28. US Army and Marine Corps, *Counterinsurgency Field Manual: US Army Field Manual no. 3-24 and Marine Corps Warfighting Publication no. 3-33.5* (Chicago: University of Chicago Press, 2007).

29. For a discussion of insurgency and counterinsurgency in theoretical and historical context, see John A. Nagl, *Learning to Eat Soup with a Knife: Counterinsurgency Lessons from Malaya and Vietnam* (Chicago: University of Chicago Press, 2005), chap. 2, pp. 15–33.

3

The Conflict Spectrum

The American way of war is shaped by four critical dimensions. First, for most Americans there is a clear distinction between the instruments of peace and those of war. The instruments of war remain dormant until war erupts, at which time the instruments of peace fade into the background, allowing whatever must be done to win. This polarization is also reflected in the way Americans tend to view contemporary conflicts. Involvement is seen as an either-or situation: the United States is either at war or at peace, with very little attention given to situations that have aspects of both conditions. That war is a continuation of politics (or policy) by other means, as Clausewitz emphasized, is as true as it is contrary to American public expectations and cultural proclivities.[1]

Second, most Americans tend to view conflicts in the world through conventional lenses, with mind-sets shaped by the US experience and by US values and norms. Issues of war and peace are seen in legalistic terms in which wars are declared and conducted between states according to established rules of law. This applies equally well to conflicts in the new era. Yet the attacks of September 11 made many aware of a new kind of war, one characterized by unconventional strategy and unconventional tactics. It is a war that does not conform to conventional mind-sets. The terms "asymmetrical," "new generation," "hybrid," and "gray zone" warfare strategies are now commonly used to denote approaches to conflict that fall outside of the boundaries of conventional armed combat, and that sometimes rely more on nonkinetic means of influence and persuasion than the actual use of force.[2] These strategies can also be a complement to warfare as traditionally understood, serving as enablers or force multipliers. The 2014 Russian invasion of Crimea, then part of Ukraine, was facilitated by various forms of hybrid warfare.

Third, the Vietnam experience left many Americans skeptical and ambivalent about the overseas commitment of US ground combat forces. This skepticism eventually reappeared with respect to the continuing US involvement in Iraq and Afghanistan that began in 2003 and 2001, respectively. Even though new generations are emerging with no memories of the Vietnam War, the Vietnam Veterans Memorial in Washington and several films of varying accuracy promise to keep the Vietnam experience alive. In addition, media coverage of US difficulties in Iraq and Afghanistan frequently made comparisons to Vietnam. There seems to be an undercurrent of caution—a Vietnam syndrome, if you will—that is latent within the body politic whenever US troops are committed to protracted unconventional wars, including counterinsurgency operations with a heavy political, social, and cultural overload such as postconflict reconstruction and nation building.[3] The Donald Trump and Joseph Biden administrations differed on many things, but they both reflected this post-Vietnam skepticism about protracted wars in their determination to withdraw US forces from Afghanistan. This was finally accomplished by President Biden in August 2021, with many American citizens and Afghans who had assisted the United States left behind and thirteen US service personnel and almost 200 Afghani civilians killed in a terrorist bomb attack in the final days. Comparisons with the chaotic US withdrawal from Saigon in 1975 were inevitable.[4]

Fourth, Americans want US involvement to be terminated as quickly as possible, with the victory providing clear decisions and final solutions. The fact that the public seeks clear and understandable solutions to complex issues compounds the difficulties inherent in policymaking. This mind-set assumes that every problem has a solution. More than three decades ago, Ernest van den Haag captured this US perspective well:

> Many Americans still are under the impression that a benevolent deity has made sure that there is a just solution to every problem, a remedy for every wrong, which can be discovered by negotiations, based on good will and on American moral and legal ideals, self-evident enough to persuade all parties, once they are revealed by negotiators, preferably American. Reality is otherwise. Just solutions are elusive. Many problems have no solutions, not even unjust ones; at most they can be managed, prevented from getting worse or from spreading to wider areas. . . . International problems hardly ever are solved by the sedulous pursuit of legal and moral principles.[5]

The consequences drive us to search for the "doable," which in turn leads to oversimplification, whether the issue is strategic weaponry, defense budgets, humanitarian issues, or unconventional conflicts. With respect to unconventional conflicts, most simplistic solutions sidestep fundamental problems and reveal a lack of understanding regarding relationships among culture, modernity, political and economic changes, internal conflicts, and less developed systems. This predilection is reinforced by the fact that many otherwise effective responses to unconventional conflicts might not fall neatly within the framework of values and norms of open systems. The most effective response to terrorist attacks against the US homeland may well involve actions that fall outside the norms of democracy—indeed, many are morally unacceptable. This does not mean that open systems are incapable of effective response. It means that open systems have difficulty in developing policies and strategies for unconventional conflicts owing to the very nature and character of those open systems.

There is increasing disagreement, however, about the appropriateness of the American way of war. Is this traditional way of war relevant in the new era? How should the US military be used in contingencies and missions short of war? Indeed, how should the United States prepare for conflicts across the entire spectrum?

Although military capability remains an essential component of the American way of war and national security, several other components are also important: diplomacy, political power, psychological strategy, intelligence, and economics. In this environment, worldwide intelligence capability takes on an increasingly important role.[6] But the fact is that in some situations none of these components can substitute for the effective use of military power. This is especially important for the United States, given its worldwide interests security objectives.

A caveat is necessary here. We do not suggest that military means should be the first or only option, but there may be times when national security and national interests require the use of military force (although this must be tempered by the nature of the conflict and the appropriate use of other instruments). The military is often the instrument of first choice, even in contingencies short of war. This was the case for the Clinton administration (1993–2001) when its security team met to hammer out US policy in Bosnia-Herzegovina. At one point, a discussion took place between United Nations (UN) ambassador Madeleine Albright and the chair of the US Joint Chiefs of Staff, General Colin Powell. As described by Powell, "The debate exploded at one session when Madeleine Albright . . . asked me in frustration, 'What's the point in having this superb military that you're always talking about if we can't use it?' I thought I would have an aneurysm. American GIs were not toy soldiers to be moved around some sort of global game board."[7] General Powell then explained that US soldiers had been used in a variety of operations other than war, as well as outright war, during recent years, but in each there was a clear political goal and the military was structured and tasked to achieve those goals. Powell was clear on the point that the military should not be used until the United States had a clear political objective.[8]

Complicating the issue, the public and some politicians tend to view war as a clear struggle between good and evil.[9] In this view, the military instrument is an implement of war that should not be harnessed except to destroy evil. This mind-set is the opposite of realpolitik and balance-of-power approaches that characterized the foreign relations of Europe's great powers from the seventeenth through the twentieth centuries. As John Spanier has written:

> Once Americans were provoked, however, and the United States had to resort to force, the employment of this force was justified in terms of universal moral principles with which the United States, as a democratic country, identified itself. Resort to the evil instrument of war could be justified only by presuming noble purposes and completely destroying the immoral enemy who threatened the integrity, if not the existence, of these principles. American power had to be "righteous" power; only its full exercise could ensure salvation or the absolution of sin.[10]

In sum, the ability of the United States to respond to situations across the conflict spectrum (discussed later) is conditioned by historical experience and the American way of war. National interests and national security policy have been shaped by the premises identified here and have influenced the way Americans see the contemporary world security environment. Yet the security issues and conflicts across the spectrum may not be relevant to US perceptions, policy, and strategy. The gap between US perceptions and the realities of the security environment poses a challenging and often dangerous dilemma for US national security policy.[11] This requires a rethinking of the nature of contemporary conflicts and the US national security posture. This rethinking, combined with the turmoil and fog of the international security landscape, is best studied by examining the conflict spectrum.

In so doing, we are aware that the focus of attention in US national security is increasingly toward the challenges of peer and near-peer competitors, like China and Russia, not to mention problems with North Korea and Iran over their nuclear programs. This kind of competition is not new and has echoes of the Cold War. Planning for conventional war is necessary because of the evolving threat, the very high stakes involved, and the massive investment in capital equipment and training required to conduct it. Planning for conventional conflicts is challenging, but simpler than planning for the myriad possibilities in unconventional war, to be discussed below. But in some ways such planning is easier, because the threat is less ambiguous than in unconventional war. We spend the most time here discussing conflicts at the lower end of the conflict spectrum because intellectual problems of conventional warfare are usually better understood and unconventional warfare will remain difficult to understand and will be problematic into the indefinite future. They are also less congenial to the American way of war and pose serious intellectual and moral challenges if they are to be waged successfully.

The Conflict Spectrum

The transformation of the US military to meet the challenges of the twenty-first century has become an indispensable part of its culture. These new directions are intended to ensure that the military remains capable across the conflict spectrum. The concept of full spectrum dominance rests on existing as well as evolving military capabilities. The US armed forces are expected to be capable of peacekeeping, humanitarian operations, disaster relief, operations other than war (OOTW), unconventional conflicts, and stability operations in addition to all the various dimensions of conventional operations and quick-reaction capabilities. All of this is to be accomplished in a strategic landscape that remains clouded in the fog of peace combined with the latent possibility of war. After 2001, that fog dissipated somewhat as increasing attention was given to homeland security.[12] Currently, however, serious questions remain regarding US political and military success, or failure, in closing the deals in Iraq and Afghanistan. Figure 3.1 shows the conflict spectrum into the twenty-first century.

The conflict spectrum is a way of showing the nature and characteristics of international conflicts. It is a useful method for assessing US capabilities and effectiveness. Contemporary conflicts are placed in various categories of intensity. The high ends of the conflict spectrum include nuclear war or large-scale interstate conventional wars; the lower ends include a variety of unconventional conflicts, including

Figure 3.1 The Conflict Spectrum

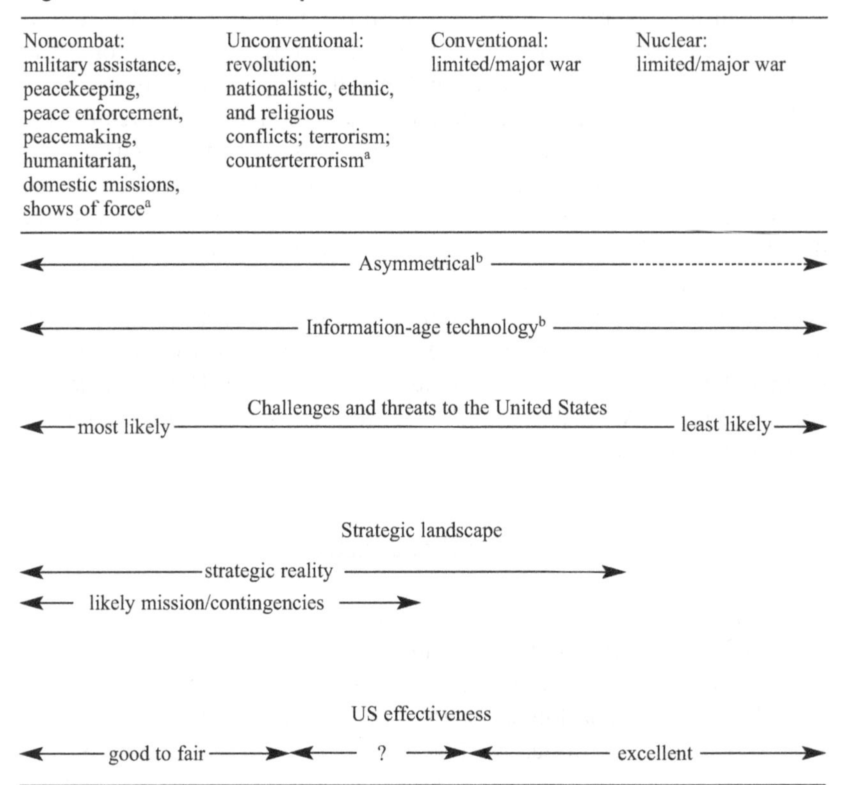

Notes: The complexity of the twenty-first-century strategic landscape and conflict characteristics is reflected in this schematic. This is even more complicated because it is possible that a variety of conflicts can occur simultaneously.

------➤ indicates an unclear impact and/or end result.

a. There is rarely a clear distinction between noncombat contingencies and unconventional conflicts. The noncombat category includes a variety of humanitarian and peacekeeping operations as well as coalition strategies and military support for UN operations. Virtually all operations other than war have the potential of developing into unconventional conflicts of one type or another. Unconventional conflicts can also take place simultaneously with conventional conflicts.

b. The asymmetrical dimension and threats emanating from information-age technology cut across all the categories of conflict.

insurgency, terrorism, cyber attack, and the use of social media for dissemination of "fake news," propaganda, disinformation, and deception. Moreover, there is disagreement in the US body politic as well as the national security system about US responses to the turmoil within other states and the international system in general.

Much attention is focused on unconventional conflicts here because of the conviction that the United States has historically shown weakness in this area. It should not detract from the reality that conventional military threats are increasing. These will be incredibly difficult to deal with, but at the conceptual level they are better understood than unconventional conflicts. Today, and for the immediate future, conflicts are likely to be multidimensional, confusing, and complex.

Three dimensions have become characteristic of this new international landscape. First, challenges to state sovereignty based on claims of self-determination by groups within a state are becoming international concerns. The UN, North Atlantic Treaty Organization (NATO), and other regional groups such as the European Union (EU) are increasingly apprehensive about intrastate conflicts, especially those that are likely to spill over into their areas. This has triggered intervention contrary to the traditional international norm of state sovereignty—the rule of the state over its own inhabitants. Some argue that state sovereignty has never been an absolute international norm. It certainly is not absolute now, as seen by NATO force involvement in Kosovo under Clinton, in Afghanistan under George W. Bush and Obama, and in Libya under Obama. The Obama and Trump administrations were also concerned about the impact of the British referendum for "Brexit" or severing its membership in the European Union. The United States has sometimes chafed at efforts of international institutions, such as the United Nations or the World Court, to impose decisions or behavioral restrictions on US national interests as seen in Washington, DC. Another example of the tug-of-war between national sovereignty and international influence is provided by the collapse of the nuclear arms reduction agreement known as the Intermediate Nuclear Forces Treaty. A marker of US-Soviet cooperation on nuclear arms reduction signed in 1987, by 2019 both the United States and Russia felt unnecessarily constrained by its restrictions while other countries built up their own missile arsenals.

Second, there is increasing attention paid to the impact of information warfare or cyber war, defined as a range of activities from criminal mischief to technologically sophisticated warfare using computer networks and a variety of communication devices.[13] Information warfare was evidently part of the US strategy in challenging Slobodan Milosevic during the Kosovo conflict in 1999 and was reportedly used by Russia in its war against Georgia in August 2008. In the latter case, a Russian cyber campaign launched coordinated distributed denial of service (DDOS) attacks against a total of thirty-eight Georgian and

Western websites over a period of three days from August 8 to 11. The targeted websites included various Georgian government offices and the US and UK embassies in Georgia. In addition, Russian political information operations attempted to shape favorable international media narrative about the war, effectively promulgating the idea that Russian actions were defensive in nature and a response to Georgian aggression. Since then, Russia has diversified its techniques for information warfare to include new technologies and strategies for shaping the international digital environment, including its much-touted Internet Research Agency in St. Petersburg, which was charged by US intelligence with interference in the 2016 and 2020 US presidential elections. In February 2022 presumably Russian intrusions into and disruption of Ukrainian cyber networks was thought to be a hybrid warfare precursor to the eventual conventional attack. In addition, President Biden accused Russia of planning false-flag operations to make it appear that Ukraine was the aggressor.

The Russian government may not always conduct these and other cyber or information operations "hands on." Unofficial, and somewhat mysterious, groups of "hactivists" or cyber criminals are suspected to have conducted some of the more sophisticated digital attacks originating from Russia (presumably with the "deniable" knowledge and approval of the Russian government). One Russian group was responsible for multiple "ransomware" attacks in the United States and elsewhere and President Biden insisted to Russian president Vladimir Putin that whether these attacks were "private" or not, the Russian government would be held responsible for not controlling them. In any case, as Ariel Cohen and Robert Hamilton have noted, Russia's war against Georgia in 2008 marked "the first time in history that Russia has used cyber war and information operations in support of its conventional (military) operations."[14] Russia's cyber depredations in Ukraine continued and were likely part of their 2022 invasion. More was to follow, and Russia along with the United States and China as well as other state and nonstate actors, would be heavily invested in various kinds of attacks on infrastructure, computer networks, and other kinds of cyber war.

Suspected or alleged attacks on governments' networks and information systems are now reported so frequently that they have begun to seem normal in international relations. The question of whether cyber attacks are equivalent to warfare in the traditional sense is a complicated one for many reasons, including the fact that many cyber strikes originate from unknown or difficult to prove sources. In addition, cyber attacks may not do physical damage unless they indirectly shut down the functioning of infrastructure such as power grids, water supplies, and emergency response networks. The US Department of Defense is studying the relationship between cyber and traditional (predigital) warfare in order to determine conditions under which it would be appropriate to respond to cyber attacks with kinetic retaliation or respond to a kinetic attack with a cyber attack. There are strong indications that the United States responded to the Iranian destruction of a US drone over the Persian Gulf in the summer of 2019 with a limited cyber attack on Iranian military installations. In addition, it has been suggested that cyber war may become a new "equalizer" of the weak against the strong—perhaps another means of asymmetrical warfare. From this perspective, proficient computer users, even those living in less developed regions, can conduct information warfare effectively against the most advanced countries. Indeed, the United States may be particularly vulnerable to this kind of attack given its industrial and military dependence on information-age technology.[15]

Third, various types of conflicts can take place in one area at any given time. That is, ethnic conflicts might be taking place while terrorism, extrastate intervention, or conventional invasions are occurring. Even more complicating is the fact that information warfare can take place against an adversary in the same conflict arena. This multiconflict scenario makes it difficult not only to pinpoint the specific adversary but also to undertake conflict resolution. The problem of mission creep also arises, whereby the specific mission assigned to the military becomes enlarged not by choice but by one's efforts to succeed. In the process, the military involvement expands uncontrollably and can lead to unacceptable consequences. In the conflict spectrum, conflicts are categorized as low intensity or high intensity primarily for policy and strategy purposes, but also in an attempt to distinguish the degree of mobilization and involvement of US forces. The intensity level should in no way be construed as representative of the actual combat area. For US military personnel (as well as for their adversaries), personal combat is high-intensity. Too often such categorization is merely a policy posture, with little relevance to the conflict environment.

US competence varies among conflict categories. At the low end of the spectrum, the United States is capable and reasonably effective in military OOTW (Operations Other Than War), stability operations, or whatever new missions short of war are labeled. Most of these military contingencies presume that operations are not likely to involve serious combat (or any combat at all). At the opposite end of the spectrum, the United States remains well positioned in nuclear weaponry and strategic forces to deter most adversaries. With the end of the Cold War, the likelihood of a major war in Europe or a nuclear exchange between major powers had appeared to diminish, but deteriorating political relations between the United States and Russia after 2007 included clashes over issues including missile defenses, Russian annexation of Crimea, destabilization of eastern Ukraine, and military intervention in Syria.[16] China's growing strategic nuclear and other long-range strike capabilities are reason for concern, including its growing potential for anti-access, area denial (A2AD) and its arsenal of ballistic missiles.[17] In any event, the United States must retain a credible deterrent to counter the use of weapons of mass destruction. Similarly, the advanced capability of the United States in conventional conflict was well demonstrated in the 1991 and 2003 wars against Iraq. In the seminal 1991 conflict, operational principles derived from a European-oriented battle scenario were fused with advanced technology, such as stealth, for long-range precision strike, command-control, and other missions.

The air war over Afghanistan in response to September 11 demonstrated a remarkable capability to use high-tech aerial weaponry to incapacitate ground-based forces deployed in difficult terrain. Still, airpower alone was not sufficient. It took US Special Operations Forces, Central Intelligence Agency (CIA), and Northern Alliance ground forces in Afghanistan to unseat the Taliban in 2001. It is in the vast middle area on the spectrum—unconventional conflicts—where the United States is at a disadvantage. Most contemporary conflicts are included in this category, ranging from revolution and terrorism to conflicts associated with coalitions of drug cartels and revolutionary groups. The tendency is to see such conflicts in terms of commando-type operations or special operations shaped by counterterrorism contingencies, but there is much more to them than this suggests.

The label *unconventional conflicts* as used here refers to conflicts that include a number of characteristics in which the US adversary employs strategy and tactics that do not correspond to US conventional force dispositions, strategy, and tactics. As a result, the United States can challenge an adversary on the ground and in the air using conventional means, but that adversary might focus on unconventional strategy and tactics, including terrorism, information warfare, and other means. In the case of information warfare, an adversary with information-age technology—a computer hacker, for instance—may be able to do significant harm to the United States, including its economy, corporations, and the US government itself. At the same time, however, US technology may have little ability to overcome a determined adversary owing to the adversary's limited dependence on information-age technology.

It is important to examine unconventional conflicts more closely because they are likely to be based on strategic cultures that do not reflect Western cultural norms or classic European-type scenarios; neither do they necessarily follow the American way of war. Moreover, competency at a minimum and, preferably, proficiency in unconventional conflicts and contingencies in the middle areas of the conflict spectrum are necessary if the US military is to be effective in those areas.[18] This will remain a difficult problem for the foreseeable future.

Unconventional Conflicts and Elite Units

The history of the United States, dating even to the pre-revolutionary period, is filled with the exploits of elite units undertaking unconventional operations. From Rogers' Rangers in the American Revolution to the First Special Service Force of World War II and the Tenth Special Forces in Korea, to the Green Berets of the Kennedy era, to the Special Operations Command of the contemporary period, the US military draws from an honored legacy of special operations and low-intensity conflict. Yet, too often this experience was placed at the periphery of classic military education, professionalism, and strategy—viewed more as curiosity than curriculum—and was generally considered tangential to real issues of war and peace.[19] Increased demands for preparing for conventional war with a near peer competitor may well result in even less attention being paid to preparing for low-intensity conflicts.

In the 1980s, a new counterinsurgency era emerged and, with it, an increased interest in special units and special operations. Resources committed to special operations increased in terms of logistics, weapons procurement, and personnel. The creation of the First Special Operations Command in 1982 was a major step in creating a permanent special operations capability within the military. In 1986, Congress provided for an assistant secretary of defense for special operations and low-intensity conflict and established a unified command for all US Special Operations Forces (Army, Navy, Air Force, and Marines). This unified command, the US Special Operations Command (USSOCOM), was established at MacDill Air Force Base, Florida, in April 1987, as a unified combatant command, commanded by a four-star general with the title of commander-in-chief. (Since the George W. Bush administration, this title, CINC, has been superseded by combatant commander, or

COCOM). USSOCOM has four service component commands: Army Special Operations Command (USASOC), Ft. Bragg, North Carolina; Naval Special Warfare Command (NAVSPECWARCOM), Coronado, California; Air Force Special Operations Command (AFSOC), Hurlburt Field, Florida; and Marine Forces Special Operations Command (MARSOC), Camp Lejeune, North Carolina; plus one sub-unified command, Joint Special Operations Command (JSOC), Ft. Bragg, North Carolina.[20] Missions assigned to USSOCOM include: civil affairs; counterinsurgency; counterterrorism; countering weapons of mass destruction; direct action; foreign humanitarian assistance; foreign internal defense; unconventional warfare; and preparation of the environment.[21] Acting Defense Secretary Patrick Shanahan, presiding at a retirement ceremony in 2019, noted that special operators make up only 3 percent of the joint force but work in more than ninety countries around the world, and added that special operations forces were the "lethal tip of our spear."[22] Although US special operations forces have expanded to some 70,000 personnel across all arms of service, early post–Cold War organizational structure largely remains in place and has been adapted for the post–September 11 world. The organization of the Special Operations Command is shown in Figure 3.2.

US special operations forces truly emerged as fully empowered within the US military command structure after September 11, 2001. US military intervention to depose the Taliban regime in Afghanistan in 2001 and the US struggle against insurgency in Iraq following the end of conventional war in 2003 both highlighted the lead role of special operations forces and their intelligence support in the global war on terror and in military counterinsurgency and counterterror operations. Previously a supporting command, USSOCOM gained authority for mission planning and a stronger budgetary position within the Pentagon as a result of its numerous post–September 11 successes and high public visibility. Ironically, some of the success stories for US special operations forces could not be told because of their highly classified nature and the extent to which intelligence and special operations were commingled in order to perform timely counterinsurgency and counterterror missions.

As USSOCOM has expanded its size and diversified its missions, one of its most important contributions has been in providing advice and training to foreign militaries, especially in states under siege from terrorists and revolutionaries. This mission of foreign internal defense is especially important in states where a large US military footprint and a US takeover of responsibility for combat missions would be unwise and counterproductive, allowing terrorists and insurgents to make the United

Figure 3.2 Special Operations Command

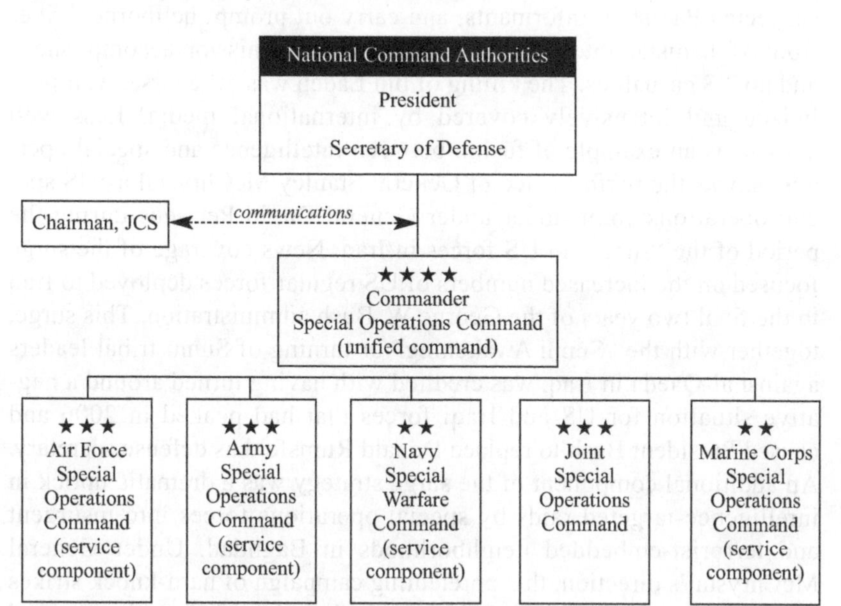

Source: Adapted from "USSOCOM Org Chart," Naval Special Warfare Command, 2011. Available at www.public.navy.mil/nsw/documents/USSOCOM_OrgChart.pdf.
Note: The stars indicate the military rank of the commander.

States the issue. Thus, for example, the United States worked with regional allies to defeat the Islamic State (ISIS) caliphate in Syria and Iraq by providing air power, intelligence, and training for friendly forces, but limiting the involvement of US ground forces in combat missions.

During the George W. Bush and Barack Obama administrations, an unprecedented convergence took place between intelligence operations and the missions assigned to the most elite unconventional warriors of the United States. Special forces became more like spies, and spies more like special warriors, in terms of their assigned roles and missions against terrorism and insurgency. The most dramatic example of mission convergence between soldiers and spies was provided by the successful raid to kill or capture al-Qaeda leader Osama bin Laden in May 2011. Then US CIA director Leon Panetta (later appointed secretary of defense) exercised overall operational command from Washington with support from the military operational chain of command, including then secretary of defense Robert Gates. Both CIA and military tactical intelligence assets were employed to locate bin Laden's hideout, ascertain his timely whereabouts, disguise US intentions from al-Qaeda and from suspected Pakistani informants, and carry out prompt heliborne strikes from Afghanistan into Pakistan and return with "mission accomplished" and no US casualties. The killing of bin Laden was, of course, well publicized and intensively covered by international media. Less well known, as an example of fusion between intelligence and special operations, was the performance of General Stanley McChrystal as US special operations commander under General David Petraeus during the period of the "surge" in US forces in Iraq. News coverage of the surge focused on the increased numbers of US regular forces deployed to Iraq in the final two years of the George W. Bush administration. This surge, together with the "Sunni Awakening" or turning of Sunni tribal leaders against al-Qaeda in Iraq, was credited with having turned around a negative situation for US and Iraqi forces that had peaked in 2006 and caused President Bush to replace Donald Rumsfeld as defense secretary. An additional component of the surge strategy was a dramatic uptick in intelligence-targeted raids by special operations forces into insurgent and terrorist-embedded neighborhoods in Baghdad. Under General McChrystal's direction, this unrelenting campaign of hard-knock strikes against cadres of resistance fighters (whether classified as "insurgents" or "terrorists" is a matter of semantics) crushed much of the military resistance in Baghdad and thus provided breathing space to the struggling and politically fractured Iraqi government. The special operations uptick within the overall surge was of necessity kept under wraps.

A variety of regular military missions short of war overlap with the role of special forces. Conventional and unconventional warriors will watch one another's progress with competitive rivalry. One report even prior to 9/11 noted: "Threatened with irrelevance by changes that allow conventional forces to conduct missions that were once its exclusive preserve, the Green Berets are refocusing on unconventional warfare."[23] This was the theme of a one-year study by the Army Special Forces as reported in *Army Times* in July 2001. "The changing strategic environment makes unconventional warfare a vital mission for the early 21st century," according to those involved in the study. The report noted, however, that "the shift will require changes in training, and could put Special Forces leaders on a collision course with superiors at US Special Operations Command."[24] By the middle of the Obama administration, special operations forces and the CIA were operating with unprecedented coordination in planning and carrying out counterterror missions, and as far as can be determined, continued this cooperation under the succeeding Trump and Biden administrations.

Conceptual Coherence

The most important aspect of developing an effective US political-military posture for unconventional conflicts is conceptual clarity and coherence. During the counterinsurgency era of the 1960s, a variety of terms came into use to provide some analytical precision to the new-found form of warfare: *insurgency, counterinsurgency, special warfare, guerrilla warfare, wars of national liberation, people's wars,* and *internal conflicts.* The 1980s saw the revival of many of these terms and added some new ones, such as *special operations, low-intensity conflict, small wars, low-level wars, operations short of war,* and *secret armies.* With the focus on combating international terrorism, *terror* and *counterterror* have become part of the special operations lexicon. In the 1990s and beyond, a variety of terms have been added, including *stability operations, peacekeeping, peacemaking, peace enforcement, humanitarian missions,* and *OOTW.* Lost are older notions of guerrilla war, insurgency, and revolution: now the tendency is to characterize many such conflicts as *ethnic, religious,* or *nationalistic.*[25]

A case in point is *revolution,* intended to achieve a strategic goal, with both strategic and tactical dimensions. It encompasses political-psychological, as well as social and economic, components and is aimed at the entire political-social order of the existing system. Revolution is a complex, multidimensional phenomenon with origins in the political-social system and a strategy based on a sweeping attack against the existing order. Revolutionary tactics range from terror, assassinations, hit-and-run raids, robbery, and kidnapping to the use of armed force in the conduct of conventional-type operations. The center of gravity of the conflict is not usually in the armed forces but rather in the political-social milieu of the existing system. The revolutionary strategy and tactics employed are shaped accordingly. As Samuel B. Griffith noted, "A revolutionary war is never confined within the bounds of military action. . . . For this reason it is endowed with a dynamic quality and a dimension in depth that orthodox wars, whatever their scale, lack. This is particularly true of revolutionary guerrilla war, which is not susceptible to the type of superficial military treatment frequently advocated by antediluvian doctrinaires."[26]

Some groups will use revolutionary strategy and tactics to achieve particular political goals that sometimes fall short of taking over the state. This is seen in ethnic, religious, and hypernationalistic conflicts, such as that in the Balkans in 1999–2000 as well as the conflicts in Iraq and Afghanistan in 2006 and beyond.

Terror and Counterterror

In this new era, cyber terrorism has become a new phenomenon. Signs already suggest that terrorist groups will try to gain strength and extend their reach by organizing into transnational networks and developing swarming strategies and tactics for destroying targets, entirely apart from whether or not they can hack into a target's computer system.[27]

In the broader scheme, there are at least four categorizations of terror: terror-qua-terror, revolutionary terrorism, state-supported terrorism, and state-sponsored terrorism.[28] All of these can have international dimensions and also be linked to other groups such as drug cartels.

Terror-qua-terror is violence for the sake of violence, combining tactical means and strategic purposes; that is, the terrorist act is perceived as an end in and of itself. The aim is to strike at the system, to gain recognition for one's terrorist group, to achieve a moral victory, to fulfill a mission in life (e.g., in a religious way), or all of these. There is very little concern about anything beyond the act or state of terrorism itself. The Islamic State, which came into international prominence in 2014, exhibited some of the preceding characteristics, but in other ways they differed from this model. ISIS sought to establish a territorial state in parts of Syria and Iraq and to use that as a nucleus for international recruitment of supporters. Although US-supported Iraqi government and Kurdish forces destroyed the ISIS caliphate in Iraq in 2019, ISIS as an international movement was very much alive in numerous countries, including Afghanistan. And the "idea" of ISIS was also transnational, propelled by the organization's relentless use of social media to spread its message and to obtain new recruits. In earlier years, factions of the Baader-Meinhof Gang, the Red Brigade, and Direct Action more neatly fit the typology of terror-qua-terror. US examples included the Symbionese Liberation Army and the Weather Underground Organization.[29]

This is also characteristic of so-called new terrorism. According to one account, "the new terrorism . . . appears pointless since it does not lead directly to any strategic goal, and it seems exotic since it is frequently couched in the visionary rhetoric of religion. It is the anti-order of the new world order of the twenty-first century."[30] However, the new terrorism of the twenty-first century also exploits contemporary technology of the Internet, including social media, to magnify its message and to construct "communities" of collaborators and anti-system revisionists. Thus, for example, neo-Nazis in the United States have been able to connect with like-minded hate groups in Europe, and Russian hackers were able to use online "crowd sourcing" techniques to create political gatherings in the United States, motivated by purposely disseminated fake stories and rumors.[31] False information on the validity of the 2020 presidential election spread online contributed to the attack on the US capital on January 6, 2021.

Terrorist acts have also been used by individuals or groups as statements against the government. This appears to be the case involving Timothy McVeigh and the bombing of the Murrah Federal Building in Oklahoma City in April 1995: "The bombing introduced many Americans to the disturbing underworld of domestic terrorism while shattering whatever illusions existed about terrorism as only an international or overseas threat."[32] Indeed, domestic terrorism was very much in the news and a subject of popular culture throughout the years of Bill Clinton's presidency, including confrontations between federal authorities and dissidents at Ruby Ridge and Waco. However, after the September 11, 2001, attacks the public and government attention quickly turned to foreign sponsored terrorism. Domestic terrorism did not go away, however. On the contrary, both the Obama and Trump years in the White House were marked by tragic shootings in schools, shopping malls, or other "soft" domestic targets across America. In early August 2019, two mass shootings within thirteen hours occurred in El Paso, Texas, and Dayton, Ohio, with some twenty-two killed in the former attack and another nine in the latter. In the summer of 2020, some peaceful demonstrations turned into riots with looting, arson, and attacks on police in cities across America, adding to a national malaise already created by the viral Covid-19 pandemic.

US government officials and experts disagreed about the causes of domestic terrorism and, therefore, about the most appropriate responses. Some called for more restrictive gun control measures, including universal background checks and banning military style weapons. Others blamed mental illness and inadequate treatment facilities and protocols for the carnage. On one hand, deaths by gang violence in major US cities such as Chicago and Philadelphia received little national media or government attention. On the other hand, the "rampage" mass shootings by demented or angry individuals in small towns and suburban communities were heavily covered by the media and provoked noisy government and media debates. The challenge was to find common causal threads among the confusing particulars of shootings across the country. For example, in some cases killers left manifestos on blogs that gave clues as to their reasons for taking such inhumane actions. In other instances, research revealed a pattern of psychological problems or social maladjustment that predisposed someone toward violence. But there were few if any candidates for a universal paradigm that could encompass a majority of, let alone all, cases of domestic terrorism. The unrelenting pattern of attacks on vulnerable domestic targets in a free society will require all institutions, including schools, churches, and other formerly safe spaces, to undergo security upgrades and preparedness for unthinkable horror inflicted by Americans against other Americans.

The September 11, 2001, attacks in New York and Washington, DC, dramatically revealed to all Americans the reach and efficiency of some international terrorist groups. They also shattered some illusions regarding international terrorism. These attacks appeared to be well organized, well financed, and well executed. Al-Qaeda attackers on September 11 created more casualties in the US homeland than any other single attack by a foreign power or group.

Revolutionary terrorism is a strategic as well as a tactical instrument of the revolutionary system: that is, terrorist tactics and strategies are designed to further the revolutionary cause. The terrorist instrument is usually under the control of revolutionary leadership, and terrorist operations are conducted so as to avoid, as much as possible, alienating the very people at which the revolution is directed. Nonetheless, in light of the international scope of revolutionary strategy, terrorists can cross over into the terror-qua-terror category in the name of revolution. In recent years, ISIS adopted some characteristic features of revolutionary terrorism and other aspects of the terror-qua-terror variety. ISIS certainly sought to overturn the existing international power structure and to destroy states or cultures that it regarded as antithetical to its mission and purpose. On the other hand, ISIS also sought to retain a territorial caliphate in parts of Syria and Iraq. Al-Qaeda perhaps fits more exactly the category of revolutionary terrorism. The attacks of September 11 were part of a global jihad against enemies of al-Qaeda, including so-called Islamic apostate regimes such as Saudi Arabia. Al-Qaeda took transnational terrorism to a new level by combining medieval fanaticism with a mastery of using modern technology for malign purposes. According to historian Max Boot:

> Bin Laden may have advocated a return to a medieval brand of Islam, but he showed a genius for using sophisticated technology and management techniques to marshal the first truly global insurgency. And he brought to his side many similar people who, like him, had little religious training but were well versed in technical subjects and familiar with the ways of the modern world. Like drug traffickers, computer hackers, and other international criminals, they represented the dark side of globalization in the twenty-first century.[33]

There may be some objection to the use of the term "revolutionary" for movements that intend to turn back the historical clock instead of moving it forward into some distant utopia or paradise. Perhaps the proper reference for movements such as al-Qaeda that find paradise in retro history would be "reactionary" instead of revolutionary. But this objection is itself subject to qualification because it is a very Western way of thinking and perhaps guilty of perceiving another civilization through eyes of our own. So "revolutionary" in this discussion refers to the scope of the ambition held by al-Qaeda and other movements with similar aspirations—revolutionary, as opposed to incremental or gradual. In that sense, the revolutionary character of Islamist radicals is noted by US scholar and expert on defense studies Eliot A. Cohen:

> The Islamist movements are vanguard organizations: that is, they view themselves as an elite among a much larger and supportive public. And they have some limited evidence to support their self-understanding. While most Muslims have a negative view of al-Qaeda, for example, percentages ranging from 5 percent in Turkey to 23 percent in Bangladesh have a favorable view of that group. Overwhelmingly a minority in absolute terms, yet percentages like these indicate that millions of the world's Muslims are at the very least susceptible to the jihadist message of al-Qaeda, let alone those of other, more successful groups.[34]

The Palestine Liberation Organization (PLO) in former years, various groups operating in the Middle East such as Hezbollah, the Vietcong during the Vietnam War, and revolutionaries in El Salvador (e.g., the Farabundo Martí Front of National Liberation and the Popular Revolutionary Army) can generally be categorized as revolutionary terrorism. Perhaps one of the clearest examples is Sendero Luminoso (Shining Path) in Peru. In early 2000, many felt that Sendero Luminoso had been all but eliminated by the efforts of Peruvian president Alberto Fujimori, yet it reemerged in 2002. In Greece, the actions of the November 17 group in 2000 may well be a version of revolutionary terror, although the group has been active on and off for decades.

Another dimension in this category is the coalition between drug cartels and revolutionary groups, as in Colombia, where the United States has been involved militarily as well as financially.[35] In this century, US pacification efforts in Afghanistan have also been confounded by relationships between regional and tribal warlords and traffickers benefiting from the lucrative drug trade in that country. This coalition in crime is mutually advantageous to the drug lords and the revolutionaries, at least in the short to middle term. The drug cartels pay protection money to the revolutionaries in order to protect drug operations in outlying areas. The revolutionaries provide protection in order to maintain a dependable financial source. This linkage has created internal turmoil and havoc, with the United States becoming increasingly involved. In the broader sense, revolutionaries can take a certain pleasure, as well as strategic advantage, in seeing Western systems, such as the United States, undermined by drug trafficking and drug use. Yet the underlying philosophical basis of revolution is contrary to the notion of long-term dependence on drug operations. From another perspective, individual revolutionaries are not immune from drug addiction. Another variant is the relationship between organized crime groups and revolutionary groups.

State-supported terrorism is difficult to identify because it is often difficult to directly link terrorist activity to states. This is a result of the states' ability to hide their support and involvement using a variety of techniques.[36] US and allied NATO operations in Afghanistan have been hampered by Pakistani military and intelligence support for Taliban fighters who cross over the Afghan-Pakistan border with impunity to sanctuaries in Pakistan. Pakistan has supported the Taliban in Afghanistan as a hedge against the uncertain outcome of the US and NATO involvement in Afghanistan and against a power vacuum that would permit increased influence by India over the political destiny of Afghanistan. The Cold War Soviet Union and China supported revolutionary movements against the United States and its allies, and the United States supported the overthrow of governments in Iran, Guatemala, Chile, and other countries as part of US "containment" strategy against the Soviet Union. The United States supported the mujahidin fighters against the Soviet occupation of Afghanistan (1979–1989).

State-sponsored terrorism is one in which a particular state actually organizes, trains, and finances terrorist groups. This includes identifying terror targets and providing recruits. Furthermore, sponsor states are able to engage in a form of psychological warfare by controlling information about their own system while gaining access to the media of open systems to broadcast their own messages. Some states engage in terrorism against their own citizens in order to maintain control or neutralize dissidents and resistance groups: totalitarian and authoritarian systems are especially noted for such activity (e.g., Nicaragua under Sandinista rule, Zimbabwe, and the former Soviet Union).[37] In the Middle East, Iraq under Saddam Hussein, Iran, Libya, and Syria have been accused of such operations. Similar accusations were made against governments across North Africa and the Middle East during the Arab Spring protests of 2011.

Some have also identified a new dimension, religious terrorism, that is linked to the new terrorism. As one observer noted: "In 1998, when Secretary of State Madeleine Albright announced a list of 30 of the world's most dangerous groups, over half were religious and included Judaism, Islam, and Buddhism. If other violent religious groups around the world were added—including the many Christian militia and other paramilitary organizations found in the United States—the number of religious terrorist groups would be considerable."[38]

The terrorist attacks on the United States in 2001 clearly illuminated the threat of religious terrorism. This has become even more pronounced and more complex in the aftermath of the US involvement in Iraq.

These labels are complicated by the fact that terrorism can be multidimensional. That is, a terrorist act can be a combination of all the categories discussed here. Add to this the possibility of cyber terrorism, and it becomes a difficult challenge for states to design effective counterterrorism strategies.

On the international level, many countries do not hesitate to define terrorist acts such as kidnapping, hostage-taking, assassination, bombings, armed robbery, and so on, as criminal in nature. Other states, especially in the developing world, resist any definition; however, that may have legal implications restricting the activities of groups fighting against neocolonial regimes. Moreover, some states see the use of terrorism as a low-risk, high-return policy affording an opportunity to strike the West, especially the United States. This lack of definition also reflects the view that one person's terrorist is another person's freedom fighter. Unfortunately, such a perspective ignores the characteristics of terrorist acts and the impact on their victims. Furthermore, this view is based on convoluted moral principles that elevate assassination and murder to humanistic ventures.[39]

Unconventional Conflicts

There is a great deal of published literature on the nature and character of unconventional conflicts, ranging from Sun-tzu and Mao Tse-tung to Che Guevara and Vo Nguyen Giap.[40] The concept now includes a variety of efforts short of taking control of the state and establishing a revolutionary government. An in-depth study of revolution and counterrevolution would require serious reading of the major works. For our purposes, it is important to identify five major characteristics that have an especially significant bearing on the ability of open systems to come to grips with unconventional conflicts.

First, unconventional conflicts are asymmetrical and, for those involved, they are "total" wars. Revolutionaries feel that they are waging a life-or-death struggle with the existing system. Terrorists often see themselves as engaged in the ultimate struggle: bringing death to themselves, if need be, to achieve their goals. For major state actors, however, involvement in unconventional conflicts is often constrained in the choice of political objectives and restricted by the outer limits of domestic popular support. More important, the nature of open systems limits strategy, tactical operations, and overall military effort.

Second, unconventional conflicts tend to be protracted, as the strategies are long-range. Those who demonstrate infinite patience and persistence are more likely to be successful: the classic example is North Vietnam, which had the staying power to outlast a technologically advanced superpower. Revolutionaries thus often adopt long-term strategies to erode the ability of the existing system to govern, and tactics and doctrines are designed accordingly. Revolutionaries do not necessarily choose a protracted conflict: they are forced into adopting such a strategy because of the initial power of the government they oppose. Some revolutionary ideologues idealize the nature of the conflict, even more than outright success. In such cases, the protracted war is used to mobilize the masses, establish a revolutionary system, and create a revolutionary myth. Contrary to popular images, many revolutionaries live difficult, not glamorous, lives, and are constantly in fear of capture or death.

Third, unconventional conflicts are also tactically unconventional. This type of conflict is not necessarily ruled by Clausewitzian principles that place the center of gravity within the armed forces of the state; Sun-tzu's formulations are more relevant.[41] The Chinese writer emphasized deception, psychological warfare, and moral influence. Combined with hit-and-run raids, assassinations, ambushes, and surprise attacks, the tactics of unconventional conflicts are difficult for conventionally trained and postured forces to counter.

Fourth, ambiguity is another characteristic of unconventional conflicts. It is difficult in revolutionary-type conflicts to separate friend from foe and to develop clear criteria for determining success. As the United States learned in Vietnam, Iraq, Afghanistan, and Syria the amount of real estate taken, weapons recovered, body count (enemy casualties), and secure areas might not be good indicators of who is winning and losing. Furthermore, the rhetoric of revolution and counterrevolution is difficult to untangle, obscuring a clear understanding of the purposes of the antagonists.

Fifth, unconventional conflicts are also characterized by their high political content and moral dimension. Although it is true that all wars are political, in unconventional conflicts operations and purposes are shaped and conducted according to political-psychological goals. For example, the deliberate sacrifice of armed revolutionary elements for the sake of a political-psychological victory is not an uncommon occurrence. The Tet Offensive by the Vietcong and North Vietnamese in 1968 is a case in point. Even though the enemy forces were decimated by US and South Vietnamese military forces, the media reports to the public back home gave the impression of a great Vietcong victory. The US and South Vietnamese forces won a major military battle but totally lost the political-psychological battle. In the long run, the latter proved to be the more important of the two, as it marked the turning point in public support for the war.[42] In addition, in this war and in other unconventional conflicts, allegations of atrocities committed by conventional militaries against suspect revolutionaries or their sympathizers have been used by insurgents to compete favorably in the political-psychological dimensions of warfare. As an example, the French lost the war in Algeria between 1954 and 1961 not mainly on account of failed military operations but mostly as a result of atrophy of popular support for the resistance to Algerian independence among Algerians and in France. This weakening of popular support was partly due to the French use of extreme measures of coercion, including torture, against Algerian civilians.[43]

Another concept related to the understanding of unconventional conflicts is that of the "gray zone." Russia under Vladimir Putin has skillfully used measures within this zone of conflict in order to advance its foreign and security interests. Gray zone activities include propaganda, disinformation, economic coercion, disguised military interventions ("little green men" in Crimea in 2014), election interference, exploitation of social media, and other tactics short of acknowledged combat between state forces. Nor is Russia alone. Other US adversaries or competitors, including Iran and China, have used gray zone techniques and tactics. Cyber war is an obvious example of gray zone measures. On the other hand, the United States is not necessarily disadvantaged in this gray zone competition. As Kathleen Hicks, deputy defense secretary in the Biden administration, explains, the greatest advantages of the United States for achieving its objectives in the face of gray zone threats are tied to "strengthening its model of governance," including

(1) the fundamental tenets of the US Constitution and civil liberty; (2) vibrant civil society, business, and media communities; (3) unparalleled alliance and partner networks; and (4) significant US government potential capacity and capability.[44]

This is an extremely important observation about the correlation between the strength of our own domestic institutions and our ability to withstand adversarial thrusts with gray zone or other attacks. Americans are often unknowingly their own worst enemies. The polarization of US political debate, and the highly partisan and emotional character of much politicking nowadays, threaten the very procedures and processes that make liberal democracy so precious and unique. An example of this is the pattern of recurring attacks on free speech from both right and left sides of the political spectrum: from flash mobs shouting down speakers on college campuses to public and private employers who punish their employees for their political opinions, even if expressed in blogs or other formats outside of the workplace. Free speech exists to protect controversial speech, not platitudes. No free speech, no democracy. The framers of the US Constitution had a reason for putting the First Amendment first. Suppressing free speech makes the US weaker, not stronger, as a society and polity. Our adversaries' greatest weapon is their ability to divide Americans into antagonistic political camps who regard one another not as "loyal opposition" but as "enemies within." The same point holds true when peaceful protests are hijacked by criminals and rioters from either end of the political spectrum. Democracy is not compatible with anarchy.

In sum, the shape and dimensions of unconventional conflicts do not easily fit into long-prevailing US notions of conflict. Moreover, the principles of warfare and battlefield conduct that are part of US military professionalism do not really focus on unconventional conflicts. US military professionals understandably seek to increase knowledge and develop skills for success in battle by defeating the armed forces of the enemy. But the center of gravity of unconventional conflicts is in the political-social milieu of the contending systems. As General Bruce Palmer noted about Vietnam: "One of our handicaps was that few Americans understood the true nature of the war—a devilishly clever mixture of conventional warfare fought somewhat unconventionally and guerrilla warfare fought in the classical manner. Moreover, from Hanoi's point of view it was an all-out, total war, while from the outlook of the United States it was quite limited."[45] US military planners recognize that forms of hybrid warfare combining conventional and unconventional methods of fighting might become more the rule than the exception in future conflicts. In the twenty-first century, the attention to humanitarian missions, as well as to a variety of peacekeeping contingencies, combined with the notion of operations other than war, broadened the concept of low-intensity conflicts and special operations, at least within many policymaking and military circles. This merely adds to the confusion already characterizing unconventional conflict concepts.

From lessons learned in a historical view of the United States and guerrilla warfare and its relevance in the current period, Anthony James Joes concluded: "Dangers lie in the path ahead. To avert or at least prepare for them, Americans need to deepen and sharpen their understanding of what guerrilla war has meant and will mean."[46] In considering involvement of the United States in future guerrilla insurgency, "the presumption should be against committing US ground forces. . . . Military victory is ephemeral."[47]

US Policy and Strategic Guidelines

Open Systems and Unconventional Conflicts

Open systems direct policy in an attempt to be decent societies based on values, norms, and an ideology that stresses respect for individual rights, justice, freedom, and equality. Underpinning this is the fact that elected officials are responsible to the people and can be removed if they fail in their duties and responsibilities. And although their goals are imperfectly achieved, the fact is that open systems develop many safeguards that protect against government interference and support individual freedom.

The environment of open systems is reinforced and perpetuated by counterbalancing forces, independent sources of information, and constitutionally protected independent political actions. The collectivity is subordinate to the individual. This is in sharp contrast to closed systems and, indeed, to the nature of most revolutionary systems, of which Roberta Goren wrote: "More often than not one sort of repressive regime has taken the place of another repressive regime. Often what has been represented as a left-wing 'liberating' regime overthrowing a right-wing repressive regime is no less reactionary than its predecessor."[48]

Many Americans view revolutions either as "glorious" affairs in which a freedom-loving people rise up against tyrants (e.g., the American Revolution) or as anticolonial affairs in one form or another. Although such views carry an implicit democratic rationalization and justification for revolution, they also reveal a misunderstanding of the prevailing security environment. More commonly, elites who initiate or co-opt revolutions are intent upon establishing rule by strong central government and may fall well short of the democratic ideal. Often, well-meaning groups here in the United States and abroad mistakenly view such revolutions through the lens of nineteenth-century liberalism. Their misperceptions and distortions can become the basis for mobilizing segments of the public against the official US political-military posture.

The very character of open systems and their value base are potentially disadvantageous in unconventional conflicts, unless US citizens and their political leaders protect and strengthen those vital institutions that make American democracy unique. Personal liberty, freedom of movement, and unfettered communications—all work to the advantage of the subversive enemy intent upon using those characteristics to its advantage, unless a strong vital center in US politics holds firm against political extremism, media sensationalism, and a Manichean view of world and domestic politics. The exploitation of social media by Russia in order to influence the US presidential election in 2016 and 2020 and possibly congressional elections in 2018 are cases in point. Russia tried to influence the elections with modern tools but in pursuit of timeless objectives: using instruments of influence other than war in order to support its foreign policy objectives. So, too, does the United States. Sovereignty is supreme power under the law: that is, a monopoly on the legitimate use of force. But legal sovereignty cannot make borders impenetrable in the age of the Internet. Cyber persuasion across national borders will be more the rule than the exception as we move further into the twenty-first century. China's concept of unrestricted warfare includes information warfare and other measures of influence, apart from the actual clash of battle.[49] Sun Tzu's dictum that the acme of skill is to win without fighting has been validated again by twenty-first century experience in the cyber realm.

Policymakers are apt to design political-military responses based on bringing overwhelming numbers or qualities of US forces to bear for short-term operations designed to solve the problem in one way or another. The result becomes ad hoc, short-term initiatives using conventional US strategies, tactics, and doctrines. Although such approaches can be appropriate in some situations, the key for most unconventional conflicts is to penetrate the political-social milieu so as to influence it more effectively (the famous adage is "winning hearts and minds").

This means that success usually goes to the side with the best people on the ground, not necessarily to the one with the largest battalions or the most sophisticated and massive firepower (such as overwhelming airpower). In the long run, solutions—if indeed there are solutions—probably require staying power over an extended period. In the conduct of the US military operation Iraqi Freedom, it became clear early on that the conflict area was much more complicated than first anticipated by the US military and political leaders.

More than fifty years ago, Hannah Arendt drew a distinction between "war" and "revolution" that seemed to forecast a bull market for revolutionary wars as against the likelihood of success in conventional interstate conflicts: "Revolution, in distinction to war, will stay with us into the foreseeable future . . . those will probably win who understand revolution, while those who still put their faith in power politics in the traditional sense of the term and, therefore, in war as the last resort of all foreign policy, may well discover in a not too distant future that they have become masters in a rather useless and obsolete trade."[50]

The American Way of War and Unconventional Conflicts

Throughout history the American way of war has been at least partly based on a moral dimension: going to war had the purpose of achieving some higher good. This usually demanded a clear identification of the enemy and its alleged evil purposes. The American way of war also makes a clear distinction between war and peace. In peacetime, nonmilitary systems prevail; in war, the military prevails. Thus the presumption was a clear separation between institutions for peace and those for war. But the line between war and peace is not as distinct in the contemporary international environment. This situation throws into disarray prevailing US notions, and it creates a basic dilemma for the US military. A condition of no war, no peace denies the US military the clear political-psychological sustenance from the body politic to engage in unconventional conflicts. At the same time, the US military must continue to maintain a credible posture for nuclear deterrence and major conventional conflicts. Open systems also require that their officials and military personnel conduct themselves within the general bounds of democratic propriety. This includes conduct on the battlefield consistent with US and Western standards of military law and ethics. Lapses from that standard will weaken political support for a war, no matter how well justified the war may be in terms of US interests.

Overall strategy and tactical operations must also fit within democratic values and norms if policy and strategy are to maintain their legitimacy. Operations based on hostage-taking, terrorism, and assassinations to erode support for a revolutionary system or penetrate and neutralize terrorist groups will generally be condemned by democratic systems. The scope and intensity of intelligence activities, both domestic and foreign, are affected by these considerations. This is extremely important given the role of intelligence in unconventional conflicts.

In responding to unconventional conflicts, US forces will usually operate on foreign soil in conjunction with host nation military forces, especially in the world outside of Europe. Supporting an existing system that counters revolutionary challenges is a difficult proposition at best, and a fundamental problem arises from the fact that US personnel socialized into the norms and values of open systems invariably have difficulty relating to and understanding those of foreign cultures. It is difficult to empathize with cultures whose view of individual worth and human rights is at variance with ours. Indeed, it is conceivable that revolutionary rhetoric can strike a more responsive chord than does the ideology of the existing system. The lack of understanding of cultures makes it difficult to understand friends as well as enemies. The same problems arise when the United States supports groups within a foreign state who seek self-determination or face genocide. Then-presidential candidate Biden stated his view that force should not be used indiscriminately, even in support of noble goals: "The responsibility I have is to protect America's national self-interest and not put our women and men in harm's way to try to solve every single problem in the world by use of force."[51] This view underlay his decision to end the US military presence in Afghanistan in the summer of 2021, even though the progress made in women's rights and civil liberties generally there was put in serious peril by his action.

Finally, the end of the Cold War and the dissolution of the Soviet Union and the Warsaw Pact changed the equation for unconventional conflicts. No longer driven by East-West ideological confrontations, such conflicts are likely to reflect indigenous issues, such as ethnic confrontation, religious freedom, and minority-group autonomy, or as Samuel Huntington asserted, "the clash of civilizations."[52] Conflicts can occur because elites or controlling groups try to gain power and impose a set of political rules and procedures derived from their version of society. Yet unconventional conflicts arising from the Cold War period may continue in one form or another, as in the Middle East, Southeast Asia, parts of Africa, and Latin America, where some conflicts are the result of drug cartel–revolutionary conspiracy. Although unconventional conflicts generally evolve from the political, social, and economic turmoil in less developed states, wherever a power vacuum exists—or where the United States is at a distinct disadvantage—regional powers, ethnic and religious groups, nationalistic revivals within states, and indigenous revolutionary groups will attempt to fill that vacuum.[53]

Thus the very nature of open systems makes it difficult to respond to unconventional conflicts, including conflicts short of war, stability operations, and the like. The increasing level of attention that the United States must devote to preparing for conventional wars against another advanced state (or a less developed state with strong military capabilities, such as North Korea) means that US strategy is properly shifting to reflect a greater emphasis on these concerns. The US record of learning from past conflicts is mixed, at best, and it would be unfortunate if more pressing concerns caused the lessons of unconventional warfare to be lost, as well. This is well argued in Sam C. Sarkesian's classic study of four "forgotten" wars: the Seminole War, the Philippine War, the Punitive Expedition into Mexico, and Vietnam.[54]

Policy, Strategy, and Military Operations

Virtually any commitment of US military forces abroad now has the potential to turn into an unconventional conflict. It is important, therefore, that US policymakers and the military be prepared for such conflicts, even if the mission is peacekeeping. The conflict spectrum (see Figure 3.1) provides a view of the conflict realities: they have clear political-military implications, especially for unconventional conflicts.[55] Responses can demand a political-military posture that is contrary to the American way of war, as well as conventional principles and doctrine. This requires training that goes beyond purely military skills. As President John Kennedy stated in response to wars of national liberation, "pure military skill is not enough. A full spectrum of military, para-military, and civil action must be blended to produce success. . . . To win this struggle, our officers and men must understand and combine the political, economic and civil actions with skilled military efforts in the execution of this mission."[56]

Although Kennedy was addressing the armed forces in 1962, his words are appropriate today. The same theme was proclaimed by President George W. Bush in the aftermath of September 11 when he successfully urged Congress to create a new Department of Homeland Security. That created the third-largest department in the national government and marked the most extensive reorganization of the government since the Truman presidency.

An effective force structure for responding to unconventional conflicts cannot be wedded to conventional hierarchy or command systems.

It requires planning, organization, and operations aimed at the political-social milieu of revolutionary-counterrevolutionary systems. It demands individuals with the requisite military skills who also understand and are sensitive to the cultural forces and nationalistic desires of foreign systems—especially those in the less developed world. They must be self-reliant individuals capable of operating for long periods in small groups isolated from the US environment. It will require patience, persistence, political-psychological sophistication, and the ability to blend in with the indigenous political-military system to prevail.

In this respect, success in revolutionary and counterrevolutionary conflicts is not necessarily contingent upon sophisticated weaponry, large numbers of troops, and massive airpower. Rather, success depends on the quality and dedication of efficient soldiers on the ground who can blend in and function as skilled political mobilizers and teachers, following the perspective of Sun-tzu.

Conclusion

What does all of this mean for US national security policy and strategy? First, in order to respond effectively to national security challenges anywhere on the conflict spectrum, policy must be based on US national interests. Although this seems obvious, it is not so clear with respect to unconventional conflicts and the less developed world. In this context, national interests and policy should include support of like-minded systems. This does not mean that support should only be extended to democratic or quasi-democratic systems. There are not many of those systems in the third world, and even in Europe some are moving in a different direction. True democratic systems need little help from the United States, and limiting US interests to that type of system is tantamount to withdrawing from the less developed world; however, this does not mean that the United States should become involved in every corner of the planet. In fact, US involvement and visible support could tarnish and undermine foreign nationalistic leaders, and in some areas US involvement would achieve little or would make matters worse.[57] US attempts to create a democratic Iraq and Afghanistan are instructive in this regard.

When compelling US national security interests are at stake, and where long-term interests can be seriously affected, US involvement should include support of states and groups who are like-minded or are the lesser of evils: this can include support of revolutionary systems.

However, the United States should never substitute one tyranny for another.[58] It is conceivable that nominally democratic systems, even certain authoritarian systems, are more susceptible to openness than revolutionary systems. The term used to describe this strategy, in which the United States is compelled to choose among undesirable options, is *suboptimizing*. Suboptimizing accepts the notion that the logic of policy and strategy is often subsumed by actual conditions that defy rational, logical, or just solutions.

Second, it follows that strategies must include a variety of options and phases (see Figure 3.1 on the conflict spectrum) that incorporate political, economic, and psychological components as well as military ones. The use of US ground combat forces must be reserved for special situations in which the area is vital to US security interests. However, if other options are effectively implemented, the use of US combat forces could be the exception rather than the rule. Strategic options should be based on support and assistance to expand the governing capacity of the existing system, to broaden its political-psychological base, and to develop a civic culture attuned to openness. Such strategies must not Americanize the conflict or the theater itself, except in unusual circumstances requiring direct involvement of conventional US combat forces.

In this regard, official or otherwise widely held theories of counterinsurgency should come with warning labels. Each generation has its own preferred theories of insurgency and counterinsurgency, or terror and counterterror. When we, the authors, were students, Mao's theory of protracted war caused fears in Washington and palpitating hearts among student radicals. In the 1970s and 1980s, various challenges to US influence in Latin America called forth demands for more effective counterinsurgency and counterterror, against Soviet-supported or otherwise reputedly Marxist revolutionaries. Fast forward to the twenty-first century, and US experience in Iraq led to rethinking of assumptions about counterinsurgency, led by General David Petraeus and his colleagues at the US Army Command and General Staff College in Ft. Leavenworth, Kansas. The impact of Petraeus and his collective of smart colonels on military thinking led to changes in practice applied by Petraeus and others in Iraq and Afghanistan.

Notwithstanding these shifting sands of prevalent thinking about counterinsurgency among US military professionals and others, it remained elusive to pin down exactly what counterinsurgency was, and under precisely what conditions it would work effectively. As Douglas Porch has shown, US approaches to counterinsurgency were often borrowed from British and French colonial experiences, and the history of those experiences written by participants was not infrequently self-serving and myopic.[59] Colonial counterinsurgency and counterterror operations were based on condescending assumptions about non-Western civilizations and were marked by brutal and insensitive treatment of subject peoples. Legendary heroes of counterinsurgency narratives benefited from grade inflation with respect to their actual accomplishments or distortion of their actual records. Thus, France "won" the Battle of Algiers although it ultimately lost the war against the National Liberation Front and affiliated groups, just as the United States supposedly never lost a battle in Vietnam, but nonetheless was forced to make a face saving exit via the declaratory policy of "Vietnamization" under President Nixon.[60] As Porch noted: "The United States failed because, like France, it inherited a strategic framework that made victory an uphill struggle, even had the opponents been less skilled and resilient than those it faced in Indochina."[61]

Third, strategies should be based on civilian-military cooperation and interservice coordination. Command structures, planning, organization, and implementation must reflect these and joint efforts. This is especially true with respect to the intelligence function. Effective intelligence requires not only military but also strategic intelligence and analysis associated with intelligence capabilities. Strategies must include counterterrorism and counterrevolutionary operations in support of existing systems and revolutions against repressive systems that promise to become harsher still. Whether such a strategy could be implemented would depend on the assessment of US national interests.

Fourth, a new realism must emerge regarding war and peace and the challenge posed by unconventional conflicts. Within military and civilian policy circles and the body politic (especially the media), there must be greater appreciation for the complexity of the conflict spectrum and a recognition of the long-term threat of unconventional conflicts. Without this new realism, it is unlikely that the necessary national will, political resolve, and staying power can be developed to respond effectively. History will not always serve up a September 11 moment that provides temporary national unity and anger directed against amorphous foes. The US war in the shadows against terrorism of various kinds continues, often invisible, unremarked and unnoticed, until the next fiasco.

Fifth, to dismiss unconventional conflicts as nonthreatening phenomena and a natural evolution of political-social turmoil is to ignore the lessons of recent history. To dismiss the susceptibility of open systems to deception, political-psychological warfare, terrorism, and totalitarian revolutionaries as unrealistic is to create a condition described by

Jean-François Revel: "Democratic civilization is the first in history to blame itself because another power is working to destroy it."[62] This may well have changed after September 11, when Americans united at least temporarily to combat international terrorism. This relates to the critical issue of staying power and national will over the long haul. Demonstrated Russian intervention in US elections may also be a catalyst for taking nontraditional security threats more seriously.

The basic dilemma for the US military grows out of these conditions, especially the fact that US norms and values do not easily match the policy and strategy required for unconventional conflicts. The traditional view of the American way of war virtually precludes a military postured for the kinds of tactical and doctrinal techniques inherent in unconventional conflicts. Fighting without appearing to fight, waging war through peaceful enterprise, using morally acceptable tactics and doctrine against adversaries who abandon any pretense of moral behavior—these are the tasks facing the US military. They become more pressing and dangerous in the long run compared to conventional war.

The capability and effectiveness of the United States to respond to unconventional conflicts are important parts of national strategy and national security policy and remain important even in the face of increasing conventional and nuclear challenges. Conflicts in the less developed world and in newly modernizing industrial states have geostrategic as well as political-military and psychological importance. Dangers include conflict escalation, regional destabilization, and the possibility of the expansion of autocratic regimes. Yet groups seeking self-determination tend to look to the United States for support in the form of resources, including political-psychological assets. In the broader national security dimension, US capability and effectiveness in unconventional conflicts are parts of deterrence.

Perceptions of US strength and political will can have a deterrent effect, and the United States needs a strong capability across the conflict spectrum, including attention to forces necessary to fight at the higher end of the conflict spectrum. US military effectiveness in unconventional, as well as conventional, conflicts sends a message about American staying power for protracted conflict in wars at all levels of conflict.[63]

Finally, the decision of whether the United States should become involved in any conflicts must be based on national strategy and national security priorities, many of which have already been identified. In several instances, US national interests are best served by avoiding military involvement and security guarantees. In other instances, US national interests are best served by political-psychological support, by traditional means of assistance (weaponry and humanitarian aid), and by intelligence operations. In any case, the United States must be prepared to respond effectively across the conflict spectrum if it is to pursue national security policy effectively.

Notes

1. Carl von Clausewitz, *On War,* edited and translated by Michael Howard and Peter Paret (Princeton, N.J.: Princeton University Press, 1976), p. 87.

2. For additional perspective, see Kathleen Hicks, "Russia in the Gray Zone," Aspen, Colorado: Aspen Institute, July 19, 2019; in Johnson's Russia List 2019 #115, July 21, 2019, davidjohnson@starpower.net and Eugene Rumer, "The West Fears Russia's Hybrid Warfare. They're Missing the Bigger Picture," July 3, 2019, https://carnegieendowment.org/2019/07/03/west-fears-russia-s-hybrid-warfare.-they-re-missing-bigger-picture-pub-79412. See also Eliot A. Cohen, *The Big Stick: The Limits of Soft Power and the Necessity of Military Force* (New York: Basic, 2016), chaps. 5, 7.

3. Marvin Kalb and Deborah Kalb, *Haunting Legacy: Vietnam and the American Presidency from Ford to Obama* (Washington, DC: Brookings Institution, 2011).

4. For a brief and excellent overview on why Afghanistan collapsed as it did, see Joseph J. Collins, "What Went Wrong in Afghanistan?" September 2, 2021. https://www.defenseone.com/ideas/2021/09/what-went-wrong-afghanistan/185061.

5. Ernest van den Haag, "The Busyness of American Policy," *Foreign Affairs* 64, no. 1 (Fall 1985), p. 114.

6. For example, see Daniel R. Coats, Director of National Intelligence, *Worldwide Threat Assessment of the US Intelligence Community,* statement for the record, Senate Select Committee on Intelligence (Washington, DC, January 29, 2019). Expert analysis on this point appears in Gregory F. Treverton, *Intelligence for an Age of Terror* (Cambridge, Mass.: Cambridge University Press, 2009).

7. Colin Powell with Joseph E. Persico, *My American Journey* (New York: Random, 1995), p. 576.

8. Ibid., pp. 576–577.

9. See, for example, Frederick H. Hartmann and Robert L. Wendzel, *Defending America's Security* (Washington, DC: Pergamon-Brassey's, 1988), pp. 26–38.

10. John Spanier, *American Foreign Policy Since World War II,* 11th ed. (Washington, DC: Congressional Quarterly, 1988), p. 11.

11. Stephen M. Walt, "How to Ruin a Superpower," *Foreign Policy,* July 23, 2020, https://foreignpolicy.com/2020/07/23/how-to-ruin-a-superpower. See also John Arquilla, *Worst Enemy: The Reluctant Transformation of the American Military* (Chicago: Ivan R. Dee, 2008).

12. See, for example, Ian Roxborough, *The Hart-Rudman Commission and the Homeland Defense* (Carlisle, Pa.: Strategic Studies Institute, US Army War College, September 2001); and Earl H. Tilford Jr., "Redefining Homeland Security," conference brief (Carlisle, Pa.: Strategic Studies Institute, US Army War College, n.d.). This was a summary of a conference sponsored by the Army War College and the Reserve Officers Association held on March 16, 2001.

13. Chris C. Demchak, "China: Determined to Dominate Cyberspace and AI," *Bulletin of the Atomic Scientists* no. 3 (2019), pp. 99–104, https://doi.org/10.1080

/00963402.2019.1604857; Gabi Siboni and Hadas Klein, "Guidelines for the Management of Cyber Risks," in *Cyber, Intelligence, and Security* no. 2 (September, 2018), pp. 23–38; David E. Sanger, *The Perfect Weapon: War, Sabotage, and Fear in the Cyber Age* (New York: Crown, 2018); P. W. Singer and Emerson T. Brooking, *LikeWar: The Weaponization of Social Media* (Boston: Houghton Mifflin Harcourt, 2018); and Nina Kollars and Jacquelyn Schneider, "Defending Forward: The 2018 Cyber Strategy Is Here," September 20, 2018, https://warontherocks.com/2018/09/defending-forward-the-2018-cyber-strategy-is-here. See also Martin C. Libicki, *Cyberdeterrence and Cyberwar* (Santa Monica, Calif.: RAND, 2009).

14. Ariel Cohen and Robert E. Hamilton, *The Russian Military and the Georgia War: Lessons and Implications* (Carlisle, Pa.: Strategic Studies Institute, US Army War College, June 2011), pp. 44–48, citation p. 44.

15. Richard A. Clarke and Robert K. Knake, in *Cyber War* (New York: HarperCollins, 2010), make this argument.

16. Angela E. Stent, *Putin's World: Russia Against the West and With the Rest* (New York: Twelve-Hachette, 2019).

17. See James Dobbins, Howard J. Shatz, and Ali Wyne, "Russia Is a Rogue, Not a Peer; China Is a Peer, Not a Rogue," October 2018, https://www.rand.org/pubs/perspectives/PE310.html.

18. Xavier Raufer, "Gray Areas: A New Security Threat," *Political Warfare: Intelligence, Active Measures, and Terrorism Report* no. 20 (Spring 1992), pp. 1, 4.

19. On the characteristics of unconventional conflicts, see Sam C. Sarkesian and Robert E. Connor Jr., *The US Military Profession into the Twenty-First Century: War, Peace, and Politics,* 2nd ed. (London: Routledge, 2006), pp. 136–140 and passim. The same authors contrast mainstream military and special operations forces in ibid., pp. 154–167.

20. Headquarters, USSOCOM, United States Special Operations Command, https://www.socom.mil; and Marine Forces Special Operations Command, https://www.marsoc.marines.mil.

21. Headquarters USSOCOM.

22. Jim Garamone, "Thomas Passes Special Operations Command Reins to Clarke," March 29, 2019, https://www.defense.gov/News/News-Stories/Article/Article/1800858/thomas-passes-special-operations-command-reins-to-clarke/.

23. Sean Naylor, "A Force to Be Reckoned With—Still," *Army Times,* July 30, 2001, p. 16.

24. Ibid.

25. Bernard Fall, *Street Without Joy: Insurgency in Indochina, 1946–1963,* 3rd rev. ed. (Harrisburg, Pa.: Stackpole, 1963), pp. 356–357. Fall pointed out years ago, "Just about anybody can start a 'little war' . . . even a New York street gang. Almost anybody can raid somebody else's territory, even American territory. . . . But all this has rarely produced the kind of revolutionary ground swell which simply swept away the existing system. . . . It is important to understand that guerrilla warfare is nothing but a tactical appendage of a far vaster political contest that, no matter how expertly fought by competent and dedicated professionals, cannot make up for the absence of a political rationale."

26. Samuel B. Griffith, ed., *Mao Tse-tung on Guerrilla Warfare* (New York: Praeger, 1961), p. 7. See also John Arquilla, *Insurgents, Raiders, and Bandits: How Masters of Irregular Warfare Have Shaped Our World* (Chicago: Ivan R. Dee, 2011).

27. See P. W. Singer and Allan Friedman, *Cybersecurity and Cyberwar: What Everyone Needs to Know* (Oxford: Oxford University Press, 2014); and John

Arquilla, David Ronfeldt, and Michele Zanini, "Information-Age Terrorism," *Current History* (April 2000), p. 179.

28. Terrorism can be placed into various categories. See, for example, Cindy Combs, *Terrorism in the Twenty-First Century,* 2nd ed. (Upper Saddle River, N.J.: Prentice-Hall, 2000).

29. Most of these groups emerged from the political and social turmoil of the 1960s and 1970s in Europe and the United States. The Baader-Meinhof Gang came out of the radicalization of the German Socialist Student Alliance. This group focused on targets in Germany, whereas the Red Brigade operated throughout Europe. The Italian Red Brigade was responsible for the kidnapping and murder of former Italian premier Aldo Moro in 1978. It was also responsible for the 1981 kidnapping of US Army brigadier general James Dozier, who was subsequently rescued by Italian antiterrorist units. Direct Action had its roots in France. The Symbionese Liberation Army emerged in the San Francisco area and was responsible for the kidnapping of heiress Patty Hearst, who subsequently disowned her family and supported the group until she was captured by authorities. The Weather Underground split off from Students for a Democratic Society and was responsible for terrorist operations against the US government and businesses.

30. On the new terrorism, see Russell D. Howard and Margaret J. Nencheck, "The New Terrorism," pp. 6–28, and Brian Michael Jenkins, "The New Age of Terrorism," pp. 29–37, both in James J. F. Forest and Russell D. Howard, eds., *Weapons of Mass Destruction and Terrorism,* 2nd ed. (New York: McGraw-Hill, 2013). See also Mark Jurgensmeyer, "Understanding the New Terrorism," *Current History* (April 2000), p. 158.

31. For pertinent examples, see David E. Sanger, *The Perfect Weapon: War, Sabotage, and Fear in the Cyber Age* (New York: Crown, 2018); and Singer and Friedman, *Cybersecurity and Cyberwar.*

32. Dennis B. Downey, "Domestic Terrorism: The Enemy Within," *Current History* (April 2000), p. 158.

33. Max Boot, *Invisible Armies: An Epic History of Guerrilla Warfare from Ancient Times to the Present* (New York: Norton-Liveright, 2013), p. 517.

34. Eliot A. Cohen, *The Big Stick: The Limits of Soft Power and the Necessity of Military Force* (New York: Basic, 2016), p. 135.

35. See, for example, Joseph R. Nunez, "Fighting the Hobbesian Trinity in Colombia: A New Strategy for Peace," in *Implementing Plan Colombia: Special Series* (Carlisle, Pa.: Strategic Studies Institute, US Army War College, April 2001).

36. Various works asserted links between the former Soviet Union and terrorism. See, for example, Roberta Goren, *The Soviet Union and Terrorism* (Boston: George Allen and Unwin, 1982). See also Claire Sterling, *The Terror Network: The Secret War of International Terrorism* (New York: William Abrahams and Owl, 1985); and Ray Cline and Yonah Alexander, *Terrorism as State-Supported Covert Warfare* (Fairfax, Va.: Hero, 1986).

37. See, for example, Arkady N. Shevchenko, *Breaking with Moscow* (New York: Knopf, 1985). Also see Andre Sakharov, "My KGB Ordeal," *US News and World Report,* February 24, 1986, pp. 29–35; Humberto Belli, *Breaking Faith: The Sandinista Revolution and Its Impact on Freedom and Christian Faith in Nicaragua* (Garden City, Mich.: Puebla Institute and Crossway, 1985); and Shirley Christian, *Nicaragua: Revolution in the Family* (New York: Random, 1985).

38. Jurgensmeyer, "Understanding the New Terrorism," p. 158.

39. Expert discussion of the various meanings of terrorism and terrorist rationalities appears in Bruce Hoffman, "Defining Terrorism," pp. 3–23, and Martha

Crenshaw, "The Logic of Terrorism: Terrorist Behavior as a Product of Strategic Choice," pp. 54–66, both in Russell D. Howard and Reid L. Sawyer, eds., *Terrorism and Counterterrorism: Understanding the New Security Environment—Readings and Interpretations,* 2nd ed. (Dubuque, Iowa: McGraw-Hill, 2006).

40. See, for example, John Arquilla, *Insurgents, Raiders, and Bandits; Selected Works of Mao Tse-tung,* abridged by Bruno Shaw (New York: Harper Colophon, 1970); *Che Guevara: Guerrilla Warfare,* translated by J. P. Morrya (New York: Vintage, 1969); General Vo Nguyen Giap, *People's War, People's Army* (Washington, DC: US Government Printing Office, 1962); and *Sun Tzu: The Art of War,* translated by Samuel B. Griffith (New York: Oxford University Press, 1971). See also Hannah Arendt, *On Revolution* (New York: Viking, 1965); and Jack A. Goldstone, ed., *Revolutions: Theoretical, Comparative, and Historical Studies* (New York: Harcourt Brace Jovanovich, 1986).

41. *Sun Tzu: The Art of War.* According to translator Samuel B. Griffith, "Sun Tzu's essays on the 'Art of War' form the earliest of known treatises on the subject, but never have been surpassed in comprehensiveness and depth of understanding" (from the foreword).

42. Stanley Karnow, in *Vietnam: A History,* rev. and updated ed. (New York: Penguin, 1997), pp. 558–561, suggests that doubts in US public opinion preceded the Tet offensive, but Tet caused a more negative media coverage.

43. Douglas Porch, *Counterinsurgency: Exposing the Myths of the New Way of War* (Cambridge, Mass.: Cambridge University Press, 2013), p. 186 and passim.

44. Hicks, "Russia in the Gray Zone."

45. General Bruce Palmer, *The 25-Year War: America's Military Role in Vietnam* (Lexington: University Press of Kentucky, 1984), p. 176. See also John Prados, *Vietnam: The History of an Unwinnable War, 1945–1975* (Lawrence: University Press of Kansas, 2009), p. 152 and passim.

46. Anthony James Joes, *America and Guerrilla Warfare* (Lexington: University Press of Kentucky, 2000), p. 3.

47. Ibid., p. 328.

48. Goren, *Soviet Union,* p. 5.

49. Timothy L. Thomas, *Three Faces of the Cyber Dragon: Cyber Peace Activist, Spook, Attacker* (Ft. Leavenworth, Kan.: Foreign Military Studies Office, 2012).

50. Arendt, *On Revolution,* p. 8.

51. Joseph R. Biden, "Face the Nation," February 23, 2020, https://www.cbsnews.com/news/transcript-joe-biden-on-face-the-nation-february-23-2020.

52. Samuel P. Huntington, "The Clash of Civilizations," *Foreign Affairs* 72, no. 3 (Summer 1993), pp. 22–49.

53. For an analysis of conflicts in the third world, see Boot, *Invisible Armies,* chap. 28 and books 6–8; Colin S. Gray, *Tactical Operations for Strategic Effect: The Challenge of Currency Conversion* (MacDill Air Force Base, Fla.: Joint Special Operations University Press, 2015); and Colin S. Gray, *Hard Power and Soft Power: The Utility of Military Force as an Instrument of Policy in the 21st Century* (Carlisle, Pa.: Strategic Studies Institute, US Army War College, April 2011). An exemplary case study is by Dodge Billingsley with Lester Grau, *Fangs of the Lone Wolf: Chechen Tactics in the Russian-Chechen Wars, 1994–2009* (Ft. Leavenworth, Kan.: Foreign Military Studies Office, 2012). See also Donald M. Snow, *Distant Thunder: Third World Conflict and the New International Order* (New York: St. Martin's, 1993). For an analysis of such conflicts in the twenty-first century, see J. Bowyer Bell, *Dragon Wars: Armed Struggle and the Conventions of Modern War* (New Brunswick, N.J.: Transaction, 1999). See also Donald M. Snow,

Distant Thunder: Patterns of Conflict in the Developing World (New Brunswick, N.J.: Transaction, 1997).

54. Sam C. Sarkesian, *America's Forgotten Wars: The Counterrevolutionary Past and Lessons for the Future* (Westport, Conn.: Greenwood, 1984).

55. See, for example, Sam C. Sarkesian, *Unconventional Conflicts in a New Security Era: Lessons from Malaya and Vietnam* (Westport, Conn.: Greenwood, 1993).

56. *Public Papers of the Presidents of the United States: Containing the Public Messages, Speeches, and Statements of the President, John F. Kennedy, 1962* (Washington, DC: US Government Printing Office, 1963), p. 454.

57. For additional perspective on this issue, see John J. Mearsheimer, "Imperial by Design," *The National Interest,* no. 111 (January/February 2011), pp. 16–34.

58. See Timothy Snyder, *On Tyranny: Twenty Lessons from the Twentieth Century* (New York: Tim Duggan, 2017).

59. Porch, *Counterinsurgency,* esp. chaps. 6, 8.

60. Christian G. Appy, *American Reckoning: The Vietnam War and Our National Identity* (New York: Viking/Penguin, 2015), pp. 221–222; and Karnow, *Vietnam,* p. 669.

61. Porch, *Counterinsurgency,* p. 222.

62. Jean-François Revel, *How Democracies Perish* (New York: Harper and Row, 1984), p. 7.

63. On military effectiveness, see Allan R. Millett, Williamson Murray, and Kenneth H. Watman, eds., "The Effectiveness of Military Organizations," chap. 1 in Millett and Murray, eds., *Military Effectiveness,* vol. 1, *The First World War* (Boston: Unwin Hyman, 1989), pp. 1–30.

4

Battling the Nuclear Genie

Nuclear weapons are unique among weapons of mass destruction or weapons of mass effect. Nuclear weapons did not go away with the end of the Cold War. A second nuclear age has given new life to the bomb in the form of new nuclear weapons states, more widely available civil nuclear technology with the potential for weaponization, and a return of Cold War–like tensions between the United States and Russia. A third nuclear age poses new challenges of nuclear modernization, deterrence, arms control, and nonproliferation.[1] Lessons learned from Cold War experience in managing US-Soviet nuclear relations are in danger of being forgotten, as are the risks of miscalculated escalation, loss of control, and failures of deterrence. As well, existing nuclear weapons states are modernizing their arsenals of ground-based, air-launched, and sea-based nuclear weapons and mating them with improved technologies for long-range reconnaissance, command-control-communications, and precision strike. In addition, the arms control regime that obtained as between the United States and the Soviet Union during the Cold War, and afterward between the United States and Russia, fell victim to a worsening in political relations between Moscow and Washington, challenges from a rising China, changes in technology and states' aspirations for nuclear modernization, and a lack of political resolve to maintain or improve existing arms control agreements that not only improved transparency and supported deterrence stability, but also served as symbolic reaffirmations of leaders' awareness that, as former US and Soviet leaders Ronald Reagan and Mikhail Gorbachev jointly affirmed: a nuclear war cannot be won and must never be fought.

A significant milestone of nuclear arms control failure was reached in August 2019 when the United States and Russia officially terminated their participation in the Intermediate Nuclear Forces Treaty of 1987.[2] This Cold War agreement was one of the linchpins of the international arms control and nonproliferation regime. It banned the development, testing, and deployment of ground-launched ballistic and cruise missiles with ranges between 500 and 5,500 kilometers. Although it applied worldwide, the focal point of the original US-Soviet agreement was the preventing of a nuclear arms race in weapons deployed in Europe by either the North Atlantic Treaty Organization (NATO) or the Soviet Union (and later Russia). By 2019 the march of history, technology, and politics had pushed the agreement toward obsolescence. Russia was accused of testing and deploying weapons precluded by the Intermediate Nuclear Forces Treaty by the Barack Obama and Donald Trump administrations. In turn, Russia accused the United States of having deployed land based missile defenses in Europe that could be repurposed for offensive strikes against European Russia. Both the United States and Russia also felt that their two sided self-denial about deploying intermediate-range missiles was leaving them at the mercy of non-treaty states with growing missile arsenals, especially China. As the treaty departed into the dustbin of history in August 2019, critics feared that door was open for a renewed arms race in Europe, including additional nuclear capable delivery systems of intermediate or shorter ranges. Indeed, the Pentagon, anticipating the demise of the treaty, was already planning new systems within formerly treaty-restricted ranges for possible deployment in Europe. NATO responsive measures would include steps to bolster deterrence and defense in Europe, including the strengthening of NATO's eastern flank (Poland and the Baltic states of Estonia, Latvia, and Lithuania): by the end of 2019, NATO planned for deployment of thirty mechanized battalions, thirty air squadrons, and thirty combat vessels within thirty days or less.[3]

The expectation of newer NATO and additional Russian intermediate-range missiles deployed in Europe overlapped with both US and Russian interest in developing and deploying nuclear weapons of lower yield (so-called mini nukes or tiny nukes), on the assumption that a wider spectrum of deliverable nuclear charges would enhance the credibility of nuclear deterrence and, in the Russian case, perhaps justify nuclear first use as a measure of "escalation for de-escalation." The latter assumption, attributed to some Russians, was that a limited nuclear first use would shock opponents into thinking twice before further escalating a large conventional war with the potential to damage Russia's vital interests. In turn, NATO, equipped with a wider range of nuclear responses, would be able to deny Russia its theory of victory by means of first or limited nuclear use, and at the same time deter additional Russian escalation.[4]

These nuanced arguments about tailored or flexible deterrence make more sense in a world without nuclear weapons than they do in a nuclear-impacted world. The first use of any nuclear weapon since the US bombing of Nagasaki will be a world historical event, shattering prior psychological expectations of a nuclear "taboo" and creating uncertainty about what lies ahead. How nuclear armed antagonists will react to this scenario in Europe is not entirely obvious, and one can imagine many scenarios. An important but unresolved question is this: Do lower levels of nuclear destruction serve as "firebreaks" to seal off the possibility of escalation to unrestricted nuclear warfare, or, to the contrary, will the use of tactical nuclear weapons serve as psychological detonators of immediate expectations of more, and greater, destruction? The answer is that, before the fact, no one knows.

Thus it remains unclear whether a post–Intermediate Nuclear Forces Treaty nuclear world will turn out to be one of stable nuclear deterrence (allowing for the possibility of small conventional wars or "hybrid" conflicts), one of nuclear jaywalking (episodic threats of nuclear use falling short of actual nuclear use), or nuclear fratricide, with unpredictable outcomes. In this regard, there is some consolation in the fact that US nuclear war-planning is no longer as straitjacketed and confined within bureaucratic "stovepipes" as was the case during the Cold War. Current US strategic planning includes flexible options for regional nuclear deterrence commingled with options for the use of conventional forces, taking into account the possible role of nuclear and conventional forces in escalation control, de-escalation, and conflict termination. The desire for improved capabilities for "tailored" deterrence supported by nuclear weapons of smaller yield is also favored by some defense experts, although others warn that the availability of weapons of smaller yield might tempt policymakers into premature nuclear first use. Nevertheless, the United States is clearly moving in the direction of composite planning that considers nuclear and conventional military forces along with other means of deterrence and defense, including cyber and space offense and defense.[5] Newer technologies, including the use of drones or hypersonic weapons for small or large attacks, will also contribute to pressures for wide spectrum military thinking about deterrence.[6]

The demise of the Intermediate Nuclear Forces Treaty, in turn, called into question whether Russia and the United States would act to extend the New Strategic Arms Reduction Talks (START) agreement of 2011 prior to its scheduled expiration in 2021. The agreement limited the United States and Russia to 1,550 operationally deployed warheads on no more than 700 deployed intercontinental launchers (intercontinental ballistic missiles, or ICBMs; submarine-launched ballistic missiles, or SLBMs; and heavy bombers) by February, 2018. The New START treaty also included provisions for monitoring and verification of one another's deployments, including on-site inspections. New START could be extended by mutual agreement without renegotiation until 2026. Once considered as an almost automatic next step, New START extension, in the wake of the Intermediate Nuclear Forces Treaty debacle and further political remonstrances between the United States and Russia, appeared precarious in the summer of 2020. It was not so much the technical issues that loomed as dark clouds over New START extension (such as US missile defenses, or Russian hypersonic weapons as countermeasures to US defenses) as it was the poisonous political atmosphere resulting from disagreements over Russia's annexation of Crimea in 2014 and subsequent destabilization of Eastern Ukraine, alleged Russian meddling in the US presidential election of 2016, US and allied economic sanctions against Russia, and allegations of Russian (as well as Chinese) cyber attacks on a variety of US targets, including probes of US electrical grids, economic infrastructure, and voting machines.[7] Fortunately from the standpoint of arms control, the administration of President Joseph Biden agreed with Russia to extend the New START agreement soon after Biden took over the Oval Office in January 2021.

Since 1945 the availability of nuclear weapons has been a critical issue in US national security policy and strategy. For much of that time the primary concern was superpower relations. Later, the acquisition of nuclear weapons by other states complicated the balance of power between the United States and the Soviet Union. With the collapse of the Soviet Union, Russia inherited the lion's share of former Soviet nuclear weapons (weapons located in Ukraine, Belarus, and Kazakhstan, as parts of the former Soviet Union, were now stranded in those new post-Soviet states, but eventually were repatriated to Russia or destroyed). The United States and Russia have continued to engage in arms control and nuclear stability discussions, but the 2019 demise of the Intermediate Nuclear Forces Treaty left New START as the only remaining nuclear arms control agreement as between Washington and Moscow.

Assessment of the nuclear future is complicated by developments in the international landscape. This landscape is ill defined and is also characterized by access to nuclear technology and weapons by international terrorist groups and rogue states. And the dispute of India and Pakistan, both of which are nuclear powers, over Kashmir highlights problems that can evolve. Combined with US involvement in the war on terrorism in the aftermath of September 11, it is no wonder that nuclear weapons remain a critical issue. In this chapter we review the evolution of nuclear weapons as a critical component of US national security. To understand the current policy and strategy, we must study the evolution of nuclear weapons policies and strategies emerging from the US-Soviet rivalry to the present period.

During the Cold War, the US homeland faced the possibility of nuclear attack, which never materialized. But on September 11, 2001, the US national security agenda changed overnight. A new context now exists for assessing threats to US national security posed by weapons of mass destruction, whether in the hands of terrorists or states. Nuclear weapons will receive specific attention here because nuclear danger remains the most geopolitically critical. This claim is supported by the decision of the George W. Bush administration to begin deploying US nationwide ballistic missile defenses in 2004 and by the Obama administration to continue research and development on improved ballistic missile defense technologies for possible future use. Are missile defenses the answer to Americans' prayers to be able to sleep more securely at night or a false security blanket? The complex relationships among US-Russian nuclear arms control, proliferation, missile defense, and homeland security provide a large challenge for US policy planners.

Nuclear Weapons Strategies

From 1945 to the end of the Cold War (1989–1991), theorists and military officers asked whether nuclear weapons had forever changed the relationship between war and politics.[8] This question remains relevant in the twenty-first century.[9] Indeed, the issue of nuclear weapons and their proliferation remains a complex and uncertain dimension of US national security.

The end of the Cold War and the demise of the Soviet Union cast further doubt on the political utility of nuclear forces and, perhaps, the threat of nuclear force. Without a global opponent and favoring regionally oriented military strategies, US military planners are uncertain as to the relevance of nuclear weapons except as a last resort. During the first decade of the present century, future president Barack Obama and a number of prestigious former policymakers in the United States and other countries called for the eventual abolition of nuclear weapons and, in the interim between now and then, for drastic reductions in the nuclear arsenals of the United States, Russia, and other acknowledged nuclear weapons states.[10]

Nuclear weapons also appeared to reverse the traditional relationship between offensive and defensive military strategies. In traditional strategy, offensive attacks were thought to be riskier than defensive stands. But the speed and lethality of nuclear weapons made offensive technology, but not necessarily an offensive *strategy,* look more imposing. Nuclear weapons that survived a first strike could be used to retaliate, and unless the attacker could protect itself against retaliation, the difference between the attacker's and the defender's postwar worlds might be politically and militarily insignificant.

The paradoxical implications of nuclear weapons for military strategy and US national security led to Cold War advocacy of nuclear strategic policies that followed one of two paths: mutual assured destruction (MAD) or nuclear flexibility. Some US officials and military thinkers favored a strategy of assured retaliation as the necessary and sufficient deterrent against any Soviet provocation, including Soviet attacks on US allies with conventional weapons. MAD required that the United States be able to absorb a Soviet surprise first strike of any size and retaliate, destroying the Soviet Union as a viable society. MAD became official US declaratory policy under Secretary of Defense Robert S. McNamara in the John F. Kennedy and Lyndon Johnson administrations.

MAD had many critics in and out of government. Some thought MAD required too much of the US nuclear arsenal; others thought that MAD would not be enough to deter a Soviet attack under some circumstances. The view that MAD was an insufficient deterrent posture for US forces gained some ground during the administrations of Richard Nixon, Gerald Ford, Jimmy Carter, and Ronald Reagan. On the other hand, although MAD was official US declaratory policy, employment policy (targeting) emphasized the destruction of enemy nuclear and conventional military forces, leadership, command-control systems, and other objectives in addition to the destruction of economic and social values.[11]

As for nuclear flexibility, under Nixon and Ford, US nuclear weapons policy was revised to emphasize options short of all-out war between the superpowers. Under Carter, Presidential Directive 59 expanded the previous guidance and called for the creation of nuclear retaliatory forces sufficiently flexible and durable to fight a protracted, or a limited, nuclear war, as might be directed by the US president in pressing circumstances. In addition, Carter's policy called for the development of nuclear command-and-control systems that would permit the United States to fight an all-out nuclear war through various phases until victory was denied to the Soviet Union.

Critics of nuclear flexibility argued that its justifications were really disguised arguments for building up nuclear arsenals well beyond the requirements of deterrence. In addition, the Soviet Union showed little apparent interest in flexible nuclear response as a means of bargaining, especially in the case of strategic nuclear weapons exploded on Soviet territory. Nevertheless, a Soviet adversary determined to exploit any relative counterforce imbalance was frequently cited by policymakers as a necessary and sufficient case for counterforce and nuclear flexibility. Although Soviet military doctrine in its political-military aspects (grand strategy) remained essentially defensive and potentially open to the concept of limited war, other aspects of Soviet military doctrine offered little in the way of encouragement. The General Staff and Politburo showed little interest in the actual limitation of nuclear strikes once deterrence had failed.[12]

Nuclear Weapons as a Deterrent

The effectiveness of nuclear weapons as a deterrent was accepted as fact by most policymakers and analysts. But carrying out war plans if deterrence failed would have been left to organizations that operated with rigid, preplanned, and detail-driven parameters, not subject to crisis control and wartime policymakers. Theorists and strategists conceived elegant ways to fight limited wars and to dominate the process of escalation, and Soviet–Warsaw Pact planners dreamed of blasting holes in NATO's forward defenses with tactical nuclear strikes during the initial phase of war. It was, for the most part, an effort to cover up mutual fears. The greatest fear was that there was no way to fight a nuclear war at an acceptable cost, making nuclear deterrence a potentially dangerous bluff.

Colin Gray was thus correct (up to a point) to insist that all nuclear deterrence strategies were tantamount to contingent nuclear war-fighting strategies.[13] Military chains of command were tasked to prepare plans for what to do if deterrence ever failed. As Cold War arsenals grew in size and complexity, some of these plans and the planning process itself took on lives of their own. The US Single Integrated Operational Plan (SIOP) for nuclear war was driven by calculations of "damage expectancy" against designated classes of targets, and damage expectancies had to be satisfied above and beyond other requirements of the war plan. US war plans were often characterized by a lack of clear policy guidance beyond vague statements to win or prevail, as if the definition were self-evident. Target planners at Strategic Air Command (now Strategic Command) in Omaha, Nebraska, were often forced to assign weapons to targets on the basis of hunches or rules of thumb. Few, if any, US presidents ever familiarized themselves with the details of nuclear war plans.[14]

The nuclear policy of the Nixon, Ford, and Carter years called for improved US counterforce capabilities: selective attacks on Soviet land-based missiles, missile submarines, and nuclear bombers as well as their supporting military command-and-control systems. None of the US nuclear policy guidance of the 1970s, however, offered any hope of protecting the US homeland in the event that deterrence failed. The Nixon administration, in signing the 1972 Anti–Ballistic Missile Treaty, made US reliance on the threat of offensive retaliation for deterrence official. Defenses were marginalized by the treaty, which constrained the size of missile defenses as well as the kinds of defenses that could be built. Both constraints were intended to preclude any possibility that either side would later deploy an effective system for the defense of its national territory.

The Strategic Defense Initiative

In March 1983, President Ronald Reagan called for the US research-and-development community to transcend deterrence based on offensive retaliation. Reagan's proposed Strategic Defense Initiative (described pejoratively by many as "Star Wars") caught his own defense bureaucrats by surprise and alarmed Moscow.[15] The technology was not available during Reagan's terms in office, or even a decade later, to provide reliable protection against a large-scale missile attack against US territory.[16] Continued deterrence through mutual vulnerability remained the only plausible option when the Cold War expired, and it remained so into the twenty-first century.

Several other strategic doctrines were considered by policymakers, experts, and critics: assured retaliation, escalation dominance, victory denial, defense dominance, and minimum deterrence.[17] In November 2001, a Texas meeting between Russian president Vladimir Putin and US president George W. Bush about nuclear weapons and the 1972

Anti–Ballistic Missile Treaty changed the context for nuclear stability by opening the door to unprecedented offensive arms reductions. In December 2001, President Bush announced that the United States would no longer be bound by the 1972 treaty. The United States and Russia agreed in May 2002 to the Strategic Offensive Reductions Treaty, limiting each side to a maximum of 1,700–2,200 operationally deployed offensive nuclear weapons by December 2012. Russia's willingness to sign this agreement pointed to the temporary post–September 11 rapprochement between Putin and Bush and to Russia's immediate decision not to let missile defenses stand in the way of US-Russian cooperation on other issues. The United States began deploying defenses in 2004 in Alaska and California. US leaders emphasized that the current and foreseeable defenses were not aimed at Russia: instead, their purpose was to deflect accidental launches and to deter attacks from rogue states with small arsenals. However, the issue of missile defenses later became a major bone of contention as between the United States and Russia, and Russian nuclear force modernization would eventually emphasize the development and deployment of new offensive strike weapons with the potential to outmaneuver or otherwise confound opposed missile defenses.

Nuclear Arms Control in the Cold War: From Confrontation to Stability

Following the October 1962 Cuban missile crisis, US and Soviet leaders perceived a mutual interest in strategic arms limitation, the avoidance of accidental and inadvertent nuclear war, and the prevention of the spread of nuclear weapons. This led to the Nuclear Test Ban Treaty of 1963 and the hotline, a direct communications link for emergency discussions between the US and Soviet heads of state. Discussions between Moscow and Washington about strategic arms limitation got under way during the latter years of the Johnson administration, continued under Nixon, and culminated in the Strategic Arms Limitation Talks Agreement of 1972 (SALT I). A year earlier, Washington and Moscow had concluded two agreements on the prevention of accidental and inadvertent war and the avoidance of unnecessary fears of surprise attack.[18]

The interim agreement on offensive arms limitation embodied in SALT I was superseded by SALT II, signed in 1979 and carried forward (although never formally ratified by the United States) until it was transformed into START during the Reagan administration. The Anti–Ballistic

Missile Treaty remained as the cornerstone of US-Soviet strategic arms limitation until the end of the Cold War. As amended by a 1974 protocol signed at Vladivostok, it limited both sides' national missile defense systems to one site of no more than 100 defensive interceptors. The United States chose to deploy its ABMs at Grand Forks, North Dakota; the Soviets, around Moscow. (The US system was closed down by Congress in the mid-1970s.)

The Anti–Ballistic Missile Treaty became a powerful symbol of affinity for advocates of mutual deterrence based on offensive retaliation. When the Reagan administration proposed its Strategic Defense Initiative in 1983, opponents argued that it would overturn the Anti–Ballistic Missile Treaty and reopen the race in offensive weapons being capped by the SALT/START process.

Even more durable than the Anti–Ballistic Missile Treaty was the Nuclear Nonproliferation Treaty (NPT), ratified in 1970 and supported by both the United States and the Soviet Union. The agreement was intended to prevent the spread of nuclear weapons and weapons-related technology. The NPT outlasted the Cold War and remains in force today, although under pressure from new and aspiring nuclear powers. The NPT was extended indefinitely in 1995 with near-unanimous approval, with the important exceptions of India, Pakistan, and Israel. All three of these states are now declared or opaque nuclear powers. The favorable climate established by the NPT extension carried forward into the 1996 multilateral agreement, a comprehensive test ban treaty (CTBT) on nuclear weapons testing, extending and deepening the impact of the original test-ban treaty and the subsequent treaties on threshold test bans and peaceful nuclear explosions. However, CTBT still lacked the ratification of the United States and other key states when President Barack Obama left office in 2017. The Donald Trump administration showed little interest in pushing the issue. The position of the Biden administration remains to be seen.

After September 11, the United States was faced with the imminent possibility of terrorist attacks using nuclear or other weapons of mass destruction, including biological, chemical, and radiological weapons.[19] Since terrorists are stateless and have no single "return address," deterrence by threat of retaliation is almost superfluous. Recognizing this, the George W. Bush administration redefined US national security policy and military strategy to emphasize the option of preemption against terrorists or rogue states planning attacks on the United States or its allies. This shift in policy and military strategy unsettled the administration's critics, who argued that a doctrine of preemption might "encourage other states to legitimize their own aggression under the guise of defensive measures."[20] Other skeptics feared that the Bush preemption doctrine was a disguised endorsement of preventive war. (Preemption is a decision to strike first against an opponent who is already planning or has set in motion a military strike. Preventive war is a decision for attack against another state whose intentions appear hostile and whose power represents a potential threat.)

Weapons Reduction

The nuclear arms control dialogue between East and West during the Cold War contributed to the reduction of political tensions in various ways. First, continuing arms control negotiations educated both sides during the Cold War about one another's strategic and defense cultures. Second, the strategic arms limitation agreements of the 1970s and 1980s (SALT/START) provided a framework that allowed both the Americans and the Soviets to avoid expensive deployments of systems that would have been militarily unnecessary and eventually obsolete in the face of improved technology. Third, cooperation between Washington and Moscow to limit the spread of nuclear weapons technology helped limit the number of nuclear aspirants during the Cold War and set a useful precedent for multilateral cooperation against proliferation afterward. US-Russian post–Cold War cooperation against the spread of nuclear weapons included the Cooperative Threat Reduction (the so-called Nunn-Lugar program) authorized by the US Congress in 1991 to encourage denuclearization and demilitarization within states of the former Soviet Union, especially the four successor states that inherited the Soviet nuclear arsenal (Russia, Belarus, Kazakhstan, and Ukraine).[21]

Efforts to limit the significance of nuclear weapons during the Cold War were complicated by the role of nuclear weapons in US and NATO strategy for the prevention of war in Europe and for the establishment of a credible defense plan if deterrence failed. Some US and European analysts and policymakers doubted that conventional deterrence was feasible; others feared that it might be. Those who doubted that conventional deterrence was feasible tended to see a viable Soviet threat of invasion, absent strong NATO military preparedness. Those who feared that conventional defense was feasible noted that a conventional war in Europe would involve very different sacrifices for Americans and Europeans. A conventional deterrent for NATO might not be as convincing as a conventional defense backed by nuclear deterrence.

NATO's willingness to settle for active-duty deployment in Western Europe of some thirty ground divisions was not forced by the economics of defense, as some politicians contended. NATO could have created conventional forces capable of credible deterrence against Soviet attack precisely because nuclear weapons made success problematic, to say the least. Nuclear weapons added a component of uncertainty and risk to Soviet, and later Russian, calculations. Yet nuclear weapons were also a curse for NATO strategy. As the numbers of tactical nuclear weapons deployed with air and ground forces multiplied during the Cold War, the problem of NATO command and control, including obtaining political approval for first use, became more complicated. Whether NATO's crisis management and policymaking process could ever have authorized nuclear release in time was widely debated. Regardless, the deployment of US nuclear weapons in Europe would link the defense of Europe and North America. In fact the United States, by spreading its nuclear deterrent outside North America, risked an attack on the US homeland in order to defend Europe from a Soviet offensive.

US-Russian Nuclear Arms Control in a New Century

Neither Mikhail Gorbachev's political career nor the Soviet Union survived the end of the Cold War, but nuclear weapons did. The Russian Federation assumed responsibility for former Soviet nuclear weapons and for continuing the process of nuclear arms control with the United States in the START negotiations. At first, US political relations with post-Soviet Russia were much improved. Under Nunn-Lugar and other projects, the United States provided funds to help Russia account for its fissile materials, transport and store its remaining nuclear weapons, dismantle obsolete or disarmed weapons and launchers, and employ displaced Russian nuclear scientists on other projects. In May 2002, as noted earlier, Bush and Putin signed the Moscow Treaty to reduce long-range nuclear weapons to a maximum of 2,200 deployed nuclear warheads for each state by the year 2012. This agreement, together with the creation of a new NATO-Russia Council for consultation on terrorism, nonproliferation, and other issues, signified for many the de facto end of the Cold War. The New START agreement negotiated by presidents Barack Obama and Dmitri Medvedev in 2010 called for even lower than SORT limits on the numbers of deployed weapons and launchers permitted for each state: a maximum of 1,550 weapons and 700 deployed launchers.[22] The New START-accountable and treaty compliant US and Russian strategic nuclear arsenals as of 2021 are summarized in Table 4.1.

After a brief honeymoon in the 1990s and early 2000s, arguments about the role of missile defenses and nuclear flexibility reappeared as points of contention and uncertainty as between the United States and Russia. Under the George W. Bush administration, the United States was thinking of missile defenses in the broader context of a reappraisal of the fundamentals of US national military strategy. According to national defense guidance and US nuclear policy statements, the former strategic nuclear "triad" of land-based, sea-based, and bomber-delivered weapons would be superseded in the new century by a "new triad" of (1) conventional and nuclear offensive strike weapons; (2) active and passive defenses; and (3) improved infrastructure, including command, control, and communications, to support US military operations worldwide, including nuclear options.[23]

In addition, the Bush administration sought from Congress approval to develop specialized mini nukes or micro nukes with low yield for use against enemy fortified bunkers and caches of WMD.[24] Nuclear and conventional weapons, according to the Bush strategy, would function as more of a seamless web of options for deterrence or, if necessary, for military action. The Bush policy shift reflected the reality of improved US conventional military forces compared to the Cold War and the information-driven character of modern weapons technology. It also responded to a new threat environment, one marked by asymmetrical threats involving weapons of mass destruction available to rogue states

Table 4.1 US and Russian Strategic Nuclear Forces, 2021

	US Strategic Nuclear Forces		
	Launchers	Warheads/Launchers	Total Warheads
ICBMs			
Minuteman III	400	1	400
SLBMs			
Trident II D-5	240	1-8	1,920[a]
Bombers			
B-52H	46	ALCM/W80-1 x 5-150	–
B-2A	20	B61-7 x 10-360/-11 x 400 B83-1 x low-1,200	–
Total bombers	66		300[b]
Total strategic offensive forces	772		2,620[c]

continues

Table 4.1 Continued

	Russian Strategic Nuclear Forces		
	Launchers	Warheads/Launchers	Total Warheads
ICBMs			
SS-18 Sickle	46	10	460
SS-19 M4 Stiletto	4	1 HGV	4
SS-25 Sickle	27	1	27
SS-27 Mod 1(mobile) Topol-M	18	1	18
SS-27 Mod 1 (silo) Topol-M	60	1	60
SS-27 Mod 2 (mobile) Yars	135	4	540
SS-27 Mod 2 (silo) Yars	20	4	80
SS-X 29 (silo) Sarmat	–	–	–
Total ICBMs	310		1189[d]
SLBMs			
SS-N-18 Stingray	16	3	48
SS-N-23 Sineva	96	4	384
SS-N-32 Bulava	64	6	384
Total SLBMs	176		816[e]
Bombers			
Bear-H6/16	55	6-16 x AS-15A ALCMs or 14 x AS-23B ALCMs	448
Blackjack	13	12 x AS-15B ALCMs or AS-23B ALCMs, bombs	132
Total air forces	68		580[f]
Total strategic offensive forces	554		2,585[g]

Sources: Hans M. Kristensen and Matt Korda, "United States Nuclear Weapons, 2021," *Bulletin of the Atomic Scientists* 77, no. 1 (2021), pp. 43–63, https://doi.org/10.1080/00963402.2020.1859865; adapted from Hans M. Kristensen and Matt Korda, "Russian Nuclear Weapons, 2021," *Bulletin of the Atomic Scientists* 77, no. 2 (2021), pp. 90–108, https://doi.org/10.1080/00963402.2021.1885869

Notes: ICBM = intercontinental ballistic missile; SLBM = submarine-launched ballistic missile; ALCM = air-launched cruise missile.

a. 240 SLBMs are on 12 deployable ballistic missile submarines (SSBNs). Two other SSBNs are in overhaul for a total of 280 launchers. Approximately 1,000 warheads are actually deployed on SLBM launchers.

b. The United States has 66 nuclear-capable bombers and about 300 warheads actually deployed at bomber bases (200 ALCMs at Minot Air Force Base and about 100 bombs at Whiteman Air Force Base). B-52H aircraft are no longer tasked with delivering gravity bombs.

c. Deployed warheads on strategic launchers are 400 on ICBMs, approximately 1,000 on SLBMs, and 300 weapons at heavy bomber bases that can be rapidly loaded into the aircraft. New START counting rules count each heavy bomber as a single warhead. The United States appears to be in compliance with New START treaty limits, with 675 deployed strategic launchers and 1,457 attributed warheads as of October 1, 2020.

d. Some ICBM launchers may be deployed with fewer than their maximum number of warheads in compliance with New START limits.

e. Some ballistic missile submarines (SSBNs) are in overhaul at any given time and do not carry nuclear weapons, so fewer than 816 warheads are actually deployed. At any given time, only 320 warheads are deployed on five operational Delta IV submarines with Sineva SLBMs. Some SLBM launchers may be deployed with fewer than their maximum number of warheads in compliance with New START limits.

f. Nuclear warheads are only assigned to 50 operationally deployed nuclear-capable bombers, although the total bomber inventory is larger. Under New START rules each bomber counts as a single warhead.

g. About 1,600 strategic nuclear warheads are deployed; slightly more than 800 on ICBMs, about 624 on SLBMs.

or terrorists. Missile defenses based on new principles, together with improved offensive and defensive forces, would provide the president with a range of options: preemption, deterrence, and defense and retaliation, if necessary.

Bush critics argued that the new triad blurred the distinction between nuclear and conventional weapons to a dangerous degree, that it would encourage nuclear proliferation, and that the past performances of missile defense technology did not point to an optimistic future. In contrast, Department of Defense technology developers foresaw a mix of conventional and nuclear strike weapons that would make the United States less dependent on the nuclear option and more capable of posing credible threats supported by realistic options. The possible assignment of Prompt Global Strike (PGS) missions to US intercontinental or transoceanic launchers continued as a point of discussion during the Obama administration and raised its head as an issue during the run-up to completion of the New START agreement. Russia remained wary of American conventional PGS systems for three reasons: the possibility of mistaking the launch of a conventionally armed weapon for a nuclear one; the possible development of a US or NATO theater-strategic conventional first-strike option against Russia's comparatively weaker nonnuclear forces; and possible threats to Russia's nuclear forces posed by US conventional PGS weapons. Partly in view of these Russian concerns, the United States agreed to count long-range, conventionally armed intercontinental ballistic missiles toward the central limits for New START.[25]

The purposes of a limited US National Missile Defense (NMD) system would be to deter and to defeat, if necessary, attacks on US territory by rogue states such as North Korea, to destroy any ballistic missiles accidentally launched at the US homeland, and to deter the use of such weapons by international terrorists and other nonstate actors. Testing of a ground-based midcourse defense (GMD) system using hit-to-kill, exoatmospheric non-nuclear interception produced inconsistent results, however. Although US NMD technology remained in its infancy, prominent Russian officials remained wary lest the United States create a first-strike capability against Russia's deterrent, backed by defenses good enough to absorb Russia's retaliatory strike. The Obama administration, aware of Russia's concerns about the viability of its nuclear deterrent, shifted the Bush plan for missile defenses in Europe to Obama's European Phased Adaptive Approach (EPAA) based on various stages of the SM-3 missile interceptor and supporting command-control, communications, intelligence, and reconnaissance systems. The Obama European missile defense plan was intended to mollify some of Russia's concerns, but Vladimir Putin's return to the Russian presidency in 2012 and Russia's annexation of Crimea in 2014, among other issues, contributed to a breakdown in US-Russian political relations and reinvigorated Russian fears of US missile defenses. Vladimir Putin's 2018 address to the Russian Federal Assembly warned, with attention getting graphics, that Russia was planning to develop and/or deploy a new generation of offensive weapons designed to defeat any missile defenses that the US might deploy.[26]

The diminished quality of Russia's conventional military forces since the end of the Cold War pressured Moscow's military planners to increase their reliance on its remaining nuclear deterrent. Several versions of Russia's post-1991 military doctrine officially declared that Russia's nuclear forces might be used not only in response to a nuclear attack but also in a variety of other contingencies, including a conventional attack on Russian territory with the potential to destabilize the Russian state. In addition to an expanded NATO and a cash-starved Russian military, Russian nervousness about the instability of its borders gave nuclear weapons a higher priority among Russian military options. Although Russia reassured the United States and its NATO allies that every precaution had been taken against a failure of its nuclear command-and-control system, the United States offered its expertise to Russia in order to prevent computer failures and other disasters that might contribute to nuclear instability. Improvements in Russia's conventional military forces after 2007 have reduced Russia's dependency on nuclear coercion, including more reliance on voluntary soldiers instead of conscripts and improved fire support from long-range, conventional weapons systems: the latter enabling more attention to the concept of prenuclear deterrence. Nowadays Russian military doctrine emphasizes the seamless connections among levels of warfare and their political, economic, psychological, and informational aspects. Particular emphasis is given to so-called gray area warfare mixing propaganda, disinformation, and deception with the use of covert action and special forces.

The Danger of Proliferation

The spread of nuclear weapons and other WMD poses new challenges for US and allied policymakers and their militaries.[27] There is a justified concern on the part of the United States and other countries about the potential spread of ballistic missiles and other long-range delivery systems. The 1998 entries of India and Pakistan into the nuclear club showed that some states value nuclear weapons as symbols of power and status, contrary to the assumptions underlying the NPT and other arms control efforts. North Korea's decisions to depart the Nonproliferation Treaty, expel United Nations (UN) inspectors, and openly declare its nuclear status have added additional uncertainties to the mix.

Two schools of thought characterize the US approach to nuclear proliferation.[28] One school, continuous with the dominant tendency in Cold War thinking, holds that any spread of nuclear weapons is inherently bad. All states aspiring to obtain nuclear arsenals should be discouraged, and all states now possessing nuclear forces (with the exception of the five permanent members of the UN Security Council and three additional de facto recognized nuclear weapons states—India, Israel, and Pakistan) should be urged to roll them back. The second school of thought sees the first approach as doomed to failure, regardless of its utility during the Cold War. This school favors selective US opposition to nuclear/WMD proliferation, based on the willingness of the state to adhere to acceptable norms of international behavior.[29] Of course, the predominant norm is nonaggression. According to this approach, the United States and its allies would not necessarily attempt to dissuade or discourage nuclear proliferation in states satisfied with the international status quo. Only those states determined to disrupt the geopolitical status quo, such as Iran or North Korea, would be targeted for counterproliferation—active measures to prevent the state from obtaining nuclear weapons and to deny it the effective ability to use them once obtained. Of course, a state's intention could change quickly, and a nuclear-capable "status quo" state could become a revolutionary "anti–status quo" state overnight. For reference, Table 4.2 summarizes the status of global nuclear arsenals as of 2020.

Since the status quo works to the US advantage in the current and near-term international system, the selective approach to nonproliferation appears hypocritical to those on the receiving end. Why, for example, should India and Pakistan have felt the past sting of US disapproval while Israel suffered nothing for its "bomb in the basement" nuclear status? In the case of US NATO allies such as Britain and France, it could be argued that the NATO defense pact gives US officials continuing contact with allied counterparts and might act, in some circumstances, as a restraint on an otherwise unilateral decision for nuclear use. But Israel is part of no such alliance, and its nuclear decisions are accountable only to its own state interests.

On the other hand, Israel, given its perceived defense predicament and small territorial size, makes a strong case for its own nuclear capacity: to deter any nuclear/WMD attack, to deny the possibility of another

Table 4.2 Global Nuclear Weapons Inventories, 2020

	Numbers of Warheads Deployed, Stored, and Retired
China	320
Russia	6,257
France	290
United Kingdom	215
United States	5,550
India	150
Pakistan	160
Israel	90
North Korea	30–40

Sources: Arms Control Association, https://www.armscontrol.org/factsheets/Nuclearweaponswhohaswhat; Hans M. Kristensen and Matt Korda, "United States Nuclear Weapons, 2021," *Bulletin of the Atomic Scientists* 77, no. 1 (2021), pp. 43–63, https://doi.org/10.1080/00963402.2020.1859865; Hans M. Kristensen and Matt Korda, "Russian Nuclear Weapons, 2021," *Bulletin of the Atomic Scientists* 77, no. 2 (2021), pp. 90–108, https://doi.org/10.1080/00963402.2021.1885869.

Note: China may have as many as 700 nuclear warheads by 2027 or 1,000 nuclear warheads by 2030. See Helene Cooper, "China Could Have 1,000 Nuclear Weapons by 2030, Pentagon Says," *New York Times,* November 3, 2021, https://www.nytimes.com/2021/11/03/us/politics/china-military-nuclear.html.

Holocaust, and to thwart adversaries who might engage in nuclear coercion.[30] Despite this logic, third world and other states aspiring to nuclear status cannot help but notice the selective application of US proliferation norms. This point, among others, has led some states to withhold their signatures from the Comprehensive Test Ban Treaty promoted by the Clinton administration and opened for signature in 1996. The proposed 2017 UN Treaty on the Prohibition of Nuclear Weapons also met with resistance from every existing nuclear weapons state, despite the near certainty that it would eventually obtain the required number of state signatories.[31]

Proponents of nuclear disarmament contend that selective nonproliferation shares the old fixation on partial instead of complete solutions to nuclear peril. The willingness to live with the existence of nuclear weapons, even in small quantities and in the hands of "reputable" states, runs the unacceptable risk of nuclear first use followed by a war of unprecedented destruction.[32] Proponents of nuclear disarmament have a respectable intellectual and political tradition going back to the very beginning of the nuclear age.[33] But they are up against the practical objection that the level of international trust necessary for verified elimination of all nuclear forces does not yet exist. And even if states trusted one another to disarm completely, the knowledge of how to build nuclear weapons cannot be disinvented, and arsenals once destroyed can eventually be rebuilt. Thus some have proposed that states could disarm weapons already assembled and ready to fire, although arsenals could be reconstituted if necessary (so-called virtual nuclear arsenals).

Conclusion

Nuclear weapons, deterrence, and nonproliferation remain critical to US national security policy and strategy as well as to those of its allies and adversaries. At the same time, the availability of related technology—and perhaps actual weapons—to states or groups willing to use them has magnified the problems of deterrence and proliferation for the United States. In addition, the technology environment for nuclear deterrence and arms control is changing rapidly. Cognitive warfare is now embedded in kinetic warfare and deterrence. Information warfare, new kinds of offensive missiles, missile defenses, and counterspace capabilities will form part of the context for future US military planning and decisionmaking.[34] Underlying these technologies will be foundational capabilities and expertise in artificial intelligence, in machine learning, in "big data" and in autonomous performance by systems of all types.[35] Even so-called "drone swarms" may be used for strategic military effects, and counter-drone swarms are also being tested.

If new technology makes antinuclear defenses viable in the coming years, how will that change the relationship between deterrence and proliferation? This depends on how good the defense technology is and who owns it. Various technologies for airborne or ground-based interception of ballistic missiles seemed more promising at the end of the first decade of the twenty-first century than they did a decade earlier. Technological innovation does not by itself, however, constitute a strategic breakthrough. Strategy involves a reactive opponent. Offensive countermeasures exist for many of the kinds of defenses that have been proposed. Among the new generations of offensive weapons posing potential threats to defenses are hypersonic weapons, already being developed and deployed by Russia, the United States, and China. Other threats to defenses are possible from attacks on satellites for surveillance, communications, command-control, and navigation by hostile satellites capable of co-orbiting within range to capture or destroy another satellite. US Department of Defense space strategy acknowledges that space is now a war-fighting domain as well as a medium used for peaceful purposes.[36] The era of offense-dominant nuclear strategy will probably continue unless space-based defenses operating at or nearly at the speed of light can be deployed and protected or some more imaginative technology for missile defenses is exploited. One possibility is the use of unpiloted aerial vehicles (UAVs, or drones) in swarms that could loiter above likely spots for missile launch and promptly attack ballistic missiles early in their flight trajectory—or even attack the missile launcher sites preemptively. Until then, we are talking about a shift not from offense dominance to defense dominance, as President Reagan hoped for, but from offense dominance to offense-defense competitiveness.[37]

If anti-missile defenses become cost-effective compared to ballistic missile offenses, then the effect will almost certainly be to dissuade interest in ballistic missile strikes, whether nuclear or not. There are other ways to deliver nuclear weapons, and one need not rely on missiles that lend themselves to interception and destruction. The terrorist attacks of September 11 were a reminder that significant destruction can be accomplished in the US homeland without missile attacks, nuclear weapons, or even regular armed forces. In addition, if antimissile defenses become good enough to make missile strikes obsolete, those defenses may also have ominous offensive (i.e., first-strike) capabilities against a variety of target sets and raise the problem of preemption to another technological level. Defenses that can intercept incoming ballistic missiles may also be able to destroy satellites, thereby disrupting communications and command and control and denying the enemy a clear picture of the battle space. Competent antimissile defenses or other space-based weapons may be the leading edge of an information warfare strategy for the twenty-first century.

Notes

1. David A. Cooper, *Arms Control for the Third Nuclear Age: Between Disarmament and Armageddon* (Washington, DC: Georgetown University Press, 2021); and Paul Bracken, *The Second Nuclear Age: Strategy, Danger, and the New Power Politics* (New York: Holt/Times, 2012). See also Brad Roberts, *The Case for U.S. Nuclear Weapons in the 21st Century* (Stanford, Calif.: Stanford University Press, 2016); and Andrew Futter, *The Politics of Nuclear Weapons* (London: Sage, 2015).

2. Justin V. Anderson and Amy J. Nelson, "The INF Treaty: A Spectacular, Inflexible, Time-Bound Success," *Strategic Studies Quarterly,* no. 2 (Summer 2019), pp. 90–122. See also Stephen Blank, "Arms Control and Russia's Global Strategy After the INF Treaty," June 19, 2019, https://www.realcleardefense.com/articles/2019/06/19/arms_control_and_russias_global_strategy_after_the_inf_treaty_114513.html; and Lawrence J. Korb, "Why It Could (but Shouldn't) Be the End of the Arms Control Era," *Bulletin of the Atomic Scientists,* October 23, 2018, https://the bulletin.org/2018/10/why-it-could-but-shouldnt-be-the-end-of-the-arms-control-era.html.

3. James Marson, "NATO Grapples With Collapse of Missile Treaty," *Wall Street Journal,* August 1, 2019. See also Tom Nichols, "Mourning the INF Treaty: The United States Is Not Better for Withdrawing," *Foreign Affairs,* March 4, 2019; William Tobey, Pavel S. Zolotarev, and Ulrich Kuhn, *The INF Quandary: Preventing a Nuclear Arms Race in Europe—Perspectives from the U.S., Russia, and Germany,* in *Russia Matters,* issue brief, January 2019, https://www.russiamatters.org/analysis/inf-quandary-preventing-nuclear-arms-race-europe-perspectives-us-russia-and-germany, and Jacob Cohn, Timothy A. Walton, Adam Lemon, and Toshi Yoshihara, *Leveling the Playing Field: Reintroducing U.S. Theater-Range Missiles in a Post-INF World* (Washington, DC: Center for Strategic and Budgetary Assessments, 2019), pp. 29–30, https://csbaonline.org/research/publications/leveling-the-playing-field-reintroducing-us-theater-range-missiles-in-a-post-INF-world.

4. See Office of the Secretary of Defense, *Nuclear Posture Review* (Washington, DC: US Department of Defense, February 2018); and Keith B. Payne, "Nuclear Deterrence in a New Age," Information Series no. 426, Fairfax, Va., National Institute Press, December 13, 2017). For related assessments of Trump nuclear posture, see Lawrence J. Korb, "Why Congress Should Refuse to Fund the NPR's New Nuclear Weapons," *Bulletin of the Atomic Scientists,* February 7, 2018, https://thebulletin.org/commentary/why-congress-should-refuse-fund-npr%E2%80%99s-new-nuclear-weapons11493; David E. Sanger and William J. Broad, "To Counter Russia, U.S. Signals Nuclear Arms Are Back in a Big Way," *New York Times,* February 5, 2018; and Hans M. Kristensen, "The Nuclear Posture Review and the U.S. Nuclear Arsenal," *Bulletin of the Atomic Scientists,* February 2, 2018, https://thebulletin.org/commentary/nuclear-posture-review-and-us-nuclear-arsenal11484.

5. Todd C. Lopez, "Nuclear Posture Review, National Defense Strategy Will Be Thoroughly Integrated," *DOD News,* June 25, 2021, https://www.defense.gov/Explore/News/Article/Article/2671471/nuclear-posture-review-national-defense-strategy-will-be-thoroughly-integrated.

6. Zachary Kallenborn,"Meet the Future Weapon of Mass Destruction, the Drone Swarm," *Bulletin of the Atomic Scientists,* April 5, 2021, https://thebulletin.org/2021/04/meet-the-future-weapon-of-mass-destruction-the-drone-swarm.

7. David E. Sanger, *The Perfect Weapon: War, Sabotage, and Fear in the Cyber Age* (New York: Crown, 2018); P. W. Singer and Emerson T. Brooking, *LikeWar: The Weaponization of Social Media* (Boston: Houghton Mifflin Harcourt, 2018).

8. For assessments of nuclear weapons and deterrence before and after the Cold War, see Futter, *The Politics of Nuclear Weapons;* Bracken, *The Second Nuclear Age;* Michael Krepon, *Better Safe Than Sorry: The Ironies of Living with the Bomb* (Stanford, Calif.: Stanford University Press, 2009); Patrick M. Morgan, *Deterrence Now* (Cambridge, Mass.: Cambridge University Press, 2003); Lawrence Freedman, *The Evolution of Nuclear Strategy,* 3rd ed. (New York: Palgrave-Macmillan, 2003); Colin S. Gray, *The Second Nuclear Age* (Boulder, Colo.: Lynne Rienner, 1999); Colin S. Gray, *Modern Strategy* (Oxford: Oxford University Press, 1999), chaps. 11–12; Keith B. Payne, *Deterrence in the Second Nuclear Age* (Lexington: University Press of Kentucky, 1996); and Robert Jervis, *The Meaning of the Nuclear Revolution: Statecraft and the Prospect of Armageddon* (Ithaca, N.Y.: Cornell University Press, 1989).

9. Keith B. Payne, *Shadows on the Wall: Deterrence and Disarmament* (Fairfax, Va.: National Institute Press, 2020).

10. See, for example, George P. Shultz, William J. Perry, Henry A. Kissinger, and Sam Nunn, "Toward a Nuclear-Free World," *Wall Street Journal,* January 15, 2008, p. A13; and Jennifer Loven, "Obama Outlines Sweeping Goal of Nuclear-Free World," *Associated Press,* April 5, 2009. For an especially insightful commentary on

this topic, see Thomas C. Schelling, "A World Without Nuclear Weapons?" *Daedalus* no. 4 (Fall 2009), pp. 124–129.

11. Fred Kaplan, *The Bomb: Presidents, Generals, and the Secret History of Nuclear War* (New York: Simon and Schuster, 2020), pp. 39–42.

12. For evidence of Soviet views on controlling and possibly terminating a major war, see Raymond L. Garthoff, *Deterrence and the Revolution in Soviet Military Doctrine* (Washington, DC: Brookings Institution, 1991), chap. 5; and Raymond L. Garthoff, "New Soviet Thinking on Conflict Initiation, Control and Termination," in Stephen J. Cimbala and Sidney R. Waldman, eds., *Controlling and Ending Conflict* (Westport, Conn.: Greenwood, 1992), pp. 65–94.

13. Gray, *Modern Strategy,* pp. 309–318.

14. Kaplan, *The Bomb.* See also Desmond Ball, "The Development of the SIOP, 1960–1983," in Desmond Ball and Jeffrey Richelson, eds., *Strategic Nuclear Targeting* (Ithaca, N.Y.: Cornell University Press, 1986), pp. 57–83. See also Richard Ned Lebow, *Nuclear Crisis Management: A Dangerous Illusion* (Ithaca: Cornell University Press, 1987), pp. 118–122.

15. Frances Fitzgerald, *Way Out There in the Blue: Reagan, Star Wars, and the End of the Cold War* (New York: Simon and Schuster, 2000), p. 198.

16. For history of US missile defense programs prior to the Strategic Defense Initiative, see Donald R. Baucom, *The Origins of SDI, 1944–1983* (Lawrence: University Press of Kansas, 1992). See also, on the politics of US missile defenses, Andrew Futter, *Ballistic Missile Defence and US National Security Policy: Normalization and Acceptance After the Cold War* (London: Routledge, 2013).

17. For an alternate perspective, see Charles Glaser, "Why Do Strategists Disagree About the Requirements of Strategic Nuclear Deterrence?" in Lynn Eden and Steven E. Miller, eds., *Nuclear Arguments: Understanding the Strategic Nuclear Arms and Arms Control Debates* (Ithaca, N.Y.: Cornell University Press, 1989), esp. pp. 113–117.

18. For an overview from SALT I to New START, see Arms Control Association, "US-Russian Nuclear Arms Control Agreements at a Glance," http://www.armscontrol.org/factsheets/USRussiaNuclearAgreementsMarch 2010.

19. The possibility of nuclear terrorism is explored in Brian Michael Jenkins, *Will Terrorists Go Nuclear?* (New York: Prometheus, 2008). See also Graham Allison, *Nuclear Terrorism: The Ultimate Preventable Catastrophe* (New York: Times/Holt, 2004).

20. Lawrence J. Korb, *A New National Security Strategy in an Age of Terrorists, Tyrants, and Weapons of Mass Destruction* (New York: Council on Foreign Relations, 2003), p. 7.

21. See Richard D. Burns and Philip E. Coyle III, *The Challenges of Nuclear Non-Proliferation* (Lanham, Md.: Rowman and Littlefield, 2015); and William J. Perry, Secretary of Defense, *Annual Report to the President and the Congress* (Washington, DC: US Government Printing Office, March 1996), pp. 63–70.

22. For text of the New START agreement, see *Treaty Between the United States of America and the Russian Federation on Measures for the Further Reduction and Limitation of Strategic Offensive Arms* (Washington, DC: US Department of State, April 8, 2010), http://www.state.gov/documents/organization/140035.pdf.

23. US Department of Defense, *Findings of the Nuclear Posture Review, Briefing* (Washington, DC: US Department of Defense, January 9, 2002). See also *DOD Briefing on the Nuclear Posture Review,* http://www.globalsecurity.org/wmd/library/policy/dod/npr.htm.

24. Robert W. Nelson, "Lowering the Threshold: Nuclear Bunker Busters and Mininukes," chap. 4 in Brian Alexander and Alistair Millar, eds., *Tactical Nuclear*

Weapons: Emergent Threats in an Evolving Security Environment (Washington, DC: Brassey's, 2003), pp. 68–79.

25. See, for example, Steven Pifer, "The Death of the INF Treaty Has Given Birth to New Missile Possibilities," *The National Interest,* September 18, 2019, https://nationalinterest.org/feature/death-inf-treaty-has-given-birth-new-missile-possibilities-81546; and Jen Judson, "U.S. Army to Fund Extended-Range Precision Strike Missile Starting in FY22," *Defense News,* June 14, 2021, http://www.defensenews.com/.

26. Vladimir Putin, "Presidential Address to the Federal Assembly," March 1, 2018, http://kremlin.ru/events/president/news/56957.

27. See Henry D. Sokolski, *Underestimated: Our Not So Peaceful Nuclear Future* (Carlisle, Pa.: Strategic Studies Institute, US Army War College, January 2016); and Burns and Coyle, *The Challenges of Nuclear Non-Proliferation,* pp. 191–210.

28. Kenneth N. Waltz, "More May Be Better," chap. 1 in Scott D. Sagan and Kenneth N. Waltz, *The Spread of Nuclear Weapons: A Debate* (New York: Norton, 1995), pp. 1–45; and Scott D. Sagan, "More Will Be Worse," chap. 2 in Sagan and Waltz, *The Spread of Nuclear Weapons,* pp. 47–91.

29. On the relationship between strategy and proliferation (or nonproliferation), see Gray, *The Second Nuclear Age,* pp. 47–78.

30. The case for an Israeli nuclear deterrent is summarized in Louis Rene Beres, *Security Threats and Effective Remedies: Israel's Strategic, Tactical, and Legal Options* (Shaarei Tikva, Israel: Ariel Center for Policy Research, April 2000), pp. 39–47.

31. United Nations Office for Disarmament Affairs, *Treaty on the Prohibition of Nuclear Weapons,* https://www.un.org/disarmament/wmd/nuclear/tpnw.

32. George P. Shultz, William J. Perry, and Sam Nunn, "The Threat of Nuclear War Is Still with Us: The U.S. Must Re-Engage with Russia to Ensure the Ultimate Weapon Doesn't Spread and Is Never Used," *Wall Street Journal,* April 11, 2019.

33. Lawrence Freedman, "Eliminators, Marginalists, and the Politics of Disarmament," in John Baylis and Robert O'Neill, eds., *Alternative Nuclear Futures: The Role of Nuclear Weapons in the Post–Cold War World* (Oxford: Oxford University Press, 2000), pp. 56–69.

34. For example, see Herbert S. Lin, "Cyber Risks in Nuclear Escalation Scenarios," Webinar presentation, Hoover Institution, Stanford University, October 27, 2021; and Herbert S. Lin, *Cyber Threats and Nuclear Weapons* (Stanford, Calif.: Stanford University Press, October 2021).

35. George Galldorissi and Sam Tangredi, "Algorithms of Armageddon: What Happens When We Insert AI into Our Military Weapons Systems," presentation to US Department of Defense Strategic Multilayer Assessment (SMA) program, April 27, 2021, https://nsiteam.com/?s=algorithms+of+armageddon.

36. Henry D. Sokolski, ed., *Space and Missile Wars: What Awaits* (Arlington, Va.: Nuclear Policy Education Center, 2021); Clementine G. Starling, Mark L. Massa, Lt. Col. Christopher P. Mulder, and Julia T, Siegel, *The Future of Security in Space: A Thirty-Year U.S. Strategy* (Washington, DC: Atlantic Council, Scowcroft Center for Strategy and Security, 2021), http://www.AtlanticCouncil.org; and US Department of Defense, *Defense Space Strategy* (Washington, DC, 2020), p. 3, https://media.defense.gov/2020/Jun/17/2002317391/-1/-1/1/2020_Defense_Space_Strategy_Summary.pdf.

37. Michaela Dodge, *Missile Defense Reckoning Is Coming: Will the United States Choose to Be Vulnerable to All Long-Range Missiles?* Fairfax, VA, Information Series no. 465 (National Institute for Public Policy, August 20, 2020), http://www.nipp.org. See also Stephen J. Cimbala, *Nuclear Weapons and Strategy: US Nuclear Policy for the Twenty-First Century* (London: Routledge, 2005), chap. 2.

5

The US Political System

In the early 1830s a young French nobleman and historian named Alexis de Tocqueville traveled throughout the United States observing its people and their government. He had much good to say about the new democracy in his 1835 book *Democracy in America,* but he also noted some problems with the way the government conducted foreign affairs: "Foreign policy," he wrote, "does not require the use of any of the good qualities peculiar to democracy, but does demand the cultivation of almost all of those which it lacks."[1]

De Tocqueville's perceptive observations remain applicable today. Democracies lack many of the qualities required to maintain an effective national security posture as well. Indeed, the very nature of democracies works against developing and implementing the strategies required for long-range success in an increasingly complicated strategic landscape.

The long-term US response to emerging regional and global challenges will be a continuing test of national will, staying power, and political resolve. As de Tocqueville noted in the case of foreign policy, "Democracy finds it difficult to coordinate the details of a great undertaking and to fix on some plan and carry it through with determination in spite of obstacles."[2] This is especially true when such plans require a long-term commitment and a degree of secrecy. The latter is especially difficult to achieve when a necessarily free press feels obligated to publish anything it can get its hands on, whatever the consequences for national security or foreign relations.[3]

At the same time, the United States confronts threats of international terrorism, unconventional conflicts, regional conflicts, nuclear proliferation, cyber attacks, and a variety of contingencies short of war. Ethnic rivalries, religious conflict, and hyper-nationalism further complicate the new security landscape. Much has changed in the international conflict arena, and fortunately some of the developments are positive. In particular, the likelihood of conflict between major powers appears remote, at least in the near term, with the exception of certain particularly sensitive flash points such as a Chinese attempt to reunify with Taiwan by force. However, an increasingly powerful Russia menacing its neighbors and invading Ukraine, as well as the rise of Russia and China as peer competitors to the United States, portend serious challenges to US national security.[4] Given the increasing military assertiveness of both competitors, they have become the primary national security challenge in the minds of US policymakers.

In addition, so-called asymmetrical wars will continue in the future, in which state or nonstate opponents attempt to offset US military advantages with innovative strategies. This became apparent during the Vietnam War and is even more true today. International terrorist groups are unlikely to confront the United States militarily, even if they could do so. Iraqi leader Saddam Hussein was unwise enough to pose a conventional military challenge to the United States. Although the conventional challenge was beaten back with relative ease, the United States had much more difficulty when the conflict became unconventional. Later phases of the second war with Iraq, as well as the war in Afghanistan, emphasized improvised explosive devices and suicide bombing rather than direct military attacks. Even a relatively powerful state such as Iran would be expected to use hit-and-run tactics against US forces rather than risk an all-out attack, unless they felt their political regime was in serious danger. Still, a military confrontation cannot be ruled out in the wake of the 2019 US abrogation of the agreement limiting Iranian nuclear weapons programs by President Trump,[5] and the difficulty of the Biden administration in getting that agreement restored.

Past experience and the nature of its political system and values make the new strategic environment difficult for the United States, which is best at conventional response, strategic deterrence, and high-tech warfare. Despite the primary need to counter threats from other industrial powers, the United States must still develop the strategy, doctrine, and tactics to mount an effective campaign against international terrorism and other threats to the US homeland. Except in clear cases of national interests, however, the nature of open systems makes it difficult to develop effective intelligence and military systems to respond to unconventional conflicts.

Why is this so? Why is it that the democratic United States finds it difficult to respond to the variety of threats to its national interests except in clear cases of overt aggression? The ambiguity of the challenges faced provides part of the explanation, but much of the answer rests with the nature and character of democracy and open systems.

US Democratic Principles

Much has been written about the meaning of democracy and its variations as practiced in the United States and elsewhere. A thorough analysis of democracy and open systems would require a philosophical and historical study that is beyond the scope of this book. Our purpose here is more modest: to touch on critical features of democracy as they pertain to national security. Initially, we need to consider the meaning of democracy, for it is in that context that US national security policy is made.

The fundamental proposition is that democracy does not adhere to a dogmatic ideological philosophy. Indeed, one of the most pervasive features of democracy is its pluralistic and pragmatic basis. Democracy is rooted in the idea of political tolerance for many views, as long as they support political equality, self-determination, and individual worth and do not foreclose the possibility of future peaceful political change. Although there are disagreements on the specific application of these concepts, they are legitimate bases for evaluating a political system. Finally, the legitimacy of a system and its leaders rests on the will of the people, a basic tenet of the US Constitution reflected in the political system it created. Put simply, the system must reflect the principles of governmental responsibility and accountability to the people.

To be sure, no political system is perfect, including that of the United States. In particular, increased political polarization in the electorate has reduced public tolerance for opposing views and those who hold them. This is dangerous for democracy, of course. Although most of the public recognizes the imperfections of democracy, they also see the merits, as did de Tocqueville: "The vices and weaknesses of democratic governments are easy to see . . . but its good qualities are revealed only in the long run. . . . The real advantage of democracy is not . . . to favor prosperity for all, but only to serve the well-being of the greatest number."[6] The Biden administration is attempting to widen the social safety net to increase the well-being of Americans who have been relatively disadvantaged economically. Democracy may not be the most efficient type of government (indeed, the Founders went to great effort to prevent government from being *too* efficient), but as Winston Churchill said in the House of Commons soon after World War II, "Democracy is the worst form of government, except for all those other forms that have been tried from time to time."[7]

Faith in democracy is tested when a country confronts difficulties. Moreover, in a society that is more respectful of diversity and becoming more multicultural, there is the possibility that the polarization and diversity politics will complicate the shaping and nurturing of democratic institutions. Many feel this is already occurring due to divisions and hyperpartisanship that makes finding consensus more difficult. Nevertheless, we believe there are fundamental principles at the root of the democratic faith that have lasted for centuries. They are ingrained in the US psyche.

This brief description of the nature of US democracy provides a starting point for an analysis of its relationship to national security in the current era. Four fundamental characteristics of democracies are relevant to the study of national security: the distribution of power, the democratic faith, multiculturalism and cultural diversity, and the messianic spirit.

The Distribution of Power

The pluralistic nature of the US political system institutionalizes the diffusion of political power. This creates an environment for power struggles among the various branches of government, within the bureaucracy, and among groups in society. This prevents one branch from becoming predominant but requires compromise to be effective.[8] Compromise is particularly difficult if a significant segment of the population views its political opponents with contempt. Power struggles can be especially severe in cases of foreign and national security policies because the issues involve vital US interests and values that are at the core of political differences in the United States. Institutionalized confrontation, the product of separation of powers, is epitomized by the struggles between the president and Congress and can become especially contentious when each branch is controlled by a different political party.

Even when one party controls the Oval Office and Congress, there are likely to be disagreements. The strong and bipartisan congressional support of President George W. Bush in the aftermath of September 11 was a response to the new threat to the US homeland, but it did not last long. The normal state of affairs in Congress and between Congress and the president is characterized more often by debate and disagreement. This was seen in 2007, when the Democratic Congress subjected the

Bush administration to serious scrutiny and undertook efforts to force an end to US military involvement in Iraq, in 2011 and 2012, when a Republican-controlled House of Representatives insisted on an agenda of limiting governmental expenditures and maintaining low tax rates, and after the 2018 election, when a Democratic House of Representatives sought to reign in President Trump and made him the first president in US history to be impeached twice.

Efforts by Congress to assert itself in the making and implementation of foreign and national security policy have led to many confrontations with the executive. These include passage of the War Powers Resolution of 1973, which attempted, without great success, to curb the war-making power of President Richard Nixon and future presidents. Despite presidential missteps over the years and furious opposition to many presidential policies, the president remains preeminent in national security decisionmaking.[9]

There is no shortage of difficult issues to resolve, guaranteeing continuing disputes between the president and Congress. These include the role of force in US international policy and the appropriate force structure and training required. Many of these decisions will be budget driven, involving as they do military pay and retirement benefits, base structure, and defense acquisition projects of great interest to members of Congress and the general public. Other issues will include the degree to which military and intelligence operations pose unacceptable risks to civil liberties, even though they may be useful to protect US interests in the new era. These concerns are receiving more public attention with increasing awareness of the capabilities of artificial intelligence, facial recognition software, and information-gathering generally.

During the administration of President Bill Clinton, battles with the Congress were numerous and often rancorous. These included US military operations in Somalia in 1993 and 1994, Bosnia-Herzegovina in 1995, and Kosovo in 1999. In 2000, the last full year of the Clinton administration, executive-legislative struggles sharpened over national missile defense (the latest version of President Ronald Reagan's "Star Wars" initiative from the early 1980s) as well as US relationships with the Russian Federation, China, and Cuba. The George W. Bush administration received a great deal of criticism from Congress over the proper response to the September 11 attacks and subsequent wars in Iraq and Afghanistan. These continued into the administration of President Barack Obama, who was criticized for vacillating military policies and setting red lines that were not enforced when crossed. The apparently ad hoc and improvisational nature of President Donald Trump's foreign and defense policies was a source of continuing criticism, both partisan and nonpartisan. The administration of President Joseph Biden did not reverse all of the policies of its predecessor, especially with respect to China, but it moved early on to repair relations with traditional allies and plan foreign and defense policies on a more strategic basis. The stage remains set for continuing power struggles between the president and Congress.

Regardless of the inherent struggles, the fact remains that the president has the lead role in foreign policy and national security, and this presidential power comes as much from politics and public opinion as it does the Constitution. As political scientist James Q. Wilson noted some time ago:

> Increasingly since the 1930s, Congress has passed laws that confer on the executive branch grants of authority to achieve some general goals, leaving up to the president and his deputies to define the regulations and programs that will actually be put into effect. Moreover, the American people look to the president—always in time of crisis, but increasingly as an everyday matter—for leadership and hold him responsible for a large and growing portion of our national affairs.[10]

As a consequence of the new security landscape, the president faces many difficult problems in both foreign and national security policies. Although some of these problems are resolved by measures short of a major war, they remain relevant to US national interests. Moreover, successful national security policy could necessitate a judicious use of nonmilitary instruments and covert operations. This requires especially effective presidential leadership in developing consensus at home for the necessity of a particular policy—a difficult prospect without bipartisan support in Congress and the public.

This institutionalized confrontation stems partly from the sharing of power. The popular notion that the US government has three separate branches with distinct powers has a corollary: the branches also share power, which allows each to influence and intervene in the affairs of the others. This gives rise to important constitutional questions and different opinions regarding the proper exercise of power.

Problems of control and responsibility evolving from power-sharing are increased by the decentralization of power within and among the various branches of government. Within the executive branch, there is usually a continuing struggle between the Departments of State and Defense as well as between the Central Intelligence Agency (CIA) and other agencies. For example, the Iran-contra hearings during the Reagan administration illuminated the conflicts between the president's national security staff and various departments. During the Clinton and Trump presidencies, problems between the Justice Department and the Federal Bureau of Investigation became well publicized. Similarly, struggles occur within Congress among committee chairs, caucus leaders, and individual members. Combined with the two-party system and frequent elections, such struggles create and nurture institutional power plays, and the decentralized federal system of the US government adds to the fragmentation. Partisan control and party machines at the state level add yet another dimension to the power equation at the national level, and the media sometimes seem like a fourth branch of government.

Historically, only the president has been considered the legal spokesperson of the United States in the international arena. The president does not have a monopoly of power to carry out policy, however, especially in today's environment. The very nature of the US system creates checks and balances that can frustrate any policy and strategy. The system has generally favored those who support mainstream policies and oppose major departures from the status quo, but this may be changing. Political developments will test whether the public remains most comfortable with mainstream politicians who focus on bread-and-butter domestic issues.

The Democratic Faith

A historical thread running through US democracy is the commitment to the free market of ideas within the body politic and the various branches of government. This supports the notion of a free press and reinforces decentralization, diffusion, and power-sharing. Linked closely to pluralism and nondogmatic philosophies, the free play of ideas is assumed to lead to the truth.

An informed and educated citizenry is essential to democracy. The citizenry must have access to information in order to exercise its will and assess the performance of government. It is a basic belief that an educated citizenry can overcome any obstacles to the functioning of democracy. The notion that information and education lead to an enlightened public and also reflects a belief in the innate goodness of people and their sense of justice, since enlightened people are thought more likely to act with wisdom and good judgment. Optimism on this point is challenged today by the concentration of influence in the largest technology companies and their communications platforms, including Facebook, Instagram, Twitter, and others. These powerful tech-information companies have been under congressional scrutiny since 2020 for their ability to monopolize access to communications and their alleged role in selecting or editing material that appears on their platforms.

Pragmatism is another ingredient of the democratic faith. This is the belief that the search for practical consequences based on common sense is important and is also a result of having an enlightened citizenry. Once problems are encountered, Americans believe they can find reasonable solutions. This pragmatic, can-do attitude pervades US society.

Finally, an important part of the democratic faith is the belief that the US system is the best one, despite its imperfections. The public believes that Americans are a decent people living in a decent society. The concern with the quality of life at the community level has a parallel with the concern with individual well-being. This is not to suggest that the culture is without prejudice and narrow-mindedness, but a concern for the individual and continuing efforts to rectify past and continuing injustices are constants in US society, strongly reinforced by the reaction to publicized incidents of police brutality against people of color. Thus far, the dissatisfaction of historically underrepresented or disadvantaged groups has been absorbed into the mainstream political process of compromise and consensus building. Conversely, there is no guarantee that future challenges to the American consensus on basic democratic values will be met successfully. Democracies are constantly under stress from destructive forces outside, or within, their national borders. The extreme polarization and partisanship in the wake of the 2020 election and President Trump's unwillingness to concede defeat, and the attack on the US Capitol building January 6, 2021, make agreement even on basic facts problematic.

Multiculturalism and Cultural Diversity

The United States is an immigrant society. Today, many immigrants are from the third world and the Far East rather than from Europe, as was the case for most of US history. Some bring a culture that differs from Anglo-Saxon and Western traditions. The result is a diverse "multicultural" system rather than an all-assimilating "melting pot." Cultural diversity lends a uniqueness, richness, and strength to American society. But some fear that multiculturalism can be taken to such extremes that the concept of Americanism is eroded in the face of cultural diversity, politicization, and polarization. There is also a fear of cultural confrontation and the continuation of age-old animosities spilling over from foreign homelands. This is complicated by the fact that some cultural precepts are difficult to change, limiting assimilation into the US mainstream. Yet assimilation into that mainstream has been the backbone of the US immigrant society. This was based not on destroying old cultures but on sharing a common language and understanding the meaning of US citizenship, heritage, and culture.

For the system to endure, it needs the support of its constituent parts and a sense of unity that transcends the particularities of race, religion, gender, sexual orientation, gender expression, or national origin. The motto on US currency—*E pluribus unum,* or "one out of many"—expresses this sentiment well.

The Messianic Spirit

Throughout US history, religion has been an important component in shaping national values and in reinforcing the messianic spirit. This is seen in the view that the United States as the leader of the West must be a moral as well as political-military leader and that moral principles should guide the behavior of governmental officials and the military. The messianic spirit is reinforced by the historic role religion has played in the evolution of the US system. The First Amendment to the US Constitution permits the free exercise of religious preference by individuals and groups and forbids the government to establish a state church. Over the years, the courts have struggled to define the exact nature of the relationship between religion and American politics; this, too, is an evolving landscape. In foreign policy, there has been some historical relationship between optimism about America's unique sense of mission in the world and the choices of American presidents with respect to military intervention. The idea of American "exceptionalism" due to the uniqueness of American values comes naturally to US politicians but grates on the nerves of some foreign observers, who reject the assumption that American values and institutions are universally applicable. Even otherwise steadfast American allies such as those in the North Atlantic Treaty Organization (NATO) can tire rapidly of a presidential style that seems preachy or presumptuous about the automatic superiority of the American way of life compared to others.[11] Some of this foreign resentment reflects other states' lack of success in cultural assimilation compared to the American experience.

For many Americans, the extension of this belief from the individual to the political results in a messianic spirit—the notion that Americans and the political system are ordained to be "the light" for other nations, lending moral weight to the democratic faith. This messianic spirit may partly explain US attempts to spread democracy wherever it thinks it can.

The Impact on National Security

These considerations—power distribution, democratic faith, multiculturalism and cultural diversity, and the messianic spirit—create contradictory forces. On the one hand, they strengthen US willingness to respond to national security challenges. On the other hand, they reveal weaknesses that may detract from national unity.

The power distribution that is characteristic of the US political system provides a basis for its legitimacy and precludes a centralized power base from forming in any one branch or individual. Furthermore, it allows inputs from many interest groups, citizens, and elected officials—the foundation of representative government. The strength created by this base of power makes the US political system and government resilient and capable of responding to mistakes, problems, and failures in a fashion not easily matched by other systems. This important point bears repetition and emphasis. The US political system is stronger, not weaker, because of the vigorous disputes and debates that occur among and within the various branches of government. The object of American government is not "efficiency" in the narrow sense, but liberty under the law. No branch of government should attempt to preempt the rights and responsibilities of the others. When, for example, presidents overreach in their attempts to use Presidential power at the expense of Congress, the legislative branch can use its inherent powers over the purse and other powers to reign in an overly ambitious president.

In developing national security policy, the president faces many constituencies and must try to build a consensus throughout the political system. This makes it difficult to design new policies and strategies that appear to challenge the prevailing democratic faith. Furthermore, maintaining a degree of secrecy and yet operating within traditional democratic parameters requires delicate maneuvering. Covert operations, for example, are often perceived as undemocratic. But the need to respond to unconventional conflicts and terrorism may require these very activities. Winston Churchill famously said that in wartime, truth must sometimes be protected by a "bodyguard of lies."

We return to the four fundamental characteristics of democracies relevant to this discussion of national security—the distribution of power, the democratic faith, multiculturalism and cultural diversity, and the messianic spirit—to see how they affect US national security policy.

The Distribution of Power

The way power is distributed creates a condition in which those responsible for foreign affairs and national security do not control all the structures and resources needed to carry out policies and strategies. The exception, of course, is when the United States faces serious threats to its existence. In such times power becomes more centralized in the executive branch. Matters of foreign affairs and national security are susceptible to political opposition within the government and public, bureaucratic foot-dragging, policy distortions, and opposition. Foreign powers may see such internal US political struggles as vacillation, weakness, and divisiveness. This has become even more complicated as information-age technology opens up the system, and foreign governments, nongovernmental organizations, business corporations, international terrorists, and individuals gain access to information on policy issues that formerly existed only within the closed realm of the bureaucratic establishment.

In the George W. Bush administration, controversy continued over China and Taiwan, NATO enlargement, national missile defense, the US role in the Middle East, and relations with Russia, and these issues remained salient into the Joseph Biden administration. The Obama administration received a great deal of criticism from those who believed that there was too much continuity with the Bush administration on national security issues. President Obama's attempts to disengage from military interventions in Iraq and Afghanistan were also the subject of great controversy. Severe criticism was leveled at President Trump for distancing himself from US allies, accommodating authoritarian figures such as the leaders of Russia, North Korea, and Saudi Arabia, and for a series of diplomatic and military policies that seemed to lack coherence. President Biden has been sharply criticized for the way in which his administration handled the US military withdrawal from Afghanistan in August 2021 and his responses to the 2022 Russian invasion of Ukraine are being closely watched.

Presidents attempt to maintain a degree of secrecy in matters of national security. In the Nixon administration much of the conduct of foreign and national security policy was cloaked in secrecy. In fact, "the administration was predisposed toward a secretive policy by its distrust of the State Department and intelligence community, by the convoluted personalities of its leaders, and by its belief that certain of its goals required extreme confidentiality and centralized direction of policy in the White House."[12] These considerations were not limited to the Nixon administration, and have been seen in all administrations since.

The Iran-contra affair, whereby the Reagan administration was trying trade arms for hostages and using the profits to support paramilitary groups opposing the Sandinista government of Nicaragua, was but one example. In the latter part of the Reagan administration, director of central intelligence William Casey was renowned for his creative efforts to limit giving information to Congress, even if it was unclassified. Casey excelled at providing Congressional testimony that avoided direct answers to difficult queries from Senators and Representatives, creating a wall of obfuscation with respect to some of the administration's clandestine operations and covert actions.

> [Casey] . . . had developed nonresponsiveness to oversight into an art form. The Senate Select Committee on Intelligence had even built a special amplifying system into its bug-proof hearing room in an effort to make Casey's muttering intelligible. What the senators did not realize was that, when he wanted to be understood, Casey spoke as clearly as John F. Kennedy. That kind of arrogance had gotten the agency involved in the Iran-contra mess in the first place.[13]

President Trump went to great lengths to limit the amount of information given to Congress on any number of subjects, but the tendency to do that is not new.

It is not unusual for a president to go to extraordinary lengths to develop support for administration policy and strategy. Proposed changes to the existing order, especially changes that promise to be innovative, involve some degree of political risk, however, both domestically and internationally. Furthermore, attempts at change invite attempts by others to derail the initiatives. This leads to the conclusion that the safest political course is to maintain existing policy and strategy and seek only incremental changes. A president who seeks new directions must be prepared to deal with a variety of power centers that can seriously challenge his or her authority. This is a problem for developing effective national security policies, especially those that require a high degree of secrecy to succeed.

There are exceptions to this generalization, of course. President Nixon's rapprochement with China was a distinct change in policy that carried national security overtones, yet it received widespread acclaim. In addition, after President Jimmy Carter's well-intentioned but ultimately unsuccessful attempt to emphasize human rights criteria in national security and foreign policy, President Reagan won praise for changing the course of the Cold War by first denouncing the Soviet Union as an "evil empire," then fostering a cordial relationship with Soviet leader Mikhail Gorbachev. President George W. Bush's speech to a joint session of Congress in the aftermath of September 11 won high praise as he spelled out a strategic course of action that was supported by a great majority of the US public. Problems of implementation—particularly the wars in Iraq and Afghanistan—eventually eroded this support. The attempted rapprochement with North Korea by President Trump began in secret, with little known about the negotiations. Similarly, the details of President Trump's private discussions with Russian leader Vladimir Putin remain under wraps.

Secret diplomacy to effect a major change in policy or strategic direction carries great political risks. Unless the effort has an immediate, positive, and visible impact on US ability to protect vital interests, various power clusters are likely to diminish the chances for success. Even the militarily successful operations in Grenada (1983), Panama (1989–1990), and the Persian Gulf (1990–1991) had their share of critics; less successful missions such as Somalia (1993), Haiti (1998), Afghanistan (2001), Iraq (2003), and Syria (2014) came under harsher criticism, not only from Congress but also from the general public.

The Democratic Faith

The democratic faith creates similar dilemmas. On the one hand, the country's commitment to individual worth, justice, and fairness strengthens the credibility of its policy and strategy; it also taps a wellspring of support from the public. On the other hand, the same commitment makes it difficult to address national security issues that require the use of force or alliances with nondemocratic systems (such as Saudi Arabia and other states in that region), although these may be advantageous to do so for both national security and economic reasons. Unconventional conflicts affect the political-social structure and make the population at large willing or unwilling participants. Moreover, involvement in unconventional conflicts requires interjecting US political-military forces into the political-social milieu of other countries, few of which may be democratic in the Western sense. The United States often has few options other than to support the lesser of evils, but having warm relations with regimes that violate human rights runs counter to US values and fosters divisiveness within our political system.

The premise that we know best makes it difficult to develop realistic perspectives on the motivation and nationalistic aspirations of other peoples. For an excellent example of this, Madeleine Albright, secretary of state during the Clinton administration, was quoted as saying that "if we have to use force, it is because we are America. We are the indispensable nation. We stand tall. We see further into the future."[14]

Many Americans, perhaps adopting the American Revolution as the frame of reference for all revolutions, fail to realize that contemporary revolutions may evolve out of considerably different circumstances. The presumption that revolutions ultimately lead to democratic systems reflects the view that the American Revolution represents the path that others should follow and also marks the primacy of hope over experience. The 2011 Arab Spring uprisings have not developed into truly democratic systems, and the prognosis for their doing so is dim. It is difficult for Americans to believe that freely elected governments may not support Western democratic values, as evidenced by the temporary resurgence of the Muslim Brotherhood in Egypt following the overthrow of President Hosni Mubarak.

The openness of US affairs, even on the most sensitive issues, provides an opportunity for adversaries to use the media as a strategic asset to further their own cause. In the twenty-first century, technology offers even greater opportunity for this information-age strategy. Such strategies include attempts to sway public opinion and elected officials and to support sympathetic interest groups. This was seen clearly in Russian hacking and social media campaigns designed to affect the 2016 and 2020 presidential elections and remains a significant concern. In addition, the level of debate within the system itself can send the wrong signals to adversaries regarding our political will. If adversaries read the US policy posture incorrectly, it could lead to dangerous strategic choices that could force a US military response. It is likely that many adversaries do not understand the United States any better than the United States understands them.

Still, the US public's commitment to the democratic faith remains a pillar of the system, and it is an inherent strength that provides the endurance to prevail over the long term. Although the democratic faith may not result in success in every venture, especially in the short term, it seems better than its alternatives.

Multiculturalism and Cultural Diversity

The cultural diversity of the United States can contribute to US national security in several ways. First, it is hard to imagine a region that does not have some representation in the US population. This makes it likely there will be knowledgeable people in the country about most everywhere in the rest of the world. Second, every country in the world has a connection with the United States by virtue of this diversity. Third, if this diversity is well integrated into the general society, different cultural groups can express their opinions and make it more likely that wise policies will emerge. To suggest that this would always be the case would be naive, of course. If, on the other hand, the diversity is not well integrated and various groups feel disconnected from the general society, these advantages will diminish.

The Messianic Spirit

The messianic spirit pervading US political culture also has strengths and weaknesses. US concern with moral and ethical issues provides strength in dealing with national security issues because its humanistic orientation strikes a responsive chord in others that is beneficial for US policy and strategy. The downside is that the belief in our messianic mission causes many foreign states and peoples to perceive Americans as self-righteous and arrogant; we ignore other cultures in trying to rule the world. Many countries, especially but not only those in the less developed world see this as imperialistic, further fueling anti-US sentiment. At the very least, US moral and ethical views can be disturbing to hundreds of millions of people, including some populations in strategically important areas whose heritage is not Judeo-Christian.[15] This provides ample opportunity for US adversaries to take advantage of differences between Americans and others.

National Security in a Democracy

De Tocqueville's assessment of democracy's ability to conduct foreign policy is especially relevant to US national security policy today. The US political system is not well suited to timely and thorough development of policy and strategy to create the most effective national security policy, and often the US system responds only at the last minute. Wise policies and strategies are most likely to occur when the security issues are clear, the adversary is identifiable, and general support exists within government and the general public. Unfortunately, many serious threats to US national security are difficult to comprehend, even though they are quite real.

The real issue, then, is how to reconcile the demands of national security with those of democracy. There are trade-offs, to be sure, but how should one balance the protection of democracy and US values, on the one hand, with the threats to national security that are not amenable to democratic processes and principles, on the other? US national security policies are constrained by all the forces and power clusters characteristic of the US political system, and most Americans expect the government to conform to democratic proprieties. Yet if success is to be achieved, something must give if the goal is to further democracy.

This is not to suggest that the United States should pursue policies that seriously undercut moral principles. But to assume that the United States must not engage in secret or covert operations in support of its national interests requires a simplistic misreading of the nature and character of the strategic landscape. Moreover, the argument that political ends can never justify covert means precludes a nuanced and calibrated approach to subtle and complex security dilemmas. Laws and procedures are ambiguous, and policies and strategies are shaped by power struggles within the US government. Put differently, as two prescient scholars noted just after the Soviet Union broke up:

> If the current trends continue, the [United States] will not be the preeminent economic or military actor in the twenty-first-century international affairs. It will not exercise global leadership or hegemony in the West, and possibly not even unchallenged hegemony in the Americas. The programme to establish a "New World Order" will have failed, if it has not already. . . . Social polarisation may lead to increased class tensions and conflict between cultural groups and to more civil disorder.[16]

Although this book focuses on policy and process, our study is pursued in the context of the US political culture. Those involved in designing policy and exercising power to influence the approval process operate within that context. To examine policy and process separate from US political culture is not only sterile, it is likely to distort the nature of policymaking and the character of the approval process.

Despite all the disadvantages open systems face in their dealings with authoritarian systems, rogue regimes, and international terrorists, in the long run democracy has the advantage. The involvement of people in the governing process, their ability to voice their views freely, and the ultimate responsibility to the people of those in office establish stability, legitimacy, and capability like no other system. Herein lies the true strength of open systems: little can prevail against the strength of people who, having examined and debated the issues, are convinced of the right policy and strategy.[17]

We next examine the institutions, offices, and individuals involved in policymaking. Similarly, we look at nongovernmental groups, the media, and the US public regarding their respective roles in national security policy and the policy approval process.

Notes

1. Alexis de Tocqueville, *Democracy in America,* edited by J. P. Mayer, translated by George Lawrence (Garden City, N.Y.: Anchor, 1969), pp. 228–229.
2. Ibid.
3. In July 2019 a British tabloid newspaper, the *Daily Mail,* printed a series of confidential cables to the British foreign office from their ambassador to the United States that were highly unflattering to President Trump and his administration. The president took great exception to his comments, and the ambassador's effectiveness as a representative being compromised, he resigned. See https://www.dailymail.co.uk/news/article-7244549/Trump-set-act-diplomatic-vandalism.html?fbclid=IwAR3E06U2tQZziqllHmHROGu1-g7ti9Yf1nvA3Xci8brMC4VXf7axZkFKLBc.
4. While it is essential to focus on the serious national security challenges that remain, one should not lose sight of progress that has been made since the end of the Cold War.
5. See John Allen Williams, "Understanding Asymmetric Warfare: Threats and Responses," *National Strategy Forum Review* (Summer 2007), pp. 23–26.
6. De Tocqueville, *Democracy in America,* p. 233.
7. Winston Churchill, speech before the House of Commons, November 11, 1947, http://wais.stanford.edu/Democracy/democracy_DemocracyAndChurchill%28090503%29.html.
8. The concepts of "separation of powers" and the resulting "checks and balances" have never been more clearly discussed than by James Madison in Number 51 of *The Federalist.* All of the "Federalist Papers" are available from the US Library of Congress. See http://thomas.loc.gov/home/histdox/fed_51.html for Federalist 51.
9. See Matthew C. Weed, *The War Powers Resolution: Concept and Practice* (Washington, DC: Congressional Research Service, March 8, 2019), https://crsreports.congress.gov/product/pdf/R/R42699.
10. James Q. Wilson, *American Government: Institutions and Policies,* 5th ed. (Lexington, Mass.: Heath, 1992), pp. 328–329.
11. European skepticism about the United States is evident in a comment by French prime minister Georges Clemenceau after World War I: "America is the only nation in history which miraculously has gone directly from barbarism to degeneration without the usual interval of civilization." See https://www.britannica.com/quotes/biography/Georges-Clemenceau.
12. Terry L. Deibel, "National Strategy and the Continuity of National Interests," in James C. Gaston, *Grand Strategy and the Decisionmaking Process* (Washington, DC: National Defense University Press, 1992), p. 48.
13. Ronald Kessler, *Inside the CIA: Revealing the Secrets of the World's Most Powerful Spy Agency* (New York: Pocket, 1992), p. xxv. See also Admiral Stansfield Turner, *Burn Before Reading* (New York: Hyperion, 2005).
14. Ray Mosley, "What Went Wrong with Pax Americana," *Chicago Tribune,* February 22, 1998, p. 1. It is not difficult to see why this is perceived as arrogance by countries that are the objects of such attention.
15. See, for example, Samuel P. Huntington, *The Clash of Civilizations and the Remaking of World Order* (New York: Simon and Schuster, 1999).

16. K. R. Dark with A. L. Harris, *The New World and the New World Order: US Relative Decline, Domestic Instability in the Americas, and the End of the Cold War* (New York: St. Martin's, 1996), p. 144.

17. See de Tocqueville, *Democracy in America,* p. 244. For a well-reasoned view of the challenges facing democracy, see Jean-François Revel, *How Democracies Perish* (New York: Harper and Row, 1984).

PART 2

National Security Institutions

6

The Presidency

Several years after he left office, President Lyndon Johnson wrote, "No one can experience with the President of the United States the glory and agony of his office. No one can share the majestic view from his pinnacle of power. No one can share the burden of his decisions or the scope of his duties."[1] Not only did Johnson capture the essence of the presidency with these few words, he put his finger on the reason why it is difficult for others to comprehend the power and responsibilities of the position.

In studying the president and the presidency, we must recognize the considerable disagreement over the power, responsibilities, and limits of the office as well as the characteristics of the most effective type of executive. Security policy is complex, and when studying the president and national security policy it must be understood that national security issues are often linked to domestic issues—although this is less so in times of crisis. This can make it difficult to isolate national security policy from other policy issues dealt with by the president.

During the Cold War, with its enormous challenges related to nuclear weapons and superpower conflicts, national security issues took on a dimension rarely known in the past.[2] With the breakup of the Soviet Union at the end of 1991, the possibility of war between major powers diminished considerably. The end of the superpower era brought with it a tendency for the US public to focus on domestic issues, at least until the September 11 attacks, but that does not make it any easier for the president. The world remains a dangerous place and the many conflicts in progress, international terrorism, and the rise of peer competitors to the United States have created a host of new challenges for national security policy. The attention of Congress to these issues waxes and wanes, but it retains the power to have a significant impact on national security policy if it chooses.

It is ironic that the responsibilities of the president in national security have become more complex even while the ability of the office to make and implement decisions has become increasingly difficult, mainly owing to the countervailing power of Congress and the unsettled international landscape. Although the impact of September 11 was significant in the short run in enhancing the leading role of the president in national security policy and focusing public attention on related issues, the lengthy US military operations in Afghanistan beginning in 2001, in Iraq beginning in 2003, and in the campaign against the Islamic State (ISIS) in Iraq and Syria beginning in 2014, eventually undermined public support for military operations abroad.

Some time ago a perceptive observer made this astute comment on presidential power: "Presidential power may be greater today than ever before. . . . It is misleading, however, to infer from a president's capacity to begin a nuclear war that the chief executive has similar power in most policy-making areas. . . . Presidents who want to be effective in implementing policy changes know they face a number of constraints."[3]

The primary purpose in this chapter is to study the president and the presidency in relation to national security policy, but this cannot be done without an understanding of the nature of the office. Below we identify some issues that help define the Oval Office and its occupant, with particular reference to national security in the contemporary period.

Evolution of the Office: An Overview

Political struggles over the nature of the presidential office are part of US history. Indeed, the disagreements began with the Constitutional Convention and the founding of the nation. Most delegates were wary of investing too much power in the executive, but they also realized that a strong executive was necessary for effective governmental action. What resulted was a series of compromises leading to the basic executive structure we see today. The acceptance of a single executive with significant powers was influenced in no small measure by the fact that delegates knew that George Washington would be the first president and he was the model for the office they created.

Washington won admiration for his fairness, honesty, and integrity as well as his leadership in the battlefield in the drive for US independence. He seemed to stand above politics and epitomize what a president should be and the role he should fulfill. Upon assuming the presidency, however, Washington commented that he felt like a "culprit who is going to the place of his execution." Other presidents have had similar feelings upon assuming office. William Howard Taft thought it was the "loneliest place in the world"; Warren Harding referred to the White House as a "prison"; Harry Truman declared that being president was "like riding a tiger. A man has to keep riding or be swallowed." While all presidents have been the subject of criticism, few have faced as much of it as Donald Trump or dealt with it with such hostility. Indeed, his background, personal history, and leadership style made him an object of great controversy even before he ran for office. President Biden is also facing a great deal of criticism as he deals with difficult foreign and domestic issues, especially regarding the chaotic withdrawal of military forces from Afghanistan in 2021. He also faced a severe crisis as Russian president Vladimir Putin sent major forces into Ukraine to topple the government there—a crisis that continues as of this writing.

Our purpose is educational and analytical rather than political, and we wish to maintain that stance when discussing a figure as polarizing as Donald Trump. Our analysis is not directed at a particular president and is not intended be political, even if it may seem so in these highly charged political times. We believe that the structure of this book and the theories we apply remain solid and provide a useful structure for an analysis of the presidency.

Donald Trump's campaign promises met with a very mixed reaction, but however appalled many people were by them, they helped propel him to victory in 2016, almost did again in 2020, and as of this writing he is considering another run in 2024. His conduct in office pleased his base, but did little to reassure his doubters. His national security policies were extremely controversial and in many ways were a sharp departure from those of his predecessors. Even more alarming to some was his improvisational and apparently ad hoc decisionmaking style. His unwillingness to seek advice from either his staff or the military, intelligence, and foreign policy establishments in government was also a source of growing concern. Losing the Republican majority in the House of Representatives in 2018 increased congressional oversight but seemed to have had little effect on his behavior.

To be fair, many of President Trump's policies may well deserve support. Ryan Evans, founder and head of the national security web magazine *War on the Rocks,*[4] speaks of "Trumportunities" and "Trumptastrophes." Reasonable people will differ on how particular policies should be categorized, but we believe the Trumportunities include pushing back on Chinese economic and military actions and attempting to get US partners to devote more resources to their own defense, and the Trumptastrophes include not using the national security establishment that worked for him effectively, sowing discord between the United States and its traditional allies, and placing partisan political considerations above sound policies.[5]

Some presidents have commented on the demands and problems of the presidency, and several have aggressively tried to expand its power. George Washington, Andrew Jackson, Abraham Lincoln, Theodore Roosevelt, Woodrow Wilson, and Franklin D. Roosevelt used their presidential powers extensively and actively, and their incumbencies are usually identified as "strong" or "expansionist" periods of the presidency. In the modern period, each president has interpreted the powers of his office expansively, regardless of political party.

The presidency, even though it was fashioned by people who had a deep mistrust of executive power, has become the focal point of national politics and the center of the national policy process. The president is the only nationally elected public official save for the vice president, and is the most recognized public leader in the country. Yet the president depends on many other political actors to accomplish political goals and implement policy. Although the president has the authority to order a worldwide alert of US armed forces and the deployment of the military, he or she can meet overwhelming resistance in more mundane matters. The president can submit programs to Congress, but cannot allocate money to them without congressional approval, a great frustration for President Trump in his attempt to expand physical barriers along the southern border. President Truman identified this anomaly when discussing the problems his successor, Dwight Eisenhower, would face upon moving into the Oval Office: "He'll sit here [tapping his desk], and he'll say, 'Do this! Do that!' And nothing will happen. Poor Ike; it won't be a bit like the army. He'll find it very frustrating."[6]

The frustration is compounded by struggles between the executive and legislative branches of government over presidential authority and initiatives—a conflict especially visible in national security policy. For example, the 1987 Iran-contra hearings, although focusing on covert operations and illegal transfers of funds, were at their core a struggle over presidential power and congressional attempts to control that power. In responding to serious and immediate threats, however, the president is accorded almost absolute power, but only for a while.

The extent of presidential power is not determined solely by the Constitution, legal grants of power, or political skill. Traditions, custom, and usage play important roles in determining the power of the Oval Office, but only if a president is willing to be constrained by them. The way in which political parties operate, the relationship of political actors to the executive office, the political climate, the functioning of the federal bureaucracy, and the nature of security threats all impact presidential powers. A realistic study of the presidency and national security thus requires an appreciation of the complex nature of the office and the presidential power base. It must also be based on the realization that the success and impact of the office depend on the personality, character, and leadership style of the president.

Theodore Sorensen, special counsel to President John F. Kennedy, put it this way:

> Self-confidence and self-assertion are more important than modesty. The nation selects its President, at least in part, for his philosophy and his judgement and his conscientious conviction of what is right—and he need not hesitate to apply them. He must believe in his own objectives. He must assert his own priorities. And he must always strive to preserve the power and prestige of his office, the availability of his options, and the long-range interests of the nation.[7]

Beginning with Andrew Jackson, who sowed the seeds of the so-called modern presidency, many presidents have used a variety of methods to expand the power of the Oval Office, including appealing directly to the people and broadly interpreting the Constitution to favor presidential power. Jackson rationalized his view of the presidency this way: "Each public officer who takes an oath to support the Constitution swears that he will support it as he understands it, and not as it is understood by others." The growth of the power of the presidency accelerated in the twentieth century to the point where Arthur Schlesinger labeled it the "Imperial Presidency."[8] But, as President George H. W. Bush learned in dealing with Congress, "Congress . . . restricted the president's conduct of the executive branch by involving itself extensively in the details of domestic and foreign policy."[9] Moreover, "it should be understood that the pendulum of power swings, only to swing back somewhat later. . . . Eventually Congress has always moved to reassert its position."[10] Presidents Bill Clinton and George W. Bush faced the same problems regarding their policies on China, Cuba, and Iraq, among others. Nonetheless, they expanded presidential powers by utilizing executive orders, a presidential prerogative that does not require congressional approval. Executive orders were used extensively by

President Obama and President Trump to advance policies that were not sufficiently popular to get through Congress, and the practice has continued under President Biden, who quickly moved to undo many of the executive orders of his predecessor.

How does the president exercise power in pursuit of US national security and national interests? What power does he have, and how does he deal with constraints and limitations on that power? What leadership style is effective in developing a coherent policy posture and strategy in national security? Any serious examination of the Oval Office and national security policy must begin with some attention to models and approaches. By first studying the broad dimension of the presidency, we design a method of linking to it the specifics of national security policy.

The Study of the Presidency

Virtually all studies of the presidency incorporate the various formal roles of the president (usually referred to as "institutionalized roles").[11] The most important institutionalized roles are chief of state, chief executive, commander in chief of the armed forces, chief diplomat, chief legislator, and party chief. The president's national security powers derive mainly from the roles of chief executive, commander in chief, and chief diplomat.

Scholars have designed several theoretical models to study the presidency.[12] Many focus on one or two important roles, showing how they affect the political system and political actors. For example, the view of the president as manager focuses on a bureaucratic model of the office. Another approach is based on personality and character and identifies the passive or active presidents and their consequences for leadership and policies.[13] The "president as great man" view assesses the officeholder according to his greatness in responding to the challenges of the time. The president's personal perception of office, which can vary from constitutional to expansionist, offers another approach. Still another approach evaluates leadership style and evolves out of studies that use personality and character to examine the president's ability to develop consensus, loyalty, and commitment in his staff and the executive branch and motivate them to pursue his policies. In addition, the "two presidencies" model was advanced to suggest that there is a domestic president and a foreign policy president.[14]

Each approach can be useful depending on the circumstances. Our study is based on two elements common to all: leadership style and personal perception of the office. We take our cue from Theodore Sorensen's view that the president's self-confidence, self-assertiveness, and philosophy are critical in the functioning of the office: "Each President has his own style and his own standard for making decisions—and these may differ from day to day or from topic to topic, using one blend for foreign affairs, for example, and another for domestic. The man affects the office as the office affects the man."[15] As Sorensen pointed out, "White House decision-making is not a science but an art. It requires, not calculation, but judgment."[16]

We believe that the character and personality of the president are critical in shaping the moral authority and power of the office. Thus there are four important components to a systematic study of the president's role in national security: (1) the president's leadership style, personality, and character as critical determinants of how the Oval Office functions with respect to national security policy and process; (2) how the president views the power and limitations of the office and his or her role in furthering its prestige and power; (3) the president's mind-set (or worldview) regarding US national interests and the international security environment and how they affect the posture the administration attempts to put into place; and (4) the president's ability to bring the first three components to bear upon the national security establishment so as to integrate its efforts to develop and implement coherent policy.

These four factors must be used to develop support among Americans, the Congress, and the federal bureaucracy. This is necessary to develop the national will, political resolve, and staying power for the implementation of national security policy and strategy. The president's effectiveness in achieving these ends rests in his or her ability to deal with all the complex issues we have identified and yet remain within the bounds of democratic proprieties.

Leadership Style

Although the national security policy process appears to be rational and clear in legal and organizational terms, in reality it is a political and, at times, a chaotic process, reflecting the power and interests of many domestic and foreign actors. The president stands at the vital center of this process. Depending upon his leadership style and philosophy and his effectiveness in exercising the power of the office, the president can minimize the influence of other political actors and guide the process to ensure that his own views and policies prevail.

The term *leadership style* is not easy to define. It refers to the president's way of doing business, which evolves from personality and character. It is the way in which the president exercises authority and power to create trust, loyalty, commitment, and enthusiasm within the administration and is crucial for successful policymaking and implementation. The term *personality* refers to the psychological and social behavior that lays the foundation for one's perceptions and worldview. The term *character* is defined as the way in which the "president orients himself towards life—not for the moment, but enduringly."[17] Put simply, personality and character fix the political behavior and shape the way the president views the world and his or her own role in it. In addition, they determine the way the president relates to subordinates, to the public, and to the nation's governing institutions.

The point to remember is that the way in which a president governs is every bit as important as the inherent power of the office. And when we talk about the powers of the presidency, we must consider three factors: sense of purpose, political skills, and character.[18] James Q. Wilson once concluded that "the public will judge the president not only in terms of what he accomplished but also in terms of its perception of his character."[19]

The leadership style that emerges from personality and character determines whether the president will follow a magisterial, bureaucratic, managerial, or corporate method of governing—or a combination of these. In the magisterial style, the president places him or herself as the authoritative head of the government. The bureaucratic style is one in which the president leads as the chief bureaucrat, with all the mindsets and perceptions that that role entails. In the managerial style, the president strives for efficiency in the administration through the close supervision advocated by managerial principles. In the corporate style, the president governs like the chairman of a large business, combining the managerial approach with commitment and loyalty.

Most presidents tend to centralize their role around a particular style, although there are elements of each in the way most modern presidents lead. Regardless of the chosen leadership style, it is implemented to establish a presence, so to speak: the president sets policy directions and saturates the administration with his views on world affairs. Yet the president must do this without frustrating the expression of alternative views and options—not an easy task.

A successful president can stamp national security with his or her personal style and outlook. If the executive is unsuccessful, then policy is likely to be incoherent and ineffective, and other domestic political actors are likely to increase their power. Short of crises or wars, gridlock between the executive and legislative branches of government on national defense and other issues is invited by a president who is either too vacuous to lead or too assertive to compromise.

Perceptions of the Office

President Johnson perceived the power of the Oval Office this way: "The source of the President's authority is the people. He is not simply responsible to an immediate electorate, either. The President always has to think of America as a continuing community. He has to prepare for the future."[20] President Richard Nixon, reaffirming the national character of the office after his 1960 loss to John F. Kennedy, stated: "The first responsibility of leadership is to gain mastery over events, and to shape the future in the image of our hopes. He must lead. The President has a duty to decide, but the people have a right to know why. The President is the only official who represents every American, rich and poor. The Presidency is a place where priorities are set and goals determined."[21]

The perceptions of office that have historically evolved and are applicable in the contemporary period cluster around four basic types:

1. The constitutionalist, or Buchanan-type, presidency.
2. The stewardship, or Eisenhower-type, presidency.
3. The prerogative, or Lincoln-type, presidency.
4. The monarchical, or Trump-type presidency.

In the constitutionalist presidency, the president views the power of the office as strictly bound by the US Constitution: the document must clearly sanction any presidential action. This narrow view of presidential authority limits the president to reacting to the policies of others, with little presidential initiative. President James Buchanan (1857–1861) provides an extreme example: he felt he was simply the custodian of the Constitution and tried to remain aloof from political battles. In 1860 he even denied that he had the power to use force to prevent the secession of southern states.

The stewardship view presumes that the Oval Office is nonpolitical, or at least nonpartisan. The president acts as an agent of the nation, supervising operations of the state machinery. Dwight Eisenhower (1953–1961) exemplified such a president. Standing aloof from party politics and political battles, Eisenhower felt that his veto power over legislative bills was the key to the presidential office. From this he took the position that the president should advise the nation, negate ill-advised legislation and ill-advised policies, and be the chief broker of the political system. This approach borrows some elements from the constitutional as well as prerogative approaches. After President Eisenhower left office, however, there was growing evidence that he was much more active and politically involved than previously believed.[22]

The prerogative view is that the powers of the presidency are exclusive rights resting in a special trust to the benefit of the nation. This approach is best exemplified by Abraham Lincoln (1861–1865), who made it clear during the Civil War that the legal limits imposed by the Constitution had to be transcended during such an existential crisis. He felt that the president was the sole representative of all the people and the office was the only institution capable of dealing quickly and decisively with major national problems. President Theodore Roosevelt (1901–1909), another prerogative president, put it this way:

> I did not usurp power, but I did greatly broaden the use of executive power. In other words, I acted for the public welfare, I acted for the common well-being of all our people, whenever and in whatever manner was necessary, unless prevented by direct constitutional prohibition. . . . My belief was that it was not only his right but his duty to do anything that the needs of the nation demanded unless such action was forbidden by the Constitution or by the laws.[23]

No president can be successful today if he ignores the prerogatives of office; the changed security landscape and new international challenges mandate such an approach. For example, in the aftermath of the terrorist attacks of September 11, the George W. Bush administration embraced a theory that assumed broad powers for the president acting in the interest of US security against imminent threats.

The special nature of the Trump administration suggests there is a fourth category, the monarchical presidency, in which the focus is on the person and interests of the president and his or her reelection. In the monarchical style, the president does not feel constrained by norms or precedents and sometimes even by laws. Personal loyalty is paramount and the president governs based primarily on what he or she believes will be of benefit personally or politically. This type of president would be driven largely by public opinion polls and the media and would come into office with no commitment to a particular type of role, other than maintaining power in office and perhaps benefiting from it as much as possible while serving. In such a presidency policy and personnel decisions would be subservient to the personal and political needs of the president and absolute loyalty to the president would be demanded.

Mind-Set and External Threats

Another key element is how the president perceives external threats, that is, the president's mind-set regarding the international security environment. Presidents often come to the Oval Office with an established viewpoint, although their views of national security matters may well evolve over the course of their presidency. Over time, a president gains experience, and this will help inform his or her strategic choices and policy implementation. Although some come to the office with more experience in foreign and national security policy than others, each has a set of beliefs that stem from public service, political involvement, and other socialization, shaped by experience with allies and adversaries alike. Some have experience only in the domestic arena, however, which may not be the best background to deal with foreign adversaries and allies.

A lack of expertise can be mitigated through the appointment of experienced and knowledgeable people to the inner circle. President George W. Bush appointed such people following his election in 2000 to compensate for his limited foreign policy experience, as did President Obama in 2009. This was true also in the early Trump administration, with the appointment of Army Lieutenant General H. R. McMaster as national security advisor and retired Marine Generals James Mattis as secretary of defense and John Kelly as White House chief of staff. For a time they provided strategic advice and structure to the administration, but President Trump soon tired of both advice and structure, and they were all replaced in a short time. This is not surprising in principle; every president surrounds himself with advisors with whom he feels comfortable.

From the end of World War II (1945) to the end of the Cold War (roughly 1989), the central US national security preoccupation was the relationship with the Soviet Union as well as the threats that Marxist-Leninist ideology posed to democracy and the West. The wars in Korea and Vietnam were fought with this as a background. The primary focus on the conflict spectrum was on nuclear and conventional conflicts, and the US fought the Vietnam War primarily as a conventional, rather than low-intensity, conflict—unfortunately for US success in the war. Each type of conflict posed a different security challenge to the United States. How the president viewed the seriousness of these conflicts, what he believed was the proper world order, and the place of the United States in that order were all critical elements in presidential performance. They remain so today.

Many observers, including those in the media, tend to oversimplify a president's mind-set by categorizing it as either hard-line (hawkish) or soft-line (dovish). Rarely does an individual's perspective fit neatly into one category or the other. True, each president does have a unique perspective about adversaries and is likely to appoint officials with compatible views to high places in the national security establishment. But the responsibilities of office, the political forces within the domestic system, and the continuities of national security policy prevent the president from implementing policies that rest solely on his own preferences and initiatives. In many cases the demands of office require compromises with a multitude of political forces, both domestic and foreign.

Finally, the term *mind-set* does not mean strict adherence to past perspectives or a dogmatic ideology. The responsibilities for US national security and protection of the homeland weigh heavily upon any president, a burden magnified by the problems of the proliferation of weapons of mass destruction (WMD) and by an uncertain world. Thus the need to reexamine US security interests and the state of US military posture is critical to national security. The uncertain world order and the changing and dangerous security landscape of the current period makes such a re-examination especially important.

The degree to which a president is able to establish US security policy and strategy is contingent upon how he or she functions in the first two elements of our framework: leadership style and perception of the office. For example, it is unlikely that a president who sees the office from a narrow constitutional perspective will be able to develop innovative policies and tackle the range of conflicts across the spectrum. Additionally, a president who cannot inspire Americans, articulate a vision for US national interests, or develop a consensus within the federal bureaucracy to support presidential policies will be unable to convince adversaries, potential adversaries, and even allies of the seriousness of US interests or of US staying power and political resolve.

The President and the National Security Establishment

To be effective the president must ensure that the national security establishment functions effectively. Thus the president's appointments to key positions are critical in determining the degree to which he will be able to move the establishment to support his or her worldview and policy directions. Appointees must be able to provide the necessary linkage between the establishment and the president. The assistant to the president for national security affairs (the national security advisor) and the secretaries of defense and state are principal actors in this respect. The individuals selected for these positions must have the president's trust and be seen by those on the National Security Council (NSC) and by the national security staff, among others, as enjoying a special relationship with the president in national security matters. Equally important, the national security advisor is a key member of the president's inner circle. Through the national security advisor, the national security staff becomes an extension of presidential power. A major problem is to ensure that this special relationship among the president, the national security advisor, and the national security staff is not seen by the secretaries of defense and state and their departments as infringing on their own prerogatives or as a threat to their roles and functions.

The behind-the-scenes influence of vice presidents is difficult to assess. Discussions of the George W. Bush presidency must include the expansive role of Vice President Richard Cheney. Contrary to historical expectations of a weak and powerless vice president, Cheney by all accounts exerted strong influence behind the scenes, particularly in the national security arena, where Cheney, a former secretary of defense, had significantly more experience than President Bush. Cheney and Secretary of Defense Donald Rumsfeld formed a powerful coalition in favor of an ambitious foreign policy and military confrontation in the Middle East. President Obama's vice president, Joseph Biden, was also more experienced in national security issues than was the president who appointed him and offered advice privately and occasionally publicly on national security issues. Such advice as President Trump's vice president, Mike Pence, offered was given privately. In contrast, President Biden has delegated two challenging areas of responsibility to his vice president, Kamala Harris: ensuring voting rights and dealing fairly with the influx of refugees and others along the southern border with Mexico. She also got international experience and exposure in the opening stage of the 2022 Russia-Ukraine crisis.

To understand the totality of the presidential role and power base, one must also consider the role of the president's spouse. First Ladies such as Hillary Clinton, Laura Bush, Michelle Obama, and to some extent Melania Trump were active in issues of particular interest to them, mainly domestic rather than foreign or national security policy. First Lady Dr. Jill Biden is concerned with education issues in general, cancer research, aiding military families, and education. In a break with tradition, she kept her paid teaching position at a community college. The degree to which their positions may have an impact on national security policy or on appointments to the law national security establishment is not publicly known.

In sum, the president's ability to control the national security establishment and to develop national will and political resolve is based on his or her leadership style, perceptions of the office, and mind-set. Even a president who sees the world in realistic terms, uses the power of the presidency aggressively, and applies effective leadership, will not always be able to develop enough public support for his or her national security policy.

The National Security Establishment

The current structure of the national security establishment was established in the aftermath of World War II. The war experience, as well as the belated recognition that a better system of unified command and control was necessary, led to passage of the National Security Act of 1947, amended by Congress in 1949, 1958, and 1986.

The 1947 act established the National Security Council, the Office of Secretary of Defense, the US Air Force, the Joint Chiefs of Staff (JCS), and the Central Intelligence Agency. For the first time, the president was provided with a principal staff member whose purpose was to give him advice and assistance in matters pertaining to national security. Although the secretary of defense was to oversee the national military establishment, it was presided over by three cabinet-level officers heading three separate executive departments: the Army, the Navy, and the Air Force.

The 1949 amendments created the Department of Defense, making it an executive department and reducing the services to military departments with no cabinet-level officers except the secretary of defense. The position of chairman of the Joint Chiefs of Staff was created to preside over the JCS, which was to remain a corporate body and principal advisor to the president and secretary of defense.

The 1958 amendments reinforced the secretary of defense by granting him or her legal authority over all elements within the Department of Defense, thereby imposing a degree of unification on the military services. Additional staff assistance was also provided. The JCS chairman was given added responsibilities over the joint staff and became a voting member of the JCS for the first time. The 1958 amendments also provided that operational commanders would report to the secretary of defense, not through the military departments and service chiefs (who would be responsible for administration and logistical support).

The 1986 Goldwater-Nichols Act provided additional changes within the Department of Defense by strengthening the position of the JCS chairman to make him or her the principal military advisor to the president, creating a deputy chairman of the JCS, requiring joint service duty for future general officers, and changing promotion criteria to encourage the best officers to serve on the joint staff.

The National Security Act of 1947 also established the basis for integrating political, military, and intelligence functions into the national security policy process through the NSC, thereby giving the president a structure for a systematized assessment of policy and strategic options. Although the move toward centralization and unification has achieved a great deal, internal problems of power decentralization and diffusion remain.

The national security establishment as it exists today is shown in Figure 6.1. An important structural evolution since 1947 has led to the emerging prominence of the national security advisor. The national security advisor is appointed by the president without approval by the Congress, and the position has been held by such important figures as Henry Kissinger, Brent Scowcroft, Colin Powell, Zbigniew Brzezinski, Condoleezza Rice, and H. R. McMaster. President Trump had several national security advisors, with very different personalities, styles, and

Figure 6.1 The National Security Establishment

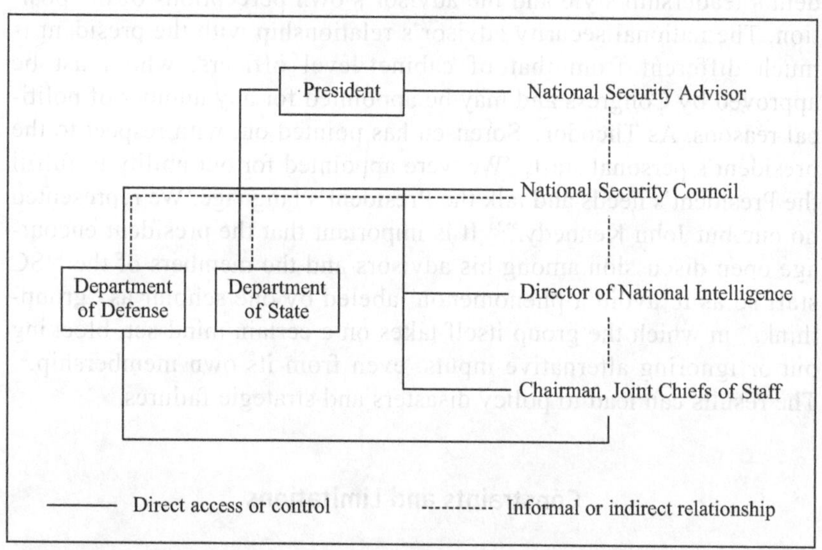

Notes: The four statutory members of the National Security Council are the president, vice president, and secretaries of state and defense. The two statutory advisors are the director of national intelligence and the chair of the Joint Chiefs of Staff. The national security advisor is an important actor in the national security policy process and usually plans and coordinates the meetings of the National Security Council.

relationships with the White House. Their continuation in office was based in large part on whether the advice they gave was consistent with what the president wanted to hear. President Biden's first national security advisor, Jake Sullivan, is widely experienced in security policy issues but not previously a national figure.

The role of national security advisor goes beyond policy coordination; indeed, some observers feel the power exercised by individuals such as Kissinger and Brzezinski undermined the role of the secretary of state. The fact that the national security advisor can provide the president with a unique view not bound by executive department perspectives makes his or her advice especially important. The national security advisor's own personality and character, perceptions of the role, and personal access to the president provide that person with a power base that translates into prominence in all areas of national security. Both Condoleezza Rice in the George W. Bush administration and John Bolton in the Trump administration were highly visible participants in the national security decisionmaking process and senior advisors to the President. John Bolton eventually left the Trump administration in 2020, publicly critical of the president and his decisions.

The extent of the NSC advisor's influence depends on the president's leadership style and the advisor's own perceptions of the position. The national security advisor's relationship with the president is much different from that of cabinet-level officers, who must be approved by Congress and may be appointed for any number of political reasons. As Theodore Sorensen has pointed out with respect to the president's personal staff, "We were appointed for our ability to fulfill the President's needs and talk the President's language. We represented no one but John Kennedy."[24] It is important that the president encourage open discussion among his advisors and the members of the NSC staff so as to avoid a phenomenon labeled by one scholar as "groupthink," in which the group itself takes on a certain mind-set, blocking out or ignoring alternative inputs, even from its own membership.[25] The results can lead to policy disasters and strategic failures.

Constraints and Limitations

The assertion that the president has the preeminent role in national security must be qualified. The first qualification is the very nature of the US political system, with its emphasis on democratic proprieties and values. A president sensitive to domestic politics and public expectations understands that the presidency is supposed to symbolize the best of the US system. Behavior and words must be in keeping with the goals and expectations of the people if a president is to be effective.

The constraints this imposes on the president create an inherent dilemma in dealing with national security. Especially in response to unconventional conflicts, some of the most effective policies and strategies stretch democratic norms and expectations. Even though secrecy and covert operations are an integral part of the response to unconventional conflicts, it is difficult for many citizens, as well as members of Congress, to accept the necessity for such operations. The view is that these operations tend to be an undemocratic and un-American way of war. Therefore the president must find ways to reconcile democratic norms with acceptable and effective strategic options.

Russia's growing interest in unorthodox political-military warfare within the "gray zone," including its efforts to influence the US presidential elections of 2016 and 2020 and the Ukraine invasion of 2022, and China's equally assertive cyber war against American public and private entities, have sparked concerns among US government officials, security experts, and pundits. Gray zone activities are those that fall between routine statecraft and direct or open warfare.[26] In addition, social media have evolved from a sidebar distraction compared to "mainstream" sources into a separate information battlefield among state and nonstate actors, including individuals and like-minded small groups with social media outreach to create transnational interest communities. As well, social media encourage Internet-based intellectual tribalism in which people are continually reinforced in their existing biases (the confirmation bias effect) instead of exposing themselves to conflicting or contrarian views. The result is to push much social media dialogue toward the extremes of distrust, including advocacy of violence, rapid spreading of fictitious or malicious messages to larger audiences, and subverting existing political authorities.[27]

Second, the current distinctions between the national security policy process and the domestic policy process are not as clear as they once were. The turmoil of the 1960s, followed by the US withdrawal from Vietnam, the Watergate affair, the impeachment of President Clinton, and subsequent events eroded public confidence in the president's ability to manage national security affairs. Neither was public faith restored by later events, such as the inability to stop nuclear weapons programs in North Korea and Iran, continued turmoil in Iraq and Afghanistan, high-seas piracy, an increasingly assertive Russia and China, a continuing terrorist threat, and the undemocratic direction of some of the Arab Spring uprisings from 2011. Although there is little a president can do about many of these issues, the public expects him or her and to have more success than is often obtainable.

The motivations behind certain national security policies and the way in which policy is made have been sensitive issues for Congress and other political actors. In these times of hacking, security breaches, and extensive public and private databases, keeping secrets has become increasingly difficult. Writing in 1998, then-senator Daniel Patrick Moynihan argued for more openness in government and for dismantling much of the secret system and apparatus. "*Analysis,* far more than secrecy, is the key to security."[28] The events of late 2001 and beyond, as well as the scope and direction of the US response, muted some of this criticism at first, but with the passage of time it has returned in full force.

In addition, the main concern of national security has qualitatively shifted. The clear purposes that characterized the Cold War era have been replaced by the challenge of a more ambiguous security environment. This began with the ascension of Mikhail Gorbachev as head of the Soviet Union. His glasnost and perestroika initiatives had the unintended result of paving the way for the dissolution of not only the Soviet empire but the Soviet Union itself. Despite their many differences, the United States and Russia agree on many international issues, as during the 1991 Gulf War, despite some last-minute maneuvering. This was also the case in the immediate aftermath of September 11 and the call by President Bush for an international response to terrorism. Some former enemies are becoming friends, and friends are becoming economic competitors. As Russia has gained in strength and confidence, however, it has been more assertive and the 2022 Russian invasion of Ukraine may presage a new Cold War era, however it eventually turns out.

Developing nations, some of which are rich with oil, are not the political-military pawns they once were. Unconventional conflicts remain difficult security problems, despite a renewed focus on preparing for the possibility of state-on-state conflict. Such conflicts are not easily understood given the American way of war, and the United States is not properly postured to respond quickly and effectively. At the same time, a newly resurgent and aggressive Russia and a more assertive China harken back to the days of the Cold War, but now with the addition of terrorism and unconventional conflict.

Third, any new president inherits his or her predecessor's budget, structures, commitments, and bureaucratic personnel and so does not have the freedom of action, at least initially, that most people would assume. Part of this constraint results from the logical reluctance to change national security dramatically early in an administration. In addition, the continuity of national security policy generally goes beyond any one president; such continuity is necessitated by the impact of US national security policy on allies and potential adversaries.

Fourth, new presidents sometimes find it difficult to deal with the national security establishment, especially the military. The character of the military profession and the general orientation of the institution, with its linkage to civilian political actors, preclude the president from becoming a completely free agent. Many items are the province of military experts and civilian specialists who develop a legitimacy from their expertise, and the president depends on them. But senior members of the military are far from apolitical in the broad sense; they are often drawn into the fray over national security policy, organizational issues, and social issues. Even President Eisenhower, whose years of military service should have provided a firm basis for national security policy (he presided over the Allied victory in Europe during World War II), encountered intense political battles. President George H. W. Bush, a decorated war hero from his Navy service in World War II, had a lot of credibility with the military.

In 1993 and 1994, President Clinton faced a serious credibility problem stemming from his avoidance of military service and his anti–Vietnam War activities. President George W. Bush got along well with the military, partly because as a member of the Texas Air National Guard he knew how to speak their language. President Obama did not have military service, but seemed attentive to military advice even if he eventually did not always follow it. President Trump expressed his great affection for the military, even though he himself did not serve and did not always seek military advice. President Biden did not personally serve in the military, although his elder son deployed to a combat zone with the Delaware Army National Guard as a member of the judge advocate general corps.

Fifth, bureaucracies in the national security policy process as elsewhere frequently resist any change that threatens their authority or budget. At times, bureaucratic loyalty overshadows policy priorities. Of central importance is the maintenance of the stature, role, and budgets of their organizations or subunits. Put simply, their perspectives are affected by bureaucratic affiliation.[29] As Henry Kissinger observed, "The nightmare of the modern state is the hugeness of the bureaucracy, and the problem is how to get coherence and design in it."[30] This has become the case in the domestic arena with the establishment of the Department of Homeland Security.

The ability of the bureaucracy to frustrate a president's national security policies is reinforced if bureaucrats are able to forge alliances with actors outside the executive office. Equally important are the so-called subgovernments and power clusters that exist within many bureaucracies, which can thwart attempts to design coherent policies.

The reality of modern US politics is that an opposition government of sorts exists within our federal bureaucracies. Staff members, attorneys, assistant division chiefs, and deputy administrators—a civil service old boys' network—stand ready to leak embarrassing information to undercut an administration. The motives may vary for this, but every administration understands that it can be sandbagged by its own people.[31] President Donald Trump was especially frustrated by what he assumed were efforts by his security and intelligence bureaucracies to slow down or block his policy initiatives and to cast doubt on the legitimacy of his election and presidency. Trump referred repeatedly to what he called a "witch hunt" investigation by Special Counsel Robert Mueller into possible collusion between the Trump campaign in 2016 and Russia, and he distrusted high officials in the Federal Bureau of Investigation (FBI) and in the US intelligence community who in his view were trying to undermine his administration.

Sixth, Congress is more assertive in national security policy through the budget process. For example, the House and Senate Budget Committees established by the Congressional Budget and Impoundment Control Act of 1974 created the House and Senate Budget Committees and established the Congressional Budget Office to give Congress its own source of expertise when evaluating budgetary issues.[32] This gives Congress a structure to examine the administration's budget and prepare alternatives. In addition, Congress has expanded its staff to improve its ability to examine national security policy and strategy.

Concerns about questionable national security activities caused Congress to step in, passing legislation to restrict intelligence activities, strengthen congressional oversight, and restrict the president's use of military force (the 1973 War Powers Resolution) and his authority to commit military assistance to other countries (the Clark Amendment with respect to Africa and, later, the Boland Amendments with respect to Central America). In 2007, Congress began a serious, but unsuccessful, effort to force the withdrawal of US forces from Iraq and in 2019 congressional attempts to limit US cooperation with Saudi Arabia in its war in Yemen were vetoed by President Trump. Congress and the country were in the mood to exert control over foreign and national security policies, regardless of the impact they had on adversaries and allies.

Although the Reagan administration marked a change—restoring the executive-legislative balance during the early 1980s—the stage was set for continuing congressional involvement in national security affairs. The changes brought on by the post–Cold War period and the concentration on domestic issues and priorities reinforced the congressional role in national security affairs. In the aftermath of September 11, this and the major role of the United States in the Middle East have changed yet again, as Presidents Bush and Obama used some broader powers to counter terrorism, both domestically and in foreign areas. President Trump regarded illegal immigration as a national security as well as a cultural issue and went to great lengths to try to expand the physical barriers on the US southern border. President Biden took office with a far different set of priorities. In the long run, national will, staying power, and political resolve will shape the scope and direction of those powers.

In any case, the need remains for flexibility and innovation—to say nothing of usable power—and thus most presidents opt for a trusted staff of advisors to conduct national security policy. For a variety of reasons, ranging from fear of media leaks to distrust of outside political actors, presidents feel more comfortable working with a select group of advisors within a structure that is under their immediate control. It is this national security establishment that is the primary structure the president uses to advance his national security policies.

Conclusion

Lyndon Johnson wrote that the president receives advice from many quarters, "but there is only one [person] that has been chosen by the American people to decide."[33] Decisions on national security were often easier to make for the president in the past. Isolationism, distance from the Old World, and military power allowed the United States relative freedom from major external threats. National security issues were relatively clear and less difficult conceptually than domestic problems. World War II and the nuclear era changed the level of threat to the US and many breathed a sigh of relief when the Cold War ended with the collapse of the Soviet Union. The new strategic landscape was complex and lacked the clarity that existed in an era of competing superpowers, and the attacks of September 11, 2001, came as a strong shock. The rise of an assertive China and the 2022 Russian invasion of Ukraine were vivid reminders of the need to refocus US defense planning toward state-on-state warfare.

The Korean War showed that beneath the nuclear umbrella the United States still requires usable conventional forces to fight limited wars and deter future ones. The evolution of regional powers, the frequency of nonnuclear conflicts, proxy wars, and the fear of direct US-Soviet confrontation reshaped the international security environment, making it less vulnerable to superpower influence and more vulnerable to the actions of smaller powers and terrorist groups.

The performance of US forces in the 1991 Gulf War seemed to overcome much of the fear associated with the earlier "Vietnam syndrome," in which there was great public aversion to US military intervention in remote areas of the world not clearly related to core interests. Indeed, in the Gulf War the commander of the coalition forces, General H. Norman Schwarzkopf, made a point of noting that US operations and command and control in the Gulf War were not like those in Vietnam. But US involvement in Somalia in 1993 rekindled visions of Southeast Asia, as did our involvement in Bosnia-Herzegovina and Kosovo later that decade. The country's response to September 11 illustrated a focus and unity not seen since World War II, but this dissipated as more questions were raised about the US role in the Middle East—especially in Iraq, Afghanistan, and Syria. Future presidents will likely have to deal with a syndrome historically every bit as constraining as was the Vietnam syndrome, especially in view of the failure of the twenty-some-year war in Afghanistan. In part, post–Cold War presidents brought this on themselves: both Republican and Democratic presidents bet on a strategy of "liberal hegemony" in which American ambitions reached beyond maintaining favorable balances of power in critical regions to the micromanagement of overseas local politics and attempts to remodel non-Western societies in America's image.[34]

The Reagan presidency, with its strengthening of the US defense posture and its perceived confidence in dealing with international security issues, reversed the pessimism of the late 1970s.

> Ronald Reagan established a pattern of leadership which his successors would be prudent to consider. He demonstrated the strength of a simple, straightforward agenda, readily explicable to the public. By concentrating his political resources on that agenda, by defining a mandate and inducing legislators of both parties to accept it, he restored the presidency as the engine that moves government.[35]

But not even Ronald Reagan could revive the earlier US supremacy: "The Reagan revolution was hampered by limitations of power inherent in the presidency and the political system, by private economic decisions, and by events abroad that lay beyond its control."[36]

National security issues in the twenty-first century, although reflecting continuities from the past, have unique characteristics, especially in the nature of conflicts and the relationship between economic strength and national security. This is complicated by the changing relationship between the domestic and national security agendas and by Internet-driven technologies for global communication and cyber war. The complexity of the issues and their undefined, fluid nature exacerbate the problem of presidential control and direction. In this context, internal struggles among government agencies leave the president vulnerable to agency biases and, in some instances, make him or her a near-captive of the bureaucracy. The pressures on the president are magnified by congressional involvement and its advocacy of policies and strategies that may be contrary to those of the administration. Add to this the interests and objectives of allies and adversaries and one must conclude that national security policy is fraught with peril and pitfalls. Yet national security is only one component of the president's total responsibility.

Another major issue is that states can have their own ideologies and conceptions of national security that are in direct contradiction to US goals. Differing strategies can challenge the Western cultural orientation of the United States.[37] Conflict in one form or another is inevitable, and US policy, even when supported by necessary resources, does not guarantee success. There are too many imponderables and uncertainties in the external environment.

According to one assessment, the weakness of the existing state system has a negative impact on leadership:

> Leaders of states in the last years of the twentieth century are weak because the nation-state, as an institution, is weak. There has been a shift of problems from the national to the global arena. . . . The amorphous challenges that have crept up in the present era are not easily countered, or conquered, by simple direct actions. Yet the only leaders in sight with vision and conviction are possessed by some form of fanatical ideology. For most, in these circumstances, muddling through is the only, even if uninspiring, style of leadership available.[38]

In summary, the posture and power of agencies within the national security establishment, the interplay of personalities between the administration and Congress, congressional power in the policy process, the politicization of national security issues, and the changed domestic and international political and security environments have bred issues far different from those of the Cold War era. Therein lies an irony: for many people, the end of the Cold War diminished the importance of national security issues compared to domestic issues. Yet global interdependence, information-age technology, international environmental issues and ecology, the new strategic landscape, and long-term threats to US quality of life have given a new impetus to the link between domestic and national security issues—a brutal reality that hit home on September 11. But it must be remembered that an increasing number of political actors, both domestic and international, affect national security and are beyond presidential control; they are constraints on the use of presidential power despite the new actions in the war on terrorism.[39]

In this environment, US national objectives and interests, as well as political-military policy and strategy, are difficult to define. This applies to adversaries and allies and political actors within the United States. US national security policy often does not have the luxury of clear-cut choices between good and evil. Rather, choices often involve living with the lesser of two or more evils. Serious and immediate threats are the exception to this general rule.

As Ernest van den Haag has written: "Just solutions are elusive. Many problems have no solutions at all, not even unjust ones; at most they can be managed, prevented from getting worse or from spreading to wider areas. Other problems are best left to simmer in benign neglect until parties are disposed to settle them."[40]

The simplifying models, perspectives, and analyses that are key to presidential performance usually fall short of the mark because there are no simple answers. The interaction among leadership style, perceptions of the office, mind-set, and the national security establishment—in the context of the domestic and international environments—preclude neat paradigms or precise model-building. The forces that affect the president's ability to exercise power and the public's expectations are difficult to integrate. Even the most respected scholars of the presidency do not agree on the power of the office, the capacity of the president to exercise this power, and the best approach to the study of the office.[41] Presidential performance in national security does not neatly follow any one rational model or specified approach.

Presidents who are most successful in the national security area have a deep understanding of the organizational dynamics and interactions within the national security establishment. Yet this should always be tempered by sensitivity to public expectations and appreciation of the system's openness. Furthermore, the president must have a realistic perception of the international scene, adversaries, and allies. Critical to all this are the character and leadership style of the individual presiding in the Oval Office.

This brings us full circle: support and consensus are contingent upon the president's leadership style and ability to set and maintain the tone of the administration. To develop coherent policy and relevant strategy requires an articulation of what the United States stands for and the national will, political resolve, and staying power to use the instruments necessary to achieve national security goals. The president's mission is to reconcile the ideals of democracy with the commitment necessary to achieve these goals in the international arena. Unfortunately, this can require the use of the military and loss of life, and only the president is in a position to lead the country to accept such sacrifices and understand why they are necessary.

In the final analysis, the uniqueness of the office, the problems of US national security, and the character of the public interact to create a distinctively US presidency. It is an almost impossible job and at the same time central to the national security of the United States.

Notes

1. Lyndon Baines Johnson, *The Vantage Point: Perspectives of the Presidency, 1963–1969* (New York: Holt, Rinehart, and Winston, 1971), preface.

2. John Lewis Gaddis, *The Cold War: A New History* (New York: Penguin, 2005).

3. Harold M. Barger, *The Impossible Presidency: Illusions and Realities of Executive Power* (Glenview, Ill.: Scott, Foresman, 1984), p. 2.

4. See http://warontherocks.com.

5. See John Bew, "Putin and the Wrecking Ball," *New Statesman* (US ed.), July 28, 2018, https://www.newstatesman.com/world/north-america/2018/07/putin-and-wrecking-ball. Bew points out that President Trump's implementation of policies was often problematical.

6. Margaret Truman, *Harry S. Truman* (New York: Pocket, 1974), p. 603.

7. Theodore Sorensen, *Decision-Making in the White House: The Olive Branch or the Arrows* (New York: Columbia University Press, 1963), p. 84.

8. Arthur Schlesinger Jr., *The Imperial Presidency* (Boston: Houghton Mifflin, 1973).

9. Sidney M. Milkis and Michael Nelson, *The American Presidency: Origins and Development, 1776–2002* (Washington, DC: Congressional Quarterly, 2003).

10. Lee Sigelman, "A Reassessment of the Two Presidencies Thesis," in Steven A. Schull, ed., *The Two Presidencies: A Quarter Century Assessment* (Chicago: Nelson-Hall, 1991), p. 60.

11. James MacGregor Burns, J. W. Peltason, Thomas E. Cronin, and David B. Magleby, *Government by the People,* national version, 18th ed. (Upper Saddle River, N.J.: Prentice-Hall, 2000), pp. 360–368.

12. See, for example, Richard E. Neustadt, *Presidential Power* (New York: Wiley, 1960), p. 9; Michael Nelson, ed., *The Presidency and the Political System* (Washington, DC: Congressional Quarterly, 1984), esp. pt. 1; and Clinton Rossiter, *The American Presidency,* 2nd ed. (New York: Mentor, 1960). See also Steven Kelman, "The Twentieth-Century Presidents," *American Democracy and the Public Good* (Fort Worth, Texas: Harcourt Brace College, 1996), pp. 460–468.

13. James David Barber, *Presidential Character: Predicting Performance in the White House,* 4th ed. (Englewood Cliffs, N.J.: Prentice-Hall, 1992).

14. See Aaron Wildavsky, "The Two Presidencies," in Wildavsky, ed., *Perspectives on the Presidency* (Boston: Little, Brown, 1975).

15. Sorensen, *Decision-Making in the White House,* p. 5.

16. Ibid., p. 10.

17. See, for example, Barber, *Presidential Character.*

18. Erwin C. Hargrove and Roy Hoopes, *The Presidency: A Question of Power* (Boston: Little, Brown, 1975), p. 47.

19. James Q. Wilson, *American Government: Institutions and Policies,* 5th ed. (Lexington, Mass.: Heath, 1992), p. 338.

20. Johnson, *The Vantage Point,* preface.

21. Richard M. Nixon, *Six Crises* (Garden City, N.Y.: Doubleday, 1962), p. 323.

22. Fred I. Greenstein, *The Hidden Hand Presidency: Eisenhower as Leader* (New York: Basic, 1994).

23. Theodore Roosevelt, *An Autobiography* (New York: Scribner's, 1913), p. 197.

24. Sorensen, *Decision-Making in the White House,* p. 291.

25. Irving L. Janis, *Groupthink: Psychological Studies of Policy Decisions and Fiascoes,* 2nd ed. (Boston: Houghton Mifflin, 2006).

26. Kathleen Hicks, "Russia in the Gray Zone" Aspen, Colorado, Aspen Institute, July 19, 2019, in Johnson's Russia List 2019, #115, July 21, 2019, davidjohnson@starpower.net. See also Eugene Rumer, "The West Fears Russia's Hybrid Warfare. They're Missing the Bigger Picture," July 3, 2019, https://carnegieendowment.org/2019/07/03/west-fears-russia-s-hybrid-warfare-they-re-missing-bigger-picture-pub-79412.

27. P. W. Singer and Emerson T. Brooking, *LikeWar: The Weaponization of Social Media* (Boston: Houghton Mifflin Harcourt, 2018).

28. Daniel Patrick Moynihan, *Secrecy: The American Experience* (New Haven: Yale University Press, 1998), p. 222.

29. Robert L. Gallucci, *Neither Peace nor Honor: The Politics of American Military Policy in Vietnam* (Baltimore: Johns Hopkins University Press, 1975), p. 138.

30. As quoted in Morton H. Halperin, *Bureaucratic Politics and Foreign Policy* (Washington, DC: Brookings Institution, 1974), p. 15. See also Morton H. Halperin, Priscilla Clapp, and Arnold Kanter, *Bureaucratic Politics and Foreign Policy,* 2nd ed. (Washington, DC: Brookings Institution, 2006).

31. Joseph C. Goulden, *The Superlawyers* (New York: Dell, 1973), p. 228.

32. US House of Representatives Historical Highlights, "Budget and Impoundment Control Act of 1974," https://history.house.gov/Historical-Highlights/1951-2000/Congressional-Budget-and-Impoundment-Control-Act-of-1974.

33. Johnson, *The Vantage Point,* preface.

34. Stephen M. Walt, *The Hell of Good Intentions: America's Foreign Policy Elite and the Decline of U.S. Primacy* (New York: Farrar, Straus, and Giroux, 2018).

35. Louis W. Koenig, *The Chief Executive,* 5th ed. (New York: Harcourt Brace Jovanovich, 1986), p. 415.

36. Ibid., p. 2.

37. See, for example, Adda B. Bozeman, *Strategic Intelligence and Statecraft: Selected Essays* (Washington, DC: Brassey's US, 1992). See also Samuel P. Huntington, "The Clash of Civilizations?" *Foreign Affairs* 72, no. 3 (Summer 1993), pp. 22–49.

38. International Institute for Strategic Studies, *Strategic Survey, 1994–1995* (London: Oxford University Press, 1995), pp. 15–16.

39. For insights into the irony of diffusion and concentration of power, see Wilson, *American Government.*

40. Ernest van den Haag, "The Busyness of American Foreign Policy," *Foreign Affairs* 64, no. 1 (Fall 1985), pp. 114–115.

41. See, for example, Thomas E. Cronin and Richard E. Neustadt, *Presidential Power: The Politics of Leadership from FDR to Carter* (New York: Wiley, 1980). Also see Edward S. Greenberg and Benjamin I. Page, *The Struggle for Democracy,* 3rd ed. (New York: Longman, 1997), pp. 402–406.

7

The National Security Council and the "Policy Triad"

The modern US president depends upon many people to formu- late and implement national security policy, coordinated at the highest level through the National Security Council (NSC). Two key presidential advisors—the secretary of state and secretary of defense—are statutory members of the NSC, along with the national security advisor. These three individuals form what we call the policy triad. The director of national intelligence (DNI) is an advisor to the NSC together with the chairman of the Joint Chiefs of Staff. The function of the Central Intelligence Agency (CIA) gives it a critical role in the national security establishment, but that agency's closed nature—the unavoidable legacy of intelligence-gathering and covert operations—means its relationship with the president, Congress, and the public is quite different. Formerly first among equals in the panoply of intelligence agency heads, the director of central intelligence now reports to the DNI along with the other intelligence agency chiefs. Nevertheless, since September 11 the CIA has increased its support for interagency cooperation against terrorism and in support of military counterinsurgency and counterterror operations, and its National Clandestine Service (NCS) remains the tip of the spear for gathering information based on human sources, also known as human intelligence (HUMINT).

The NSC and its staff are primarily advisory units: even though recommendations can be made by the NSC and approved by the president, their interpretation and implementation rest mainly with the State Department, Defense Department, and the intelligence community. It is not surprising that within the NSC the views of operational departments often clash. In addition, national security advisors and their staffs develop increased influence as presidents rely on them for filtering information, brokering interagency conflicts, and sorting out policy agendas. As the US Commission on National Security/21st Century concluded, "The power to determine national security policy has migrated toward the [NSC] staff. The staff now assumes policymaking and operational roles, with the result that its ability to act as an honest broker and policy coordinator has suffered."[1] On the other hand, this assessment must be qualified by the recognition that the president and his style of management ultimately determine how effective the NSC can be.

Aside from their advisory functions as statutory members of the NSC, the secretaries of state and defense also play significant roles in the national security establishment as cabinet members and department heads. Furthermore, the Departments of State and Defense have substantial links to Congress and are involved in a variety of formal relationships with other countries. Their perspectives thus reflect many influences. State and Defense rank as the bureaucratic players with the most potential clout in foreign and national security policymaking, but they are not equals. The size of the Department of Defense in people and budgets makes it first among equals in the national security firmament, as do the time urgency and national visibility of defense missions.

The State Department, although small in size compared to the Defense Department, nevertheless provides the indispensable diplomatic background for the implementation of defense and national security policies. When the diplomatic table is well prepared prior to the outbreak of hostilities, the US chances for success rise in proportion. An example of diplomatic preparedness as a prelude to military success was provided by the diplomacy of then secretary of state James Baker under President George H. W. Bush in 1990 and 1991. Before the Gulf War, the United States isolated Iraqi leader Saddam Hussein from political support, even from Middle Eastern countries not necessarily friendly to the United States but equally or more worried about Iraq's aggression against Kuwait.

In contrast, the Donald J. Trump administration arrived in Washington, DC, in January 2017 with a president and White House inexperienced in foreign policy and international relations. Trump's first secretary of state, businessman Rex Tillerson, gutted the organizational chart of the State Department and left ambassadorships unfilled in important countries such as Saudi Arabia and South Korea. In addition, President Trump preferred an improvisational style of management that leapfrogged his own cabinet department heads as well as the permanent civil service. Trump's views on relations with Russia were also at odds with many members of Congress and professional staff in the Departments of State and Defense. Distrust between President Trump and the US intelligence community was sparked by Trump's views on the efficacy of US intelligence in the war against Iraq in 2003 and by his ongoing distrust of the Federal Bureau of Investigation (FBI) leadership and its investigations into his presidential campaign in 2016. On the other hand, the Joseph Biden administration, which assumed office in January 2021, offered a more traditional approach to staffing cabinet departments with experienced personnel, including the Department of State.

Whereas the secretaries of state and defense bring their own worldviews and operational methods, the national security advisor advances the president's perspective and performs an advisory role. This person also sets the agenda and coordinates the activities of the national security staff in support of the NSC.

It is the power and relationships of these three—the two secretaries and the national security advisor—relative to one another and to the president that define the direction of US national security policy. The policy triad is the core around which the policy process revolves (see Figure 7.1).

Figure 7.1 The Policy Triad

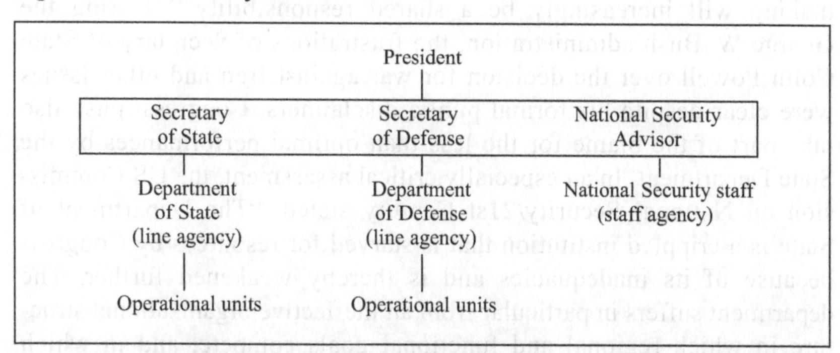

Notes: The secretaries of state and defense each wear two hats: staff advisor to the president and operational department head. Thus their perspectives on national security are usually conditioned by the capability of their departments to implement policy and strategy. The national security advisor, however, has no operational units; the national security staff is just that—a staff agency. The national security advisor and national security staff attempt to provide a presidential perspective and to stand above department issues. Perceptions, mind-sets, and responsibilities differ between the two secretaries and particularly between the two secretaries and the national security advisor.

The Department of State

The secretary of state is the president's primary advisor on foreign policy and the operational head of the department responsible for its conduct. The department is organized along two broad lines: functional and geographic. Country desks operate under assistant secretaries responsible for a particular region (e.g., the Nigeria country desk would fall under African affairs). The functional areas, such as intelligence and research and politico-military affairs, cut across geographic boundaries. In 1993 and 1994, plans were implemented to change the organization to reflect the foreign policy of the Bill Clinton administration and to simplify the department's burdensome structure. This included creating the Office of the Secretary to bring together separate groups within the department and to clarify the reporting process. Efforts were also made to give more power to undersecretaries of state.[2] In addition, programs were undertaken to improve security at US embassies and consulates worldwide, and these efforts continued into the George W. Bush and Barack Obama administrations.

Despite such efforts, some have concluded that during the Clinton and George W. Bush administrations the department's role in foreign and national security diminished. According to one assessment of the Clinton policymaking process: "The secretary of state in the Clinton administration, like other recent administrations, will continue to be prominent, but will not dominate policy formulation, instead policy making will increasingly be a shared responsibility."[3] During the George W. Bush administration, the frustrations of Secretary of State Colin Powell over the decision for war against Iraq and other issues were clear despite his formal public disclaimers. Congress must also take part of the blame for the less than optimal performances by the State Department. In an especially critical assessment, the US Commission on National Security/21st Century stated, "The Department of State is a crippled institution that is starved for resources by Congress because of its inadequacies and is thereby weakened further. The department suffers in particular from an ineffective organizational structure in which regional and functional goals compete, and in which sound management, accountability, and leadership are lacking."[4] The organization of the State Department is shown in Figure 7.2.[5] It is notable that President Obama's first secretary of state, Hillary Clinton, attempted to increase the overall budget for State during her first months in office and succeeded in raising the State Department's budget by 7 percent. Regardless the administration, the Department of State's

Figure 7.2 The Department of State

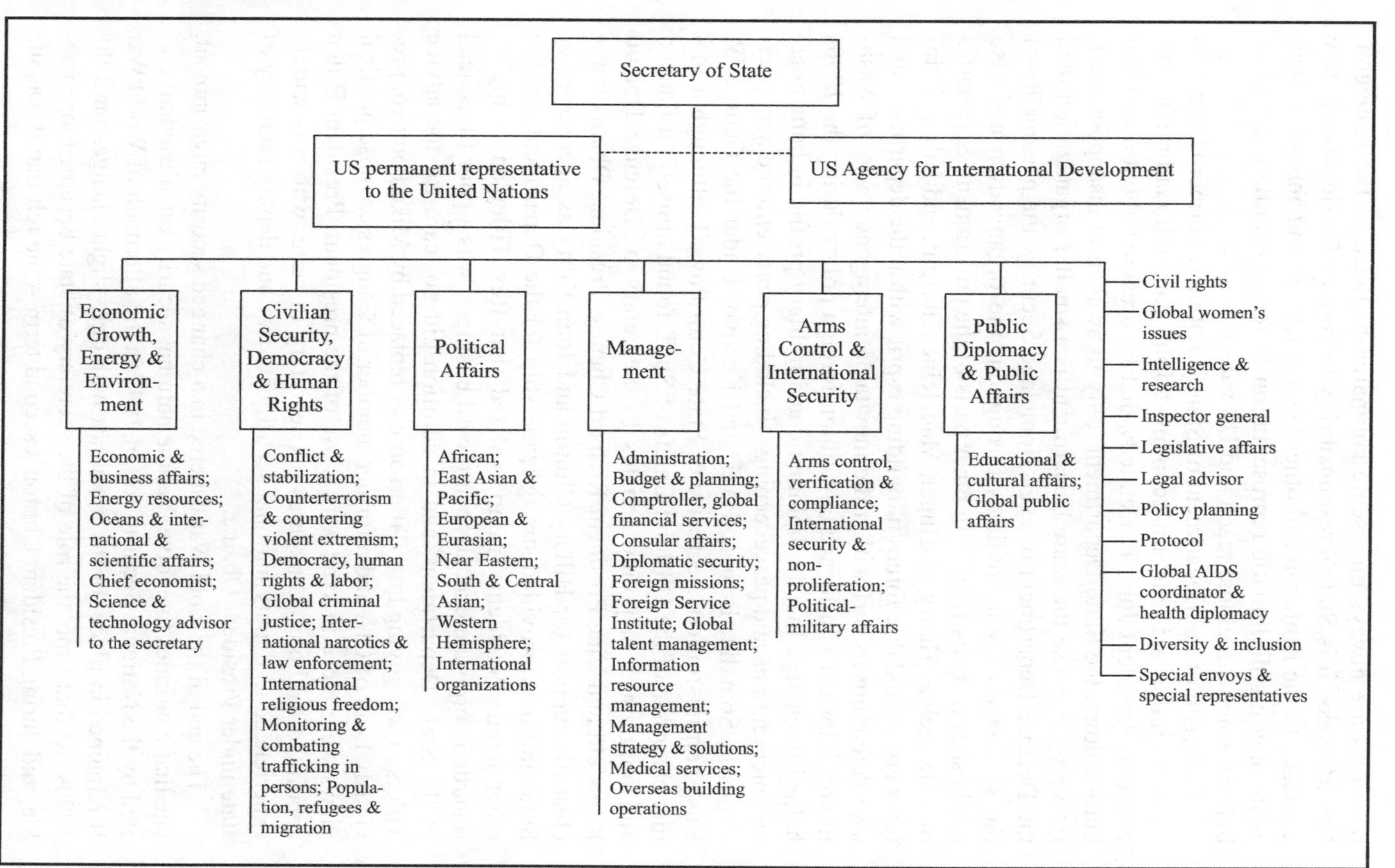

Source: Adapted from US Department of State, August 2021, https://state.gov/wp-content/uploads/2021/09/DOS-Org-Chart-August-2021.pdf.

resources are always minuscule compared to those of the Pentagon. Nevertheless, it is State's responsibility to convey Presidential policy guidance to the Pentagon and other US agencies acting abroad, as well as its ongoing diplomatic representation in foreign capitals that symbolizes American global reach and US values.

Criticism of the Department of State is not new. Former US ambassador to India and Harvard economist John Kenneth Galbraith once wrote to President John F. Kennedy that he dreamed about a fire that burned down the State Department, only to awaken in disappointment. However, because the State Department is a smaller organization than the Defense Department or even Homeland Security, the personality of the secretary of state radiates throughout the organization and also establishes its bona fides with other parts of the government. Secretaries of state such as Henry Kissinger, Madeleine Albright, and Hillary Clinton were formidable forces in building esprit within the department and also defending its turf within the inevitable interagency wars of Washington. Albright's influence on Clinton foreign policy during the 1990s helped to enlarge the US diplomatic and military profile in humanitarian interventions and peacekeeping—including peace enforcement operations in Somalia, Bosnia, Haiti, and Kosovo. Under the George W. Bush administration, Secretary of State Colin Powell, although a former chairman of the US Joint Chiefs of Staff, found himself outflanked and outgunned in policy debates by Secretary of Defense Donald Rumsfeld and Vice President Richard Cheney. President Obama made clear his support for Hillary Clinton and John Kerry as Secretaries of State and both provided strong leadership for the Department. On the other hand, President Trump realized that Rex Tillerson, a highly regarded corporate executive in the oil business, was a poor fit as head of the State Department, and in addition did not care for the advice Tillerson was giving him. Tillerson was replaced by Mike Pompeo, previously head of CIA and a former member of Congress, whose political views were more in sync with those of the president. President Biden appointed Anthony Blinken as his secretary of state, who was widely experienced as deputy national security advisor and deputy secretary of state under President Obama.

The end of the Cold War ushered in a changed security environment, one that continues to evolve. Yet the national security establishment created by the National Security Act of 1947 in the aftermath of World War II remained in place for the most part with only slight changes until the 1990s. At that time, the role of the secretary of state became more visible, and during President Clinton's second term some felt that it was the dominant player in the policy process. At the same time, the decline of serious threats and the drawdown of the US military—combined with deep cuts in the defense budget—crimped the scope and power of the secretary of defense. Thus, whereas national security issues were a driving force in US policy during the Cold War, the 1990s favored international economics and democratization in the form of "engagement and enlargement." In the twenty-first century and after September 11, the focus has once again been on national security policy, with Secretary of Defense Donald Rumsfeld taking center stage behind only the president and vice president. This further marginalized the Department of State. In 2006, Rumsfeld resigned and Robert Gates became secretary of defense. Many noted that Rumsfeld's resignation was due primarily to the US military problems in Iraq. But foreign policy and national security policy are increasingly intertwined. Former secretary of state Condoleezza Rice and defense secretaries Rumsfeld and Robert Gates became involved in policy issues that fused foreign policy and national security issues.

At the same time, other players took the stage in foreign policy and national security. Arms-control initiatives and agreements with the Soviet Union and, later, Russia, as well as US relations with China, Russia, and the Middle East, are many of the primary concerns of the secretary of state. Yet all this has a huge impact on national security issues. Peace in the Middle East, primarily a foreign policy matter, also has national security implications. And in the 1990s, US involvement in Somalia, Haiti, and Bosnia-Herzegovina intermingled foreign and national security policies. The US involvement with the North Atlantic Treaty Organization (NATO) in Kosovo beginning in 1999 also shows the close relationship between foreign and national security policies.

Other State Department activities beyond the realm of national security nevertheless influence the department's perspective on security matters. The secretary of state is responsible for consular services, aid to US citizens overseas, diplomatic missions, and US embassies—the focal point of the US presence in many countries. Although ambassadors can deal directly with the president, their main contact is the secretary of state. The embassy staffs are typically given the responsibility to coordinate all official US activities in the host country. This country-team concept, first formalized by President Dwight Eisenhower, was an attempt to bring consistency to US activities overseas through centralized control (although US military operations were not included in this concept).

The primary bureaucratic mind-set within the Department of State is deeply involved with traditional diplomatic and consular tasks, embedded in traditional notions of courtly, courteous, Old World diplomacy. The focus is on negotiations and compromise. The department's organizational behavior and internal mind-sets stem from these institutional roots, imposing a template of education and socialization that produces foreign service officers and department employees quite different from their counterparts at Defense. The result can be serious disagreements between the two secretaries. At times the departments work at cross-purposes, although accord is usually reached in clear cases of national security and when crises help forge a consensus. One of the hallmarks of President Obama's national security team was the high level of personal respect that existed between his first defense secretary, Robert Gates, and the secretary of state, Hillary Clinton, contributing to interdepartmental cooperation on a number of key issues, including Iraq and Afghanistan.

The State Department has been criticized as a bureaucracy committed to stability, the status quo, and self-protection of its organizational integrity and autonomy. According to some, this has produced bureaucratic inertia and burdensome procedures that allow little room for initiative and innovation. The US Commission on National Security/21st Century had this to say about the reorganization of the Department of State:

> The President should propose to the Congress a plan to reorganize the State Department, creating five Under Secretaries with responsibility for overseeing the regions of Africa, Asia, Europe, Inter-America, and Near East/South Asia, and redefining the responsibilities of the Under Secretary for Global Affairs. . . . The Secretary of State should give greater emphasis to strategic planning in the State Department and link it directly to the allocation of resources through the establishment of a Strategic Planning, Assistance, and Budget Office.[6]

Several studies have also shown that the department has difficulty in formulating long-range policies that link to domestic political concerns as well as foreign policy issues. As Christopher Shoemaker concluded, "This rather important deficiency stems both from the department and from historical proclivities of the Foreign Service."[7] He went on to note that "bureaucratic power within the State Department is normally vested in the regional bureaus, which, despite their staffing by seasoned professionals, are virtually unable to come to grips with the development of long-range policy."[8]

Nor is this all. Some critics have charged that the US foreign and defense policy establishment embraced a misguided grand strategy of "liberal hegemony" from the end of the Cold War to the beginning of the Trump administration. Liberal hegemony committed the United States to a policy of expanding American influence and liberal values, including democratic governments and market capitalism, regardless of the variations in cultures and foreign policy priorities of other states, and supported by overcommitment to the use of force in overthrowing contrarian regimes. Harvard political scientist Stephen M. Walt argued:

> It was the height of hubris for Americans—who are, after all, only 5 per cent of the world's population—to believe that they had discovered the only workable model for a modern society and the only possible blueprint for a durable and peaceful world order. It was naïve for them to think they could create stable and successful democracies in deeply divided societies that had never been democratic before. It was positively delusional to assume that this objective could be achieved rapidly and at low cost. It was unrealistic to believe that other states would not be alarmed by America's efforts to reshape world politics and to assume further that opponents would not devise effective ways to thwart US designs.[9]

That being said, the US State Department was responsible for one of the few examples in US history of farsighted policy planning, anticipating both the end of the Cold War and, as well, the US diplomatic and military grand strategy that would be required for eventual success. In 1947, as head of the policy planning staff in the State Department, George F. Kennan authored (anonymously, although insiders knew his identity) an influential article entitled "The Sources of Soviet Conduct." Against the grain of much contemporary argument that took the Soviet challenge to the West as primarily military, Kennan interpreted the Kremlin's challenge as mainly political and economic and argued that it should be met by "long-term, patient but firm and vigilant containment."[10] Kennan's prescription emphasized steadiness of purpose and military strength sufficient to deter Soviet aggression across the major fault lines of the East-West divide, but it also cautioned against political rashness and military adventurism that would be counterproductive for US interests. Critics on the right favored "rollback" of Soviet power by force if necessary; the left preferred a more accommodating US stance toward Soviet ambitions and power. Kennan was vindicated when the Cold War ended with the political collapse of the Soviet Union and without the outbreak of World War III in Europe.

The Department of Defense

In contrast to the 200-year tradition in the Department of State, the Department of Defense is a relatively new organization, established after World War II by the National Security Act of 1947 and its amendments in 1949 and 1958. Major problems faced the secretary of defense in trying to centralize defense policy and changes were taking place in the late 1980s and throughout the Clinton and George W. Bush administrations (1993–2008). The Department of Defense is still involved in changes in response to information-age technology and the changing international security landscape. More changes were implemented in the George W. Bush, Barack Obama, and Donald J. Trump administrations. The organization of the Department of Defense is shown in Figure 7.3.

The president's responsibility as commander in chief of the military is usually exercised through the secretary of defense and the operational elements of the Department of Defense. Like the secretary of state in the foreign policy area, the defense secretary performs dual functions in defense policy: the primary advisor to the president on defense policy and the head of the operational elements of the department. In the advisory capacity, the secretary focuses on the types of forces and manpower levels needed to effectively pursue the goals of US national security policy. Among the secretary's concerns are the composition of forces, weapons acquisition, training, planning, and operational implementation. The operational arms of the Department of Defense are highly visible: coercive military forces. Their symbolic—as well as substantive—role plays a great part in the perceptions of allies and adversaries regarding US capability. Therefore, the views of the secretary of defense are important in shaping policy.

The structure of the Office of the Secretary of Defense (as distinct from the service branches) includes several functional units supervised by assistant secretaries (for example, regional affairs). In 1993, Secretary of Defense Les Aspin changed the department's structure to overlap with certain responsibilities associated with the Department of State, such as promoting democracy and human rights. In early 1994, Secretary of Defense William Perry, who succeeded Aspin, reduced this overlap by shifting the department toward traditional roles and concerns. In addition, the service branches, as well as the Defense Intelligence Agency (DIA) and National Security Agency (NSA), operate or collect information that has a direct bearing on the function of the State Department.

The management styles of defense secretaries vary according to their temperaments, prior professional experience, and presidential relationships. Former US senator William Cohen became Clinton's third secretary of defense in 1996 after Perry had refocused the department on its traditional missions and stabilized what many felt was disarray created by Clinton's first defense secretary, Les Aspin. Aspin's "Bottom-Up

Figure 7.3 The Department of Defense

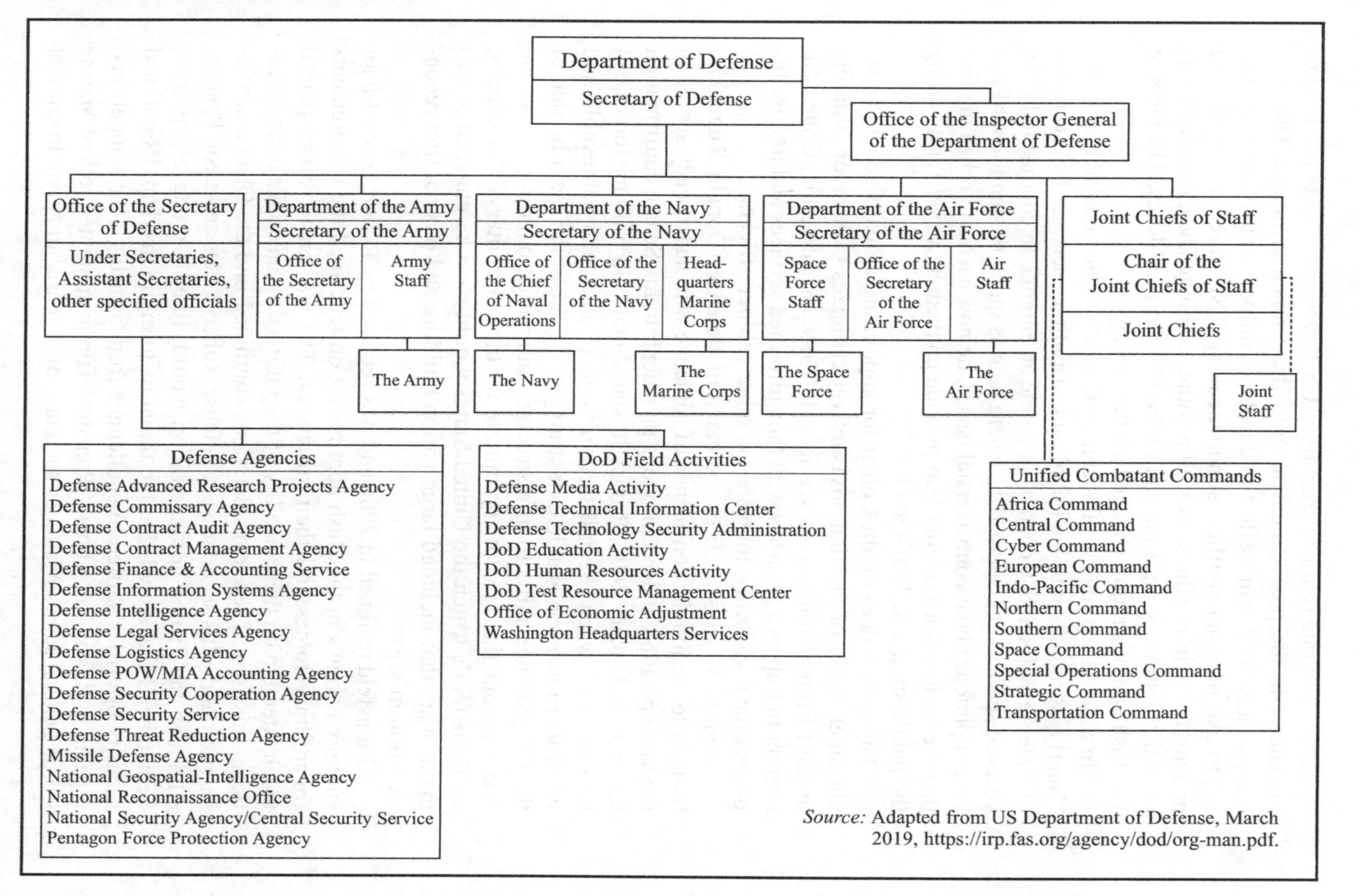

Source: Adapted from US Department of Defense, March 2019, https://irp.fas.org/agency/dod/org-man.pdf.

Review"—a strategy for restructuring US forces and a response to contingencies in the post–Cold War era—drew much criticism. This was compounded by Aspin's effort to develop a military capable of responding to two regional conflicts simultaneously.[11] Nonetheless, the need to prepare for twin conflicts, as well as other contingencies, remained part of military strategy until the end of the Clinton presidency. The George W. Bush administration changed all this yet again.

In a June 1998 review of Secretary Cohen's tenure, one author wrote: "Nearly 18 months into his initial tour of military service, US Secretary of Defense William S. Cohen has become an enigma. In contrast to his 24 years of service on Capitol Hill, where he was quick to step forward and voice opinions about controversial issues, he now seems to relish the fact that the legions surrounding him provide convenient cover for avoiding the public spotlight. That's too bad."[12]

The author does credit Cohen for appointing a qualified and capable inner circle of civilian advisors yet criticizes Cohen for isolating himself from the uniformed services. In 1999 and into 2001, Cohen did attempt to improve relations with the uniformed services and to raise his international visibility compared to other cabinet officers.

George W. Bush's first secretary of defense, Donald Rumsfeld, brought to the job the confidence of someone who had already served in the same position under a previous president. In his Senate confirmation hearings in January 2001, Rumsfeld stated that "US military forces can best be used when the military mission is clear and achievable and when there is a reasonable exit strategy. . . . When the main burden of the US presence shifts to infrastructure and nation building . . . we are into missions that are not appropriate for the US military."[13] Ironically, on Rumsfeld's watch the United States would soon be engaged in two major wars that included large nation building and postconflict reconstruction missions.

Rumsfeld resigned in 2006 and was replaced by Robert Gates. Major strategy reviews in the Bush administration redefined the requirements for military forces and their structures. The war on terrorism placed much attention on the capability of the military to undertake nonconventional missions. These considerations continued into the Obama administration under his secretaries of defense Robert Gates and Leon Panetta. Although the Trump administration would, like its predecessors after September 11, acknowledge the threat of foreign terrorism, its national security priorities shifted to challenges from aspiring peer competitors, especially Russia and China.[14] President Trump appointed retired Marine Corps general James Mattis as secretary of defense, an appointment that was reassuring to those concerned with Trump's lack of experience. Trump did not care for Secretary Mattis's advice and none of his future Secretaries of Defense had an equivalent stature. President Biden appointed retired Army general Lloyd Austin as his secretary of defense, whom he knew and respected from his days as vice president.

The Quadrennial Defense Review

Questions raised by many outside the Department of Defense regarding US military capability led to legislation that established the Quadrennial Defense Review (QDR). This legislation directed the secretary of defense and the JCS to conduct a defense review and provide a report by 1997 and every four years thereafter. QDR 1997 reflected the status quo, retaining the focus on two major regional conflicts and conventional forces.[15] In 2000, Steven Metz wrote, "Consensus is emerging that this QDR should be strategy driven rather than budget driven like the QDR 1997."[16] That appeared to be the case with QDR 2001, as Secretary Rumsfeld spelled out a new direction, shifting focus from two major regional conflicts, establishing guidance for transforming the military, budget guidelines for 2003, and so-called terms of reference for the military (these included reassuring friends and allies of US commitments, warning adversaries of US resolve, deterring threats and counter coercion, and defeating adversaries if deterrence fails). The 2006 QDR made it clear that the war against terrorism was likely to go on for a number of years and required the use not only of the military but also of US intelligence, economic, and diplomatic institutions.[17]

The 2010 Quadrennial Defense Review of the Obama administration noted that the Department of Defense must balance resources and risks among four priority objectives: prevailing in today's wars; preventing and deterring conflict; preparing to defeat adversaries and succeed in a wide variety of contingencies; and preserving and enhancing the US all-volunteer force.[18] The specific challenges that the Department of Defense is required to deal with in the future, according to the 2010 QDR, included the following: (1) defeat al-Qaeda and its allies; (2) support a national response to attacks on, or natural disasters in, the United States; (3) defeat aggression by adversary states, including those armed with advanced anti-access capabilities or nuclear weapons; (4) locate, secure, or neutralize weapons of mass destruction, key materials, and related facilities; (5) support and stabilize fragile states threatened by terrorists and insurgents; (6) protect US citizens in harm's way overseas; (7) conduct offensive operations in cyberspace; and (8) prevent human suffering due to mass atrocities or large-scale natural disasters abroad.[19] The breadth and depth of these expectations are emphasized in another section of the QDR as follows:

> In the mid- to long term, US military forces must plan and prepare to prevail in a broad range of operations that may occur in multiple theaters in overlapping time frames. This includes maintaining the ability to prevail against two capable nation-state aggressors, but we must take seriously the need to plan for the broadest possible range of operations—from homeland defense and defense support to civil authorities, to deterrence and preparedness missions—occurring in multiple and unpredictable combinations.[20]

The QDR was replaced in 2018 by the classified National Defense Strategy, which had a similar purpose.[21] Soon after his inauguration President Biden issued his own strategic guidance.[22]

September 11 focused attention on homeland defense, with the establishment of the cabinet-level Office of Homeland Security. Former Pennsylvania governor Tom Ridge was appointed director of homeland security, and he immediately began coordinating agencies' activities for purposes of counterterrorism. The Office of Homeland Security was replaced by the Department of Homeland Security (DHS) in 2003. President Obama appointed Janet Napolitano as his secretary of homeland security. Under President Trump the Department had a higher profile in border protection and deporting undocumented persons already in the country. His first DHS secretary was retired Marine Corps general John Kelly, who soon became President Trump's chief of staff, and who was replaced by his deputy, Kirstjen Nielsen. She was in office for approximately two years, and the position was unfilled during the rest of the Trump presidency. President Biden appointed Alejandro Mayorkas, former deputy Homeland Security secretary. The organizational chart of the department is shown in Figure 7.4.

The nature of the military profession, as well as the education and socialization of civilian officials and employees, shapes the institutional posture of the Department of Defense. Logically, the posture leans toward the military solution in responding to national security issues. In turn, there is an orientation within the department to ensure adequate staffing levels, resources to develop sophisticated weaponry, and satisfactory compensation and benefits for service personnel. This fits hand-in-glove with the effort to develop a skilled military that can effectively perform in war. Increasingly, these efforts have expanded to include operations other than war, unconventional conflicts, and political-military situations that in the past were primarily Department of State concerns.[23]

Figure 7.4 The Department of Homeland Security

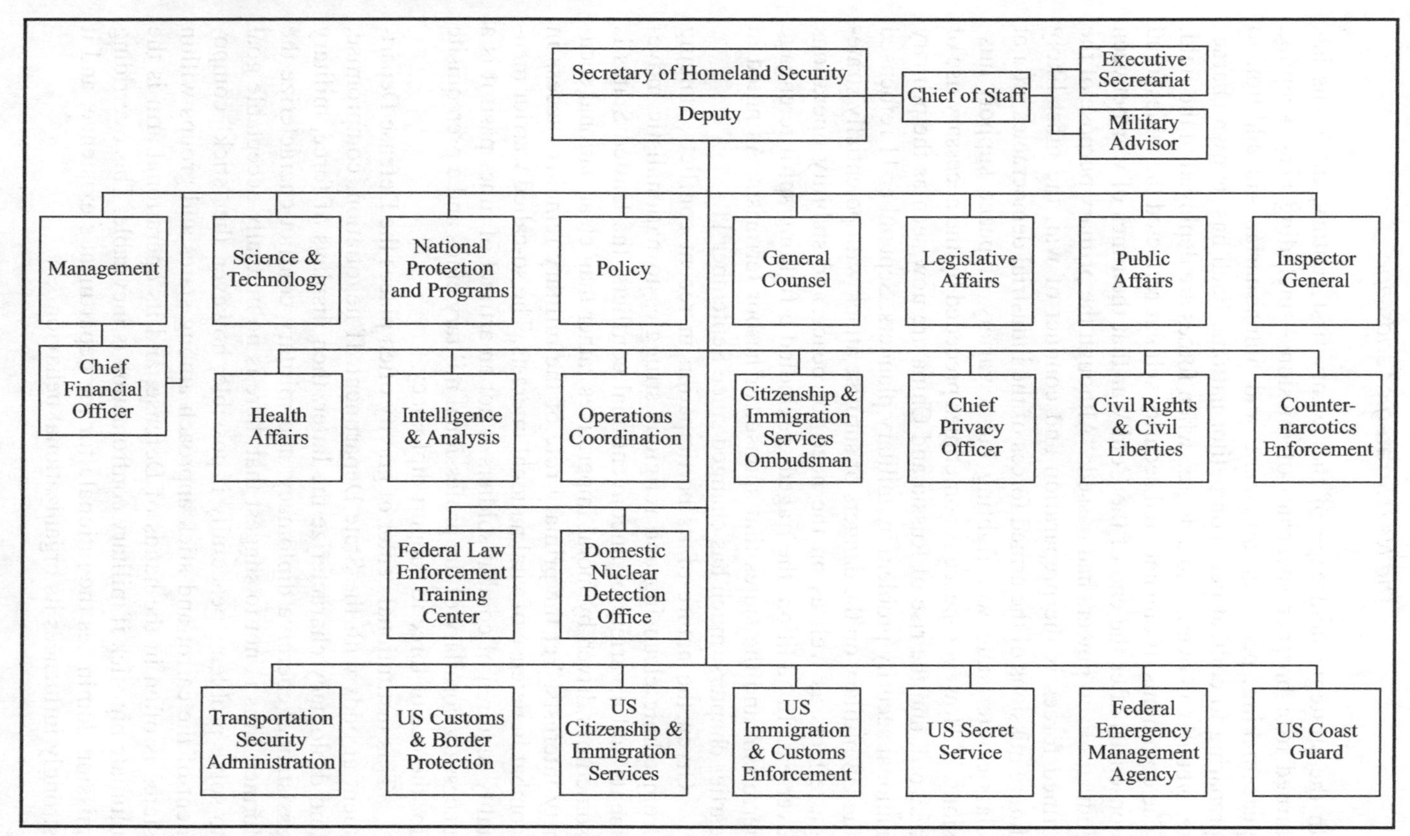

Source: US Department of Homeland Security, April 2012, available at http://www.dhs.gov/xlibrary/assets/DHS_OrgChart.pdf.

The New Strategic Landscape

In the strategic landscape of the twenty-first century, the focus has turned to a broader spectrum of missions—including peacekeeping, peacemaking, peace enforcement, and humanitarian—in addition to preparing to combat peer competitor nations. Each has its own character, yet all can overlap to a degree when forces are deployed in the field. The comments of one national security scholar reflected the widespread consensus after the end of the Cold War that the times of war between industrialized powers had passed: "Although the primary purpose of the armed forces is the preparation and conduct of war, the most likely future missions of the armed forces of the industrial democracies are not classic inter-state war fighting but a variety of peace support missions."[24] However, peace is sometimes preceded by the necessary use of a sword, and the rise of Russia and China are now seen as the primary national security problem by military planners. September 11 refocused the US military on the dangers of surprise attack with potentially strategic effects as well as on the need for peace and stability operations overseas. In addition, the tragedy rekindled a fighting spirit to engage terrorism and the states that sponsor or harbor terrorists. As noted in earlier chapters, much has changed since September 11.

Given the nature of its participation in recent conflicts, ranging from ethnic cleansing and religious struggles to nationalistic movements and a variety of unconventional conflicts, the United States is sometimes driven by moral imperatives rather than clear national security interests. Yet the primary role of the military remains success in combat in defense of vital national interests. The so-called warrior mentality required of combat soldiers is not an artifact of times past; it is a necessary corollary of the professional military ethos and a prerequisite for the use of force to support diplomacy.

This institutional center of gravity often places the Defense Department at odds with the State Department. If negotiation, compromise, and diplomacy characterize the latter, then displays of force, military assistance, coercive diplomacy, and military conflict characterize the former. This is not to suggest that there is no mutually acceptable goal to solve problems peacefully if possible; however, the "stick" component of the carrot-and-stick approach among states and groups within states is often in the hands of Defense, and its operational arm is the ultimate big stick if military confrontation is inevitable. This overriding mission dominates the rationale for the department's existence, and it strongly influences its organizational behavior.

Colin Powell's views on the use of the military, expressed to UN ambassador Madeleine Albright when he was JCS chairman, were quoted earlier in Chapter 3. He clearly believed that the use of military force should follow, not precede, the setting of clear political goals.[25]

After the end of the Cold War, the likelihood of global conventional war was considerably reduced for a time. Many within the Department of Defense questioned the commitment of US forces outside existing treaty areas. This had been the general view since the end of the Vietnam War. The Gulf War of 1991 (Desert Storm) was an exception, although many within the military initially favored the extensive use of sanctions. Operation Desert Storm was planned and analyzed as a Middle Eastern version of a European scenario in which US forces, based on traditional strategic and operational concepts, brought sophisticated weaponry to bear against a clearly identifiable and conventionally postured adversary. Notwithstanding US success in Desert Storm, military caution and reluctance to engage in large scale, conventional wars of choice in non-Western societies and cultures are conditioned by the fear of involvement in situations without clear political objectives or with problems that might not be resolved by military force.

Equally important is the underlying concern that involvement in ambiguous operations will lead to political opposition at home and the undermining of military capability.[26] But today many continue to contemplate the use of military force in response to international terrorism. Although much of the action has focused on special operations and relatively casualty-free air operations, follow-on conventional forces have become part of the overall strategy, and major battles involving US ground forces have ensued. For example, the US invasion of Iraq in 2003 to depose the regime of Saddam Hussein was unexpectedly followed by a protracted Iraqi insurgency against the occupying force of the United States and its coalition partners. Given the challenges of nation states, more traditional war-fighting missions are being emphasized now.

Because there is no sharp delineation of responsibility and power in several policy areas between the Departments of Defense and State, there is often disagreement on policy preferences. This is magnified when the departments are headed by strong personalities with their own views on the international security environment and their own policy agendas. If sufficiently strong, one secretary can dominate his or her department's approach. In any case, traditional models of national security policy, relationships among cabinet officers, and the power and responsibility of departments rarely go by the book.

The National Security Advisor

President Eisenhower was the first president to create the position of special assistant for national security affairs, later to be retitled assistant to the president for national security affairs, and now commonly known as the national security advisor. According to an authoritative study of US foreign policy, by the mid-1970s the national security advisor had become a "second secretary of state."[27] This was not the original intent. In the administrations of Harry Truman and Dwight Eisenhower, the role was simply to "arrange meetings of the [NSC] and manage the paperwork."[28] During the early years of the NSC, the secretary of state was the prominent figure in national security policy.

Several factors helped to elevate the national security advisor to prominence by the 1970s. President John F. Kennedy, frustrated by the bureaucratic inertia at the State Department, sought a more streamlined and responsive system for foreign policy advice and implementation. Furthermore, the relatively quiet days of the Eisenhower administration were replaced by the US-Soviet confrontations over Berlin, the Bay of Pigs, the Cuban missile crisis, and the beginning of US involvement in Vietnam. President Kennedy, wanting more direct control over US foreign policy and strategy, appointed McGeorge Bundy as assistant to the president for national security affairs. The combination of Bundy's strong personality, expanding US involvement overseas, the increased complexity of national security issues, and the difficulty in developing flexibility and responsiveness within existing departments placed the NSC and the national security advisor in a more prominent policy position.

The post assumed its greatest importance under Presidents Richard Nixon, Gerald Ford, and Jimmy Carter; Henry Kissinger occupied the position, followed by Brent Scowcroft and Zbigniew Brzezinski. Strong-minded, covetous of their prerogatives, and enjoying direct access to the president, Kissinger and Brzezinski expanded the power of the position and tried to impose their personal policy preferences on the national security establishment. Under Carter, there was considerable friction between Brzezinski and Secretary of State Cyrus Vance, leading to the latter's resignation in the wake of the failed 1980 attempt to rescue US diplomats held hostage in Iran.

Since then, there has been considerable debate regarding the proper role of the national security advisor. During the first term of Ronald Reagan, there was an attempt to restrict the role to coordinator and expediter rather than initiator and innovator. Some argued that this downgrading allowed lesser personalities to occupy the position and may have led to some of the excesses revealed in the 1987 Iran-contra hearings. The argument is that a leader with a strong personality is needed to ensure the proper functioning of the national security staff. This is true whether the advisor has a high-profile style, such as Kissinger and Brzezinski, or a low-profile style, such as Scowcroft.

In 1993, for the first time in twelve years, the Democratic Party controlled the Oval Office and Congress. President Clinton appointed Anthony Lake as national security advisor and Sandy Berger as deputy national security advisor. Lake later resigned and was replaced by Berger, who remained in that position until the end of the Clinton administration. In 2001, Condoleezza Rice was appointed national security advisor in the George W. Bush administration. She was the first woman to hold the post and was a national security and defense intellectual. Rice was challenged in her role of coordinator of national security policy by the continuing friction between Colin Powell's perspective on the use of force and the relationship between force and policy, on one hand, and that preferred by the phalanx of Dick Cheney and Donald Rumsfeld, on the other. When Powell chose not to continue into Bush's second term, he was succeeded as secretary of state by Rice. Presumably Rice got along somewhat better with the duo of Rumsfeld and Cheney than did Powell—at least, open friction was less obvious.

Most important, the role and power of the national security advisor depend upon the president. Even though the NSC was established by Congress, the way in which it is used is almost entirely in the president's hands. In any case, the national security advisor is in a position to make a major impact on national security policy. This is true for several important reasons. First, that person is the president's personal confidant, as the post is filled without Senate consent, as required for cabinet secretaries. Second, the office is located inside the White House, so the person has direct access to the president and usually sees him every working day. This fact alone creates a perception of power not accorded other officials. Third, the national security advisor is not tied to any agency, which grants a degree of flexibility that secretaries usually cannot match. In addition, the national security advisor is not encumbered by operational responsibilities, and the NSC staff is relatively small compared to the huge bureaucracies at State and (especially) Defense.

Finally, the national security advisor represents the president's views, in contrast to the organizational tendencies represented by the secretaries of state and defense. Thomas Donilon was President Obama's national security advisor in 2011, working effectively with

Secretary of State Hillary Clinton and Defense Secretary Robert Gates. President Donald Trump first appointed Lieutenant General Michael Flynn, formerly head of the Defense Intelligence Agency (DIA), as his national security advisor. But Flynn ended up under FBI investigation for allegedly false statements to them about his contacts with Russia and was accused of misrepresenting them to Vice President Mike Pence. Under a cloud, he was replaced by Lieutenant General H. R. McMaster, a serving military officer and noted author of widely read works on defense policy and military affairs.[29] McMaster was one of three high-level appointments in the early Trump administration held by serving or retired military officers; the others were James Mattis as secretary of defense, and John Kelly as White House Chief of Staff. All three would leave the administration before the end of President Trump's term of office. President Biden's national security advisor is Jacob (Jake) Sullivan, an experienced policy advisor from the Obama administration.

It is apparent that a strong personality in this position can strongly influence national security policy. Furthermore, the intermingling of foreign policy, national security, and domestic issues enhances the role. That person is in a position to synthesize these policy issues and bring to bear a perspective that is closely linked to the president's perceptions of office and mind-set. This presidential confidant is appointed by, works for, and owes allegiance to the president. President Donald Trump, after several false starts with appointees who were capable professionals but unable to adjust to the hurly-burly style of policymaking characteristic of the administration, appointed former United Nations (UN) ambassador John Bolton to the position of national security advisor. Much more than a policy expeditor, Bolton had not disguised his support for an assertive US defense and foreign policy, and he was probably an influential force in convincing President Trump to repudiate the 2015 Iran nuclear deal signed by the Obama administration. Eventually Bolton departed the Trump administration and became very critical of the president's policies and management style.

The National Security Council

It is useful to examine the organization and function of the NSC in detail, especially with respect to the policy triad and the NSC system. Figure 7.5 illustrates the organization of the NSC, the relationship of the NSC staff, and the procedures for making recommendations and input.

Figure 7.5 The National Security Council

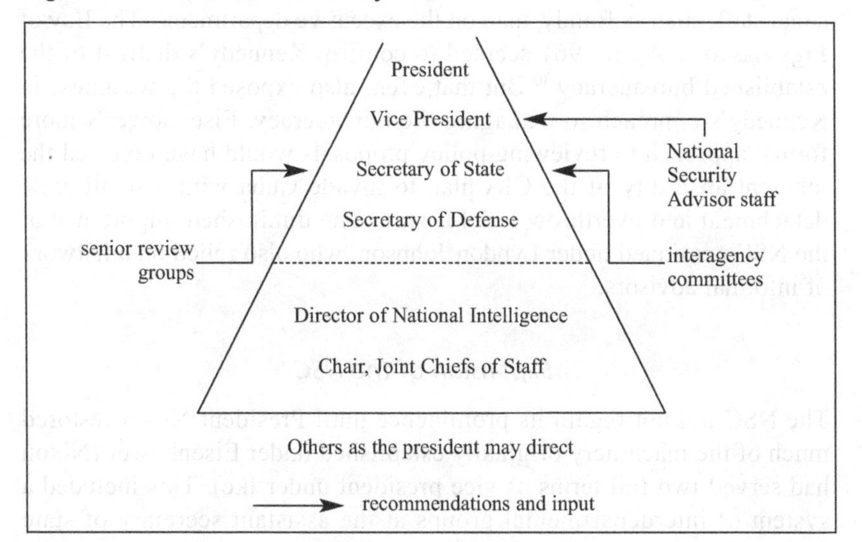

The president determines how the NSC is used—even whether it should be used. Each president from Harry Truman to Donald Trump has shaped the NSC and used it according to his leadership style and perceptions of the office. Truman, the first president to interact with the newly established NSC, made it clear that he considered the NSC to be a strictly advisory body and did not regularly attend its meetings until the Korean War. He maintained that a president could not abdicate his responsibilities in foreign and national security policy to a structure such as the NSC.

President Eisenhower possessed unmatched military experience and established a formal and systematic method for using the NSC. Besides creating the post that would eventually become known as national security advisor, he set up a system of committees to serve the NSC. As many scholars agree, Eisenhower's preference for complete staff work and well-defined procedures gave the NSC and its staff previously unknown prominence in national security. Following military procedures, Eisenhower wanted options exhaustively discussed within the NSC, with clear recommendations presented to him identifying the best policy.

President Kennedy, however, preferred a smaller and more streamlined staff and eliminated much of Eisenhower's machinery. Kennedy's appointment of McGeorge Bundy reflected the importance he attached to integrating foreign and defense policies. His disdain of formal procedures and rigid bureaucracy made him rely more on the NSC staff and his personal staff, such as Bundy, than on the executive departments. The Bay of Pigs fiasco in April 1961 seemed to confirm Kennedy's distrust of the established bureaucracy.[30] But that event also exposed the weakness in Kennedy's approach to managing the bureaucracy. Eisenhower's more formal approach to reviewing policy proposals would have revealed the inherent absurdity of the CIA plan to invade Cuba with a small exile detachment and overthrow Fidel Castro. The diminished importance of the NSC continued under Lyndon Johnson, who also relied on a network of informal advisors.

Expansion of the NSC

The NSC did not regain its prominence until President Nixon restored much of the machinery originally established under Eisenhower (Nixon had served two full terms as vice president under Ike). This included a system of interdepartmental groups at the assistant secretary of state level and senior committees at the undersecretary and deputy secretary levels to examine and coordinate policies on defense, intelligence, covert operations, and other national security matters; Vietnam also received special attention. This structure continued, for the most part, under President Ford but changed under President Carter, who retained the procedures but reduced the committees to two and gave more authority to the executive departments. Nonetheless, the strong personality and views of National Security Advisor Brzezinski under Carter tended to prevail, placing the NSC staff in a powerful position vis-à-vis the executive departments.

The Reagan administration strengthened the NSC committee system and established interagency groups at the assistant secretary level to focus on defense policy, foreign policy, and intelligence. Problems surfaced, however, as Reagan's personal staff became involved and made decisions outside the purview of the established structure. Friction between the national security advisor and the secretaries of state and defense affected several issues, leading to the resignation of Secretary of State Alexander Haig. Turnover was high: the Reagan administration disposed of national security advisors wholesale, including Richard Allen, William Clark, Robert McFarlane, John Poindexter, and Frank Carlucci.

In the George H. W. Bush administration, the roles of the national security advisor and his deputy were strengthened, as each chaired an important interagency committee within the NSC. The appointment of Brent Scowcroft to the position of national security advisor saw a new model of professional competency, management skill, and bureaucratic legerdemain unseen before or since. This internal restructuring was expected to provide better coordination and sharply define policy options. In late 1991 and into 1992, Bush's national security establishment was put to the test by a range of new issues emerging that did not fit easily into the Cold War perspective. Many of these issues continued through subsequent administrations to this day.

The Twenty-First Century: The NSC and Its Staff

In 2001, President George W. Bush and his inner circle made an early impact on the NSC and staff:

> The first weeks of President George W. Bush's tenure as Commander-in-Chief portend significant changes in the national security arena. Old hands in dealing with both the domestic and international ramifications of defense affairs—Cheney, Rumsfeld, Rice, and Powell—are contributing their collective expertise in fashioning the national security construct that will frame the political boundaries of America's military-engagement policies during the next four years. . . . A top-to-bottom scrub of the National Security Council has already occurred, and about one-third of the NSC positions have been eliminated. . . . The size of the professional staff at the NSC doubled under the former president, from about 50 to more than 100 full-timers.[31]

The NSC staff is an arm of the NSC itself. As such it not only performs administrative duties but also is involved in policy-related functions. Muddling the policy process is the fact that the national security advisor lacks a legal basis for operational action. "The National Security Council (NSC) should be responsible for advising the President and for coordinating the multiplicity of national security activities. . . . The NSC Advisor and staff should resist the temptation to assume a central policymaking and operational role."[32] The effectiveness of the staff is limited if it becomes too large and unwieldy, yet it must be large enough to respond to the tasks at hand. The real source of power in the NSC staff rests in its relationship to the president. As such, it has a great degree of bureaucratic freedom and flexibility and reflects presidential views.

After reviewing the history of the NSC staff until 1987, the Tower Commission concluded:

> What emerges from this history is an NSC staff used by each President in a way that reflected his individual preferences and working

> style. Over time, it has developed an important role within the Executive Branch of coordinating policy review, preparing issues for Presidential decision, and monitoring implementation. But it has remained the President's creature, molded as he sees fit, to serve as his personal staff for national security affairs. For this reason, it has generally operated out of the public view and has not been subject to direct oversight by the Congress.[33]

The Iran-contra hearings of 1987 brought the role of the NSC staff into full view as Congress attempted to uncover violations of law by members of the NSC staff, specifically Marine Lieutenant Colonel Oliver North, in using the profits from weapons sales to Iran to get around Congressional strictures and provide funds to help overthrow the government of Nicaragua.[34] As intelligence historian John Prados concluded, "Ultimately it is the President's responsibility to keep his house in order and banish the conflicts among unruly subordinates. Presidents have compiled a rather poor record in this regard."[35] This was no less true during the Clinton administration. The George W. Bush administration was a comparatively top-down operation with message discipline superior to that of Clinton or Reagan on national security issues: however, interdepartmental conflicts were nevertheless played out and overlapped on occasion with the NSC decisionmaking process. The Barack Obama administration in foreign and defense policy featured the president's preference for apparent tidiness and cohesion, consistent with the "no drama Obama" character of his decisionmaking style. Nevertheless, conflicts developed over Afghanistan policy between National Security Advisor James L. Jones and the State Department's point man for Afghanistan, former diplomat Richard C. Holbrooke. Holbrooke's abrasive style annoyed Jones and his staff, who responded with attempts to restrict Holbrooke's travel to the region and isolate him from key meetings in order to diminish his influence over President Obama.[36]

The NSC and its staff are a well-established structure with statutory authority. It has proven to be a useful and important structure when used as an advisory agency that is allowed to analyze policy and strategy options. The way the NSC and its staff function in any administration is complex and subject to personal preferences, mind-sets, and leadership styles—including those of the policy triad. To be sure, the president is central, and the NSC cannot substitute for his or her central role. Yet as one authority concluded: "The decision-making process is always problematical and often controversial. The process requires the interaction of people with differing personal, political, and institutional perspectives on policy issues. By design and evolution, the system promotes rivalry among the branches of the government and within those branches as policy questions move toward resolution."[37]

Expectations for drastic improvement in the US policymaking process with respect to national security are confounded by the matrix of domestic and foreign policy inputs that collide with one another to frustrate policymakers and military commanders. The experience of the Obama administration in this regard is instructive. Obama's liberal supporters were sometimes dismayed by the gap between his campaign promises about national security issues in 2008 and the actual policies adopted by his administration thereafter. Obama decided against the hopes of some supporters not to close the US detention facility at Guantanamo Bay, Cuba; to double down on US commitments to the war in Afghanistan with a military "surge" of some 30,000 additional troops; to maintain the Bush timetable for withdrawal of US combat forces from Iraq; and to adjust, but not to cancel, a plan for missile defenses deployed in Europe against the threat posed by Iran or other sources in the Middle East.

In addition, Obama's attorney general, Eric Holder, decided, for the most part, against criminal investigations of alleged CIA abuses under Bush in coercive interrogations (making exceptions for several especially egregious cases). These Obama "disappointments" in terms of the expectations of his followers were "déjà vu all over again" for the White House. Obama's predecessor George W. Bush campaigned on the need for a "humble" foreign policy and argued against assigning US armed forces to nation-building missions. September 11 and military campaigns in Afghanistan and Iraq connected the Bush administration and its successor to new agendas of threat assessment, policymaking, deterrence, and war, including transnational conflicts against nonstate actors from terrorists to cyber hackers.

Need for Change

The US Commission on National Security/21st Century concluded that changes in the world since the end of the Cold War have not been accompanied by any major institutional changes in the executive branch of the US government. Reorganizations of the US government since September 11, including the creation of the Department of Homeland Security, require amendment of this statement. But the commission's statement still underscores some perennial problems in national security policymaking. According to the commission, deficiencies exist that only a significant organizational redesign can remedy. Most troublesome is the lack of an overarching strategic framework guiding US national security policymaking and resource allocation. Clear goals and priorities are rarely set. Budgets are prepared and appropriated as they were during the Cold War.[38] Time will tell whether the increased concern with dealing with peer competitors improves US strategic thinking.

All this must be qualified by the realities of the political process and the number of actors involved in the national security system. Examining the role of the power centers in the US political system and their role in foreign and defense policies, Roger Hilsman wrote: "The president and his staff in the White House constitute the most powerful of these power centers, but the presidency is far from being all powerful. . . . [A] political system composed of multiple power centers gives a veto to a relatively small coalition of those power centers who oppose some new initiative."[39] Nonetheless, the policy triad in concert with the president forms a formidable power center that is at the core of national security policy and is the driving force behind the national security establishment. It becomes even more formidable when one party controls both the Oval Office and Congress. As Hilsman concluded, "Policy is made through a political process. Power is an element in politics. But power diffused can lead to evil as surely as power concentrated. Herein lies the irony."[40]

Conclusion

Whatever was intended by the framers of the Constitution, the presidency has evolved into the principal focus of US national politics. This inescapable centrality of the presidency includes national security and defense policies. Some presidents have taken to these responsibilities with more enthusiasm or expertise than others: but the national security "buck" stops on the presidential desk, whatever the personality of the office holder. This was illustrated clearly by the reaction to the rapid departure of the US from Afghanistan in August of 2021, when many US citizens and Afghan allies were left behind in the initial evacuation and a terrorist explosion killed thirteen American service members and many more Afghan civilians waiting to be evacuated. Whatever advice President Biden relied upon from the policy triad and the intelligence community, his presidency will be judged by how he responded to that advice.

This overview of the primary structure for advising the president on national security provides the basis for several conclusions:

1. The president by law and of necessity has the central role in national security policy.
2. The informal policy triad is critical in shaping national security policy. The interactions among the secretaries of state and defense and the national security advisor strongly influence the way policy is formulated and the kind of advice given to the president.
3. All of the primary players on the NSC have their own constituencies and bases of power. Strong personalities can impose personal policy preferences on the NSC and the system, at times making it more difficult for the president to consider the full range of options.
4. Those in the policy triad and on the NSC can mobilize in a variety of ways to oppose policy options they disagree with. Adding to the problems of coordination and consensus is the fact that the secretaries of state and defense have different organizational goals and may well view problems through different lenses.
5. The NSC does not have an operational arm. If its recommendations are approved by the president, they must be implemented through the two departments—the Departments of State and Defense—and at times through the DNI and its component agencies, especially the CIA. When the interplay of power relationships within the NSC and the policy triad gives rise to multiple advocacy, the president will find it difficult not to become involved in the policy formulation process.

Strong personalities can pose a dilemma for any president, but appointing people who simply parrot the president's version of national security—the "groupthink" mentality—can lead to dangerous weaknesses in national security policy. Honest disagreement can be valuable, but at the same time some fundamental meeting of the minds must be expected. Though the president cannot constantly interject himself into the policy formulation process to resolve differences among his top personnel, reconciling disagreements is the responsibility of the president. For this reason, presidential leadership is critical to effective national security policy formulation and execution.

Notes

1. US Commission on National Security/21st Century, *Road Map for National Security: Imperative for Change,* final draft report (February 15, 2001), p. 47.

2. See, for example, John M. Goshko, "State Department Reorganizes Ranks," *Washington Post,* February 6, 1994, p. A8.

3. James M. McCormick, *American Foreign Policy and Process,* 3rd ed. (Itasca, Ill.: F. E. Peacock, 1998), p. 390.

4. US Commission on National Security/21st Century, *Road Map,* p. 47.

5. Ibid., p. 57.

6. Ibid., p. 54.

7. Christopher C. Shoemaker, *The NSC Staff: Counseling the Council* (Boulder, Colo.: Westview, 1991), p. 42.

8. Ibid.

9. Stephen M. Walt, *The Hell of Good Intentions: America's Foreign Policy Elite and the Decline of U.S. Primacy* (New York: Farrar, Straus, and Giroux, 2018), p. 89.

10. Stephen E. Ambrose and Douglas G. Brinkley, *Rise to Globalism: American Foreign Policy Since 1938,* 9th rev. ed. (New York: Penguin, 2011), pp. 95–96.

11. John G. Roos, "First Glimpse of Bottom-Up Review Was a Nice Job of Packaging, Anyway," *Armed Forces Journal International* 132, no. 3 (October 1993), pp. 17–18.

12. John G. Roos, "Another Bridge Too Far; It's Time for the SecDef to Fight Some Close-In Battles," *Armed Forces Journal International* 144 (June 1998), p. 2.

13. Vince Crawley, "Bracing for Change," *Army Times,* January 22, 2001, p. 8.

14. Daniel R. Coats, Director of National Intelligence, "Worldwide Threat Assessment of the US Intelligence Community," statement for the record, Senate Select Committee on Intelligence, January 29, 2019, pp. 24–26. See also James Dobbins, Howard J. Shatz, and Ali Wyne, "Russia Is a Rogue, Not a Peer; China Is a Peer, Not a Rogue" (Santa Monica, Calif.: RAND, October 2018), https://www.rand.org/pubs/perspectives/PE310.html.

15. See Steven Metz, ed., *Revising the Two MTW Force Shaping Paradigm* (Carlisle, Pa.: Strategic Studies Institute, US Army War College, April 2001).

16. Steven Metz, *American Strategy: Issues and Alternatives for the Quadrennial Defense Review* (Carlisle, Pa.: Strategic Studies Institute, US Army War College, September 2000), p. ix.

17. US Department of Defense, *Quadrennial Defense Review Report* (Washington, DC, 2006), p. 9.

18. US Department of Defense, *Quadrennial Defense Review Report* (Washington, DC, 2010), p. v.

19. Ibid., p. 15.

20. Ibid., p. vi.

21. Mara Karlin, "How to Read the 2018 National Defense Strategy" (Washington, DC: Brookings Institution, January 21, 2018), https://www.brookings.edu/blog/order-from-chaos/2018/01/21/how-to-read-the-2018-national-defense-strategy.

22. Joseph R. Biden, "Renewing America's Advantages: Interim National Security Strategic Guidance" (Washington, DC: White House, March 2021), https://www.whitehouse.gov/wp-content/uploads/2021/03/NSC-1v2.pdf.

23. Sam C. Sarkesian and Robert E. Connor Jr., *The US Military Profession into the Twenty-First Century: War, Peace, and Politics* (London: Frank Cass, 1999), esp. chap. 9–10.

24. Christopher Dandeker and James Gow, "Military Culture and Strategic Peacekeeping," in Erwin A. Schmidl, ed., "Peace Operations Between War and Peace," *Small Wars and Insurgencies* 10, no. 2 (special issue, Autumn 1999), p. 58.

25. Colin Powell with Joseph E. Persico, *My American Journey* (New York: Random, 1995), p. 576.

26. See Alan Ned Sabrosky and Robert L. Sloane, *The Recourse to War: An Appraisal of the "Weinberger Doctrine"* (Carlisle, Pa.: Strategic Studies Institute, US Army War College, 1988).

27. John Spanier and Eric M. Uslaner, *American Foreign Policy Making and the Democratic Dilemmas,* 6th ed. (New York: Holt, Rinehart, and Winston, 1993).

28. Ibid.

29. For example, see H. R. McMaster, *Dereliction of Duty: Lyndon Johnson, Robert McNamara, the Joint Chiefs of Staff, and the Lies That Led to Vietnam* (New York: HarperCollins, 1997).

30. Although President Kennedy publicly accepted the blame for the Bay of Pigs failure, privately he placed much of the blame on the poor planning and cumbersome procedures in the military and the CIA, as well as the staff procedures of existing agencies.

31. John G. Roos, "A New Beginning," *Armed Forces Journal International* (March 2001), p. 2.

32. US Commission on National Security/21st Century, *Road Map,* p. 50.

33. *President's Special Review Board* (Washington, DC: US Government Printing Office, February 26, 1987), generally referred to as the Tower Commission Report.

34. See, for example, *Report of the Congressional Committees Investigating the Iran-Contra Affair with Supplemental, Minority, and Additional Views,* 100th Cong., 1st sess., House Report no. 100-433 and Senate Report no. 100-216 (Washington, DC: US Government Printing Office, 1987), esp. pp. 36–51.

35. John Prados, *Keepers of the Keys: A History of the National Security Council from Truman to Bush* (New York: William Morrow, 1991), p. 561.

36. Rajiv Chandrasekaran, "The War Within the War Cabinet," *Washington Post,* June 25, 2012, http://global.factiva.com/hp/printsavews.axpx?pp=Print&hc=Publication.

37. Donald M. Snow, *National Security: Defense Policy in a Changed International Order,* 4th ed. (New York: St. Martin's, 1998), p. 100.

38. US Commission on National Security/21st Century, *Road Map,* p. x.

39. Roger Hilsman with Laura Gaughran and Patricia A. Weitsman, *The Politics of Policy Making in Defense and Foreign Affairs: Conceptual Models and Bureaucratic Politics* (Englewood Cliffs, N.J.: Prentice-Hall, 1993), pp. 340, 343.

40. Ibid., p. 349.

8

The Military Establishment

The military establishment is a critical operational arm of the national security system. It is essential to understand how it is organized and its relationship to the president and other political actors. Military success requires highly skilled and competent individuals at all levels in the military hierarchy. Furthermore, the education, socialization, and mind-sets of military professionals are important in shaping the military establishment and in determining its ability to pursue the goals of US national security policy. This in turn has made the president and the national security establishment heavily dependent upon the military for sound advice.

The US military establishment has gone through several important changes in organizational structure since the end of World War II, and these changes can be expected to continue. The US military must adjust to a variety of political and social forces that have affected its structure and missions. It must also adjust to an ill-defined and ambiguous national security landscape. These changes reflect challenges across the conflict spectrum and complicate strategy, force planning, and training.

Over the past few years, the main focus of US national security policymakers has shifted from unconventional, or "irregular" warfare to high intensity conventional and even nuclear warfare against rising nuclear powers such as Iran and North Korea and increasingly powerful peer competitors, such as Russia and China. This does not mean that the capabilities for unconventional warfare will erode, but the bulk of attention and financing will be devoted to preparation for high intensity warfare against states with capabilities similar to those of the United States.

This was made clear by then–secretary of defense Mark Esper at a talk to the students and faculty of the US Naval War College on August 7, 2019, when he noted, "Many of you spent most of your career fighting irregular warfare. But times have changed. We are now in an era of great-power competition. Our strategic competitors are Russia and China."[1] Over the years the concept of national security expanded to include military participation in humanitarian and peacekeeping missions, combating international terrorism, and possibly domestic missions that would have been unthinkable before September 11, such as border control. Despite this, the US military is transforming its forces and training to respond to rising peer competitors such as China and Russia. The lower-level challenges will still remain, so the military will need to remain flexible and prepare for multiple contingencies.

The US military will need to become more flexible and agile to meet current and emerging threats. This envisions changes in strategy, doctrine, weaponry, and training. We will explore these issues by looking at the command-and-control structure of the military, in particular the multiservice unified combatant commands who do the actual fighting. Although the president is commander in chief of the military, Congress also has an important role in funding and influencing military actions, including the size and composition of military forces. We then discuss the norms of the military profession that affect how wars are prepared for and waged. A crucial issue for policymakers is how to decide when the use of force is justified. President Reagan's secretary of defense, Caspar Weinberger, set out some conditions that should be met before US ground troops should be committed. Although these are almost forty years old, they still provide useful standards for initiating military actions. Finally, we look at the military's role in the policy process in the national level. Although the military resolutely avoids participation in partisan political issues, they are "political" in the sense that they try to shape decisions about military funding, structure, training, and actions.

The Command-and-Control Structure

The military establishment's focus of power shifted to the secretary of defense when that office was given control of the military departments in 1949; changes in 1986 further expanded the secretary's power, as well as that of the chairman of the Joint Chiefs of Staff (JCS). The Joint Chiefs of Staff comprises the uniformed military service chiefs of the

Army, Navy, Marine Corps, Air Force, and Space Force, plus the head of the National Guard Bureau. In addition, it includes the chairman and vice chairman of the JCS.[2]

Several reference points need to be reviewed with respect to the structure of the Department of Defense (see Figure 7.3 in the previous chapter). First, the secretaries of the military departments (Army, Navy, and Air Force) have no operational responsibilities; revisions to the National Security Act of 1947 downgraded their executive department status to that of military departments (1949) and later removed them from the chain of command (1958). The primary responsibilities of the service secretaries are in administrative and logistical areas: manpower, procurement, weapons systems, service effectiveness, military welfare, and training responsibilities, among other duties. A strong personality in the service secretary's office can have an influence in shaping the posture and operational capability of the service, but he or she will still lack operational responsibility.

Second, the role of the JCS chairman has been strengthened. Formerly, the JCS was a corporate body, and the chairman served principally as spokesperson for joint decisions. In addition, the chairman had little control over who served on the joint staff from the services, whose commanders also rotated the chairman's functions among themselves in his absence. The position of chairman was in some ways symbolic rather than substantive, but strong chairmen, such as Admiral William Crowe and General Colin Powell, were able to have significant impact on policies.

The 1986 Defense Reorganization Act (the Goldwater-Nichols Act) made the JCS chairman the primary military figure in the defense establishment. It gave him direct access to the president and assigned him responsibilities not only for strategic planning, but also for a range of other matters (including budget assessments and readiness evaluations), affording him a more direct relationship with the commanders in unified combatant commands. He now has direct control over assignments to the joint staff. Most important, he is officially the "principal military advisor" to the president, the secretary of defense, and the national security council. The position of vice chairman of the JCS was also created by this legislation, relieving the chairman of some detailed responsibilities. Also, the chairman is no longer required to accept joint staff members nominated by the various services.

In short, the chairman of the JCS is now the most important member of that body, responsible only to the secretary of defense and the president. The chain of command and control of the operational arm of the military runs directly from the commander in chief to the secretary of defense through the JCS chairman to the commanders of the unified combatant commands (see Figure 7.3). Note that the chairman of the JCS is not technically in the chain of command, although he facilitates communication between the secretary of defense and the unified combatant commanders.[3]

Third, a new officer specialty was created by Congress.[4] This joint specialty provides for a lifetime career path for officers qualified as staff officers in joint staff positions. The objective is to have a pool of officers from all services qualified to serve on joint staffs, although it was not intended to create a general staff corps on the old German army model. The program has led to important changes within the profession and in the functioning of the joint staff, but officers are still promoted primarily on their record of success in positions of command. A career based on service on joint staffs is unlikely by itself to result in promotion to a flag officer—that is, general or admiral. Still, service on joint or combatant command staffs is important in fulfilling the joint qualification requirement for promotion to flag officer or command of joint forces.

Unified Combatant Commands

A unified command, as the name suggests, is a joint service operational responsibility. Commanded by a four-star officer from one service, all services are generally represented in each command. There are currently eleven unified commands, with specific details set forth in the classified biennial Unified Command Plan (UCP).[5] Seven of these have responsibility for US military operations in defined regions of the world (roughly the United States, South Asia and the Pacific, Europe, Latin America, Africa, Southwest Asia, including the waters adjacent to these areas), and President Trump directed the addition of US Space Command in December, 2018:[6]

- USAFRICOM: United States Africa Command
- USEUCOM: United States European Command
- USCENTCOM: United States Central Command
- USNORTHCOM: United States Northern Command
- USSOUTHCOM: United States Southern Command
- USINDOPACOM: United States Indo-Pacific Command
- USSPACECOM: United States Space Command

Four of the nine unified commands are defined by their function:

- USSOCOM: United States Special Operations Command (special operations)
- USSTRATCOM: United States Strategic Command (global strike warfare, space, missile defense, combating weapons of mass destruction)
- USTRANSCOM: United States Transportation Command (movement of military forces)
- USCYBERCOM: United States Cyber Command (cyber defense and offense)

Until the 1986 Defense Reorganization Act, the commanders of unified commands had little control over what units were assigned to their command. This was the responsibility of the various services, whose component commanders had to depend on their own services for resources. The composition of forces was determined by each service. Thus unified commands reflected a mix of doctrines, equipment, and missions as determined by the services.

Now the commanders of unified combatant commands have been given more power in budget matters pertaining to their commands, hiring and firing authority over subordinate commanders, and direct access to the secretary of defense and JCS chairman, bypassing the respective services. The president, in turn, can give the JCS chairman primary responsibility for overseeing the activities of these commands. The 1986 Defense Reorganization Act established the JCS chairman as the spokesperson for the commanders of the combatant commands, especially in operational requirements. Commanders now have the authority to act as real commanders, independent from control of their respective services. The changes resulting from the 1986 act have had a positive impact on jointness and in the operational direction of the various commands, although there will sometimes be friction between the combatant commanders and the services over the amount and type of support the services can provide the commanders.

It is also the case that in the new strategic environment, interservice rivalries remain as each service tries to protect its turf in terms of missions and budgets. This remains a characteristic of the military establishment as each service attempts to prepare for twenty-first-century warfare and the new strategic landscape. Congress tried to override interservice squabbles and problems of command and control by imposing a strengthened secretary of defense and JCS chairman, as well as a strengthened command system, on the military establishment, but internal rivalries remain. The number of assistant secretaries of defense reflects the variety of matters under the responsibility of the secretary of defense. Indeed, the scope of activities and the level of resources required to maintain and expand them created a vast managerial complex—perhaps so complex as to preclude an efficient military system.

The inherent rivalries leave the Defense Department vulnerable to politicization of its operational arms and hamper development of coherent policy and feasible options. With the inauguration of George W. Bush in 2001, a new national security team was put into place and, with it, a revised defense organization under Secretary of Defense Donald Rumsfeld. This included a system designed to operate in several unconventional environments, particularly to combat international terrorism.

Secretary Rumsfeld's goal was to transform the military into a twenty-first-century force designed to respond to threats and potential threats. This encompassed changes in the military system that focused on doctrine, training, and weaponry and incorporated information-age technology. The initial actions in Afghanistan in 2001 and Iraq in 2003 seemed to offer support for Rumsfeld's vision of lighter and more mobile forces at the expense of numbers, but the need for quantity as well as quality of forces soon became apparent as the military became bogged down in both wars. Rumsfeld's replacement by Robert Gates in 2006 resulted in a reappraisal of the specifics of military transformation in view of the US experience in Afghanistan and Iraq.[7] These reappraisals continued in subsequent administrations, and given their impact on preferred types and levels of forces and their employment, the process can be very contentious. Still, service doctrines can change in the face of new challenges.[8]

Congress: Guns and Butter

The role of Congress does not stop at legislation to restructure the military. Executive-legislative skirmishes over constitutional roles have already been discussed, but the division of authority over military appraisal and allocation also impacts the formulation of strategy and its operational implementation. This is reflected in guns-and-butter issues: How much should be spent for defense, and how much for nondefense issues, such as building a wall on the southern border of the United States? The president is the commander in chief of the military, but Congress has the power of the purse. The weapons acquisition process, which greatly affects the military establishment's performance, is but one example. These struggles magnify the political dimension of the military. In addition, Article 1, Section 8, of the US Constitution gives Congress the power to "make Rules for the Government and Regulation of the land and naval Forces." This increases the importance of the house and senate armed forces committee, which exercise close supervision over all aspects of the military.

Congressional hearings on strategy, resource allocation, and military performance strike at the heart of the military establishment, requiring operational commanders at all levels to become sensitive to the political nature of their responsibilities. The desire in Congress for more explicit military recommendations and more control over military commitments in contingencies short of war means it struggles with the president, the secretary of defense, the chairman of the JCS, individual service chiefs, and any number of high-level operational commanders.

This was true even before US forces became engaged in Iraq. In 1993 and 1994, these struggles were especially visible over the reduction of the military and the shrinking defense budget. In the aftermath of September 11, budget allocations for defense increased, especially for the wars in Iraq and Afghanistan and to improve homeland security. Those budget increases could not be sustained in the face of the fiscal retrenchment required by the condition of the US economy after 2008. A shrinking budget makes it even more urgent to understand the forces necessary to support US national security strategy. Table 8.1 presents the relative allocations for several defense categories averaged from 2001 to 2017, showing the relatively large amounts allocated for operations and maintenance, military personnel, and procurement.[9]

Directly and indirectly, Congress has increased its oversight of the military establishment. Congress brings along a variety of political considerations, from personnel issues to weapons procurement to base

Table 8.1 Department of Defense Budget Allocation, 2001–2017

	Average Percentage of Select Defense Military Accounts (Rounded)
Operations and maintenance	41
Military personnel	24
Procurement	19
Research, development, test, and evaluation	12
Military construction	2

Source: Congressional Research Service, Library of Congress, "Defense Primer: The National Defense Budget Function," March 17, 2017, https://sgp.fas.org/crs/natsec/IF10618.pdf.

closure decisions. Its judgments on budget allocations affect the structure of the military establishment, procurement decisions, and personnel issues. Ultimately congressional decisions affect the ability of the United States to pursue its vital interests.

Civilians employed by the Department of Defense are an important element in the relationship between the military establishment and other political actors. In 1990, there were more than 1 million civilians working for the department, not counting the more than 1 million employed in defense-related industries. During the Bill Clinton administration, there was a reduction in the number of civilians employed and in the money devoted to defense industries. By 2000, the number of civilians working for the Department of Defense had dropped to about 700,000. Those employed in defense-related industries had also been reduced.[10] As of 2021 the number of civilian employees was some 759,000.[11]

Close links with the Department of Defense provide a channel for civilian attitudes and mind-sets to penetrate the military profession. Conversely, military attitudes and mind-sets penetrate the civilian component. This interpenetration is greatest at the higher levels, as all appointments at the assistant secretary level and above are political and normally civilian, including many retired military. A more noticeable degree of separation exists at lower levels, where Department of Defense civilians are rarely present in operational units. There is, however, a growing tendency to use private contractors to perform formerly military functions. Contracted personnel can be very cost-effective in the long run, because whatever their daily rate, the Defense Department does not incur longtime pension or health-care obligations to them. The degree to which civilians should perform military functions is controversial, and it can be difficult to hold them legally accountable for misbehavior.

The dual influences in the military establishment simply confirm the long-standing norm that civilian control of the US military is a well-established fact in law and reality. Equally important, the military profession has accepted this as a basic premise. Over the years, civilian rule has slowly but surely permeated weapons acquisition, force composition, strategic options, and command-and-control issues. It is in such areas that the president and the military establishment face some of the most serious disagreements.

Force Restructuring and Composition

The size and composition of US military forces are determined by civilian leaders based on several factors, including the level of resources available, technological developments, and the perceived threat. One can never be sure what forces will be required or what the intentions are of potential adversaries, a problem complicated by the very long lead times required to develop new systems. Given the inability to demonstrate conclusively which forces may be needed in the future and given also the vested interests of those who benefit from current procurement, change comes slowly in this area. Serious strategic thought is required in developing force structure, of course. Former chief of staff of Army General Gordon Sullivan and Lieutenant Colonel James Dubik noted that force structure and budgetary decisions must be made with an awareness of the missions required and the resources to fund them:

> American political leaders expect the military to contract in both size and budget, contribute to domestic recovery, participate in global stability operations, and retain its capability to produce decisive victory in whatever circumstances they are employed—all at the same time. . . . International and domestic realities have resulted in the paradox of declining military resources and increasing military missions, a paradox that is stressing our armed forces. The stress is significant.[12]

Much of the US force structure is a "legacy force" designed for the Cold War. It is a force unmatched for large scale high-intensity combat. With the demise of the Soviet threat and the subsequent rise in unconventional and terrorist threats, several attempts were made to match forces to new missions. Now, with the increasing power of peer competitors such as Russia and China, some old missions have become increasingly relevant.

Another milestone was, of course, the terrorist attack on the United States on September 11, 2001, and the military response to the challenges posed by both irregular military forces and terrorists. The Bush administration entered office in 2001 skeptical of the importance of nation building, but opted for a program that combined counterinsurgency (COIN) and counterterrorist (CT) operations in both Afghanistan and Iraq. Experience there reminded the United States of the difficulty of trying to win the "hearts and minds" of a population with very different beliefs, historical experiences, and interests.

President George W. Bush's first defense secretary, Donald Rumsfeld, designed his battle plans for both the Iraq and Afghanistan wars to validate his views on the transformation of the US military into a lighter, more agile, and more lethal force—one less dependent on quantity of forces and more dependent on quality and capabilities. As a consequence, there were not sufficient forces on hand to control the societies after their governments were toppled, let alone try to transform them into something more to US liking. This proved impossible to accomplish and future decisions on force structure and modernization will reflect the revived interest in preparing for conflict between great powers.

The Military Profession

If the command-and-control structure is an important ingredient in the formulation of national security policy, the character of the military profession, including its values, norms, and mind-sets, is equally so.[13] How military professionals perceive threats, assess the capability of the military instrument, and develop professional skills and capabilities is a significant determinant of the advice they give. In addition, professional behavior has much to do with the perceptions of Congress, allies, and adversaries regarding US military capability and effectiveness.

Six elements are paramount in shaping the character of military officers: (1) the profession has a defined area of competence based on expert knowledge; (2) there is a system of continuing education designed to maintain professional competence; (3) the profession has an obligation to society and must serve it without concern for remuneration; (4) it has a system of values that perpetuate its professional character and establish and maintain legitimate relationships with society; (5) there is an institutional framework within which the profession can function; and (6) the profession has control over the system of internal rewards and punishments and is in a position to determine the quality and quantity of those entering the profession.

One factor distinguishing the military profession is captured in the US Military Academy motto, "Duty, Honor, Country." Its sole client is the state, including the people of the state. The military professional accepts the doctrine of "ultimate liability": one must be prepared to give one's life for the state. Other professions, such as peace officers and firefighters, also are prepared to give their lives for others, but the characteristic unique to the profession of arms lies in its primary purpose: to win the nation's wars. Because of the enormous power resting with those who bear arms in defense of the state, the military profession is subject to constraints not found in other callings. The military cannot publicly or formally engage in political partisanship to secure better wages, conditions of employment, or operational commitments. It must comply with the policy decisions of civilian officials even if it does not agree with them. Institutional and professional loyalties as well as professional motivations preclude expressing public outrage at orders from above. The military must accept the decision to destroy the enemy with all of the power at hand once the political leadership decides it shall be done.

Professional career patterns and success depend on the ability to work within the system and follow the established path. Professional values and mind-sets, which preclude military professionals from stepping outside the system, also socialize them into an institutional perspective that supports a particular set of relationships with society. Perhaps most important, there is a system of teaching that establishes the way military professionals view themselves, their institution, society, and the outside world. While other professions follow somewhat similar patterns, nonmilitary professions move in a civilian mainstream that has relatively undefined boundaries.

The military profession, in contrast, tends to be a more self-contained community, socially, legally, economically, and intellectually—all of which reinforces the primary professional purpose of success in battle. To be sure, civilians sometimes find it difficult to appreciate the lifestyles and mind-sets thereby engendered. This lack of understanding is at the root of the negative and sometimes antagonistic views that military professionals hold about self-styled military experts, especially in the media, Congress, and academia. Many military professionals see such critics as uninformed about the military, or even hostile to it, which helps perpetuate the self-contained community mind-set of the military and magnifies differences between the military and society in general.

As US society has become more sensitive to issues of diversity and inclusion, there is a greater expectation that the military should reflect society more closely in its norms and interactions. Despite the greater concern for preparing for major war, there remains an expectation that the military must be more society-sensitive in dealing with noncombat operations such as peacekeeping, humanitarian, and constabulary missions. Meanwhile, the military has been dealing with important personnel issues, including gender, sexual orientation, sexual identity, sexual assault, morale, recruitment, and postcombat stress. The Biden administration added to this list the concern that extremist political perspectives might penetrate military organizations and distort their priorities.

The military perspective casts the external environment and the requirements for military effectiveness in a mold that may differ considerably from the view of elected officials and the public. It can also differ from views within the executive branch. Evolving in no small part from the Vietnam experience, military professionals have become sensitive to the political dynamics created by military commitments. As a result, military leaders tend toward extreme caution about military involvement in situations likely to generate domestic political opposition, a concern expressed eloquently in 1976 by General Fred Weyand, chief of staff of the Army:

> Vietnam was a reaffirmation of the peculiar relationship between the American Army and the American people. The American Army really is a people's Army in the sense that it belongs to the American people who take a jealous and proprietary interest in its involvement. When the Army is committed the American people are committed, when the American people lose their commitment it is futile to try to keep the Army committed. In the final analysis, the American Army is not so much an arm of the Executive Branch as it is an arm of the American people. The Army, therefore, cannot be committed lightly.[14]

Note, however, that "commitment" has its own spectrum. For example, US Army and Marine Corps combat troops deployed overseas have a distinctly different connotation for the public than do US Navy personnel sailing aboard vessels in international waters.

Once forces are committed, most military professionals expect to employ the resources at hand to prevail quickly, believing it unconscionable not to use the most effective weaponry and tactics to subdue the enemy. Furthermore, the principles of war stress that overwhelming firepower and resources should be centered on the enemy's weakest point to achieve victory. Military professionals seek overwhelming superiority, not simply adequacy, in the belief that such superiority will reduce casualties and be the most economical in achieving success. Political fireworks often accompany the use of the military in conflicts short of war.

Homeland security is another mission that complicates the role of the military. Military professionals recognize the need for homeland security but are concerned about command and control and the coordination required among the active military, reserve forces, domestic security forces, and emergency and medical teams.[15] This issue will become even more complicated and broader in scope as the most recent department, the Department of Homeland Security, works out its missions and relationships with other national security agencies. Since the Posse Comitatus Act of 1878, there have been restrictions on the use of active duty military forces in domestic law enforcement. By and large, these restraints are supported by the military leaders, who have no wish to be involved in unpleasant, messy, and highly controversial actions on US soil.

Finally, military professionals expect to be provided clear policy goals and operationally precise tasks, tailored to support an identifiable strategy. Mainstream military posture is shaped best to respond to these dimensions. Contingencies of an unconventional nature are more likely to be policy incoherent, strategically obscure, and operationally muddled.

The Weinberger Doctrine

The Weinberger Doctrine was set forth by secretary of defense Caspar Weinberger in 1984 to spell out the conditions under which US ground combat troops should be committed. This doctrine was designed to ensure that the use of military force would be as a last resort and would be effective if employed. Events in Somalia in 1993 seemed to confirm the relevance of these approaches. During the Somalia mission, the killing of eighteen US soldiers in the failed attempt to capture the warlord Mohammed Farrah Aidid in Mogadishu led critics to point out that a lack of clearly defined military objectives, inadequate support, and mission creep were fundamental causes of military failure.

The Weinberger Doctrine includes the following elements:[16]

- There should be no overseas commitment of US combat forces unless a vital national interest of the United States or important US ally is threatened.
- If US forces are committed, there should be total support—resources and manpower to complete the mission.
- If committed, US forces must be given clearly defined political and military objectives, and the forces must be large enough to be able to achieve these objectives.
- There must be a continual assessment between the commitment and capability of US forces and the objectives, which must be adjusted, if necessary.
- Before US forces are committed, there must be reasonable assurance that Americans and their elected representatives support such commitment.
- Commitment of US forces to combat must be the last resort.

General Colin Powell later spelled out his view on the use of force: "Have a clear military objective and stick to it. Use all the force necessary, and do not apologize for going in big if that is what it takes. Decisive force ends wars quickly and in the long run saves lives. Whatever threats we faced in the future, I intended to make these rules the bedrock of my military counsel."[17]

The massive use of military force at the center of gravity of the adversary remains a clear military principle. How this can be applied to operations other than war, stability operations, and unconventional conflicts, including international terrorism, remains contentious. The US military is unmatched in its ability to destroy regular military forces, but what is the center of gravity in a conflict that is rooted in political, ethnic, or religious considerations? These problems would resurface in a big way in both Iraq and Afghanistan.

The Military and the Policy Process

The structure of the Department of Defense and the nature of military professionalism are the primary determinants of the military's role in the national security policy process. That role has more to do with administrative and operational considerations than with the serious formulation of national strategy, in which the military is in a distinctly secondary position. The civilian leaders determine national objectives, and the military develops military strategies to achieve them.

The multiservice perspective necessary for effectiveness is difficult for military professionals to maintain, despite the increased emphasis on "jointness" in career development. Career success still runs through the respective services. Even the members of the JCS, with their dual role, are sorely tempted to tend to their service responsibilities first. The same is true of joint staff officers, and although their performance on the joint staff and their relative efficiency are determined by the JCS, service perspectives and career considerations tend to erode a joint perspective. Service parochialism and professional socialization instilled over a long career are difficult to overcome by assignment to a joint staff.

The drive in the military toward high technology, reflected in sophisticated battlefield weaponry, electronic warfare, and the evolution of an intricate organizational defense structure, has tended to shape the military establishment along the lines of civilian corporate and managerial systems. Further, great efforts have been made within senior service schools to develop officers capable of dealing with such complex conditions. As a consequence, military managerial capability has become an important factor in military efficiency and personal career success. But not everyone agrees with this approach. Many military professionals and critics argue that the real need within the military services is for leaders and warriors. According to this view, emphasis on management erodes the ability of the military services to command and lead operational units successfully in battle. Reliance on technology shifts the focus away from the psychological-social dimensions of human behavior, thereby reducing competence in the art of leadership.[18]

Notwithstanding its coveted aloofness from politics, the military has been dragged into the fray on many fronts, including weapons acquisition, budgets, and relations with Congress. Furthermore, contemporary conflicts at all levels of the conflict spectrum include important political considerations.[19] As some are inclined to argue, the military professional must begin functioning long before the opening salvos of war and operations other than war. Indeed, wars and conflicts in the late twentieth and early twenty-first century have been characterized by inextricably interconnected political and military factors. Cases in point include the monitoring of the Iraqi military and the Kurds in northern Iraq after the first Gulf War, the post–September 11 Iraq and Afghanistan wars, involvement in Bosnia-Herzegovina and Kosovo after the Bosnian War, concern over political developments in Russia, the China-Taiwan issue, and the US intervention in Syria and assistance to Saudi Arabia in its war in Yemen, among others. For military professionals to be successful, they must acquire political as well as military skills. At a minimum, military professionals must be able to deal with the political dimensions of conflict, at least to understand the political dimensions of problems they may be asked to handle militarily. In the twenty-first century, such considerations are no longer distinct from the operational environment.

Many observers fear that attention to politics not only detracts from developing military skills but also exposes the profession to politicization, harming professional integrity and autonomy. Their solution is to maintain a clear separation between politics and the military profession. The traditional concept of civil-military relations and an apolitical military associated with democratic systems is based on this separatist view. Others argue that there is no such thing as a completely apolitical military and that the US military is no exception. How these matters can be reconciled with the political character of conflicts and the military's role in a democracy remains a persisting dilemma for the military profession.

In this perspective, the military must inform both elected officials and the public of their views regarding the military system and the use of force. Sam C. Sarkesian suggested that the military profession cannot function as a "silent order of monks."[20] He noted in another context,

"The military profession must adopt the doctrine of constructive political engagement, framing and building a judicious and artful involvement in the policy arena. A politics-savvy military profession is the basic ingredient for constructive political engagement."[21]

These characteristics of the military establishment do not allow simplistic views of its role in the national security policy process or in the formulation of strategy. On one hand, the views of the military cannot be ignored; on the other, the character of the military institution, the nature of the military profession, and the way the US political system works preclude a lead role for the military. Research on the matters examined here reveals the following:

> The combination of domestic and international issues promises to weigh heavily on the military profession and its strategic and doctrinal orientation. These issues strike at the core of the professional ethos. Even more challenging is that the military profession must be prepared to respond to a variety of national security challenges with diminished resources and considerably fewer personnel, and it must do so even in the face of skepticism within the body politic regarding issues of national security. Exacerbating all of this is that few politicians and academic commentators in the new era have had any real military experience. Although military experience is not the sine qua non for serious examination and analysis of national security and the military profession, without that experience it is difficult to design realistic national security policies or to understand the nature and character of the military profession.[22]

How the president and the national security establishment incorporate the military establishment into the policy process while maintaining the character of the profession and not violating the norms and expectations of a democratic system is a problem facing every president. The challenge cannot be met without an understanding of the nature and character of the military establishment and its professionals. That understanding is also essential for the study of national security.

In thinking about the military establishment, one must not lose sight of its purpose. As one author observed,

> [T]he purpose of military forces is to fight and win our nation's wars. . . . All other considerations are subordinate to this purpose and the peacetime preparation to accomplish it. . . . The military should not be viewed as a social welfare agency, a human rights organization, a giant petri dish for social experimentation, an engine driving the economy, or a conduit for defense contractors to reach the public purse.[23]

These considerations should not preclude the expectation that the military will fight as humanely as possible and avoid noncombatant casualties to the extent that can be done. It does mean that one should not lose sight of the purpose for which military forces are raised and used.

Notes

1. See https://usnwc.edu/News-and-Events/News/Secretary-of-Defense-Esper-Tells-US-Naval-War-College-Students-His-Focus-is-Great-Power-Competition. Video at https://youtu.be/JQ_gDd27IjA.

2. See Congressional Research Service, "Defense Primer: the Department of Defense," December 20, 2018, https://crsreports.congress.gov/product/pdf/IF/IF10543.

3. The Congressional Research Service notes (ibid.) that "although the Chairman plans, coordinates, and oversees military operation involving U.S. forces, neither the Chairman nor the JCS has a formal role in the execution of military operations—a role instead assigned to the unified combatant commanders."

4. A *specialty* is a primary or secondary career pattern for which officers may qualify by virtue of performance and education, among other considerations.

5. US Department of Defense, "Combatant Commands," https://www.defense.gov/Our-Story/Combatant-Commands.

6. Executive Office of the President, "Establishment of US Space Command as a Unified Combatant Command," December 18, 2018, https://www.federalregister.gov/documents/2018/12/21/2018-27953/establishment-of-united-states-space-command-as-a-unified-combatant-command.

7. It is noteworthy that as these words are written in 2021 the United States still has some forces in Iraq and only ended the war in Afghanistan by beating a hasty retreat in August of this year.

8. See the 2019 planning guidance of the incoming commandant of the Marine Corps, General David H. Berger, which urged a rethinking of Marine Corps doctrine and amphibious shipping requirements in view of the increased dangers from precision targeting by defensive forces. See https://www.marines.mil/Portals/1/Publications/Commandant's%20Planning%20Guidance_2019.pdf?ver=2019-07-17-090732-937.

9. Congressional Research Service, "Defense Primer: The National Defense Budget Function," March 17, 2017, https://crsreports.congress.gov/product/pdf/IF/IF10618.

10. William S. Cohen, Secretary of Defense, *Report of the Secretary of Defense to the President and Congress, 2001,* appendix C, http://www.nti.org/e_research/official_docs/dod/2001/101DOD.pdf.

11. Congressional Research Service, "Defense Primer: Department of Defense Civilian Employees," January 12, 2021, https://sgp.fas.org/crs/natsec/IF11510.pdf.

12. General Gordon R. Sullivan and Lieutenant Colonel James M. Dubik, *Land Warfare in the 21st Century* (Carlisle, Pa.: Strategic Studies Institute, US Army War College, February 1993).

13. For a detailed study of US military professionalism, see Sam C. Sarkesian and Robert E. Connor Jr., *The US Military Profession into the 21st Century: War, Peace, and Politics* (New York: Routledge, 2006). Also see Don M. Snider and Gayle L. Watkins, "The Future of Army Professionalism: A Need for Renewal and Redefinition," *Parameters: US Army War College Quarterly* 30, no. 3 (Autumn

2000), pp. 5–20. What many consider classics on military professionalism and civil-military relations are Morris Janowitz, *The Professional Soldier: A Social and Political Portrait* (New York: Free Press, 1971); and Samuel P. Huntington, *The Soldier and the State: The Theory and Politics of Civil-Military Relations* (New York: Vintage, 1964).

14. General Fred C. Weyand, "Vietnam Myths and Realities," *CDRS Call* (July/August 1976), reprinted in Harry G. Summers, *On Strategy: The Vietnam War in Context* (Carlisle, Pa.: Strategic Studies Institute, US Army War College, January 17, 1981), p. 7. General Weyand was the last commander of the Military Assistance Command Vietnam and supervised the withdrawal of US military forces in 1973.

15. See, for example, Lieutenant Colonel Antulio J. Echevarria II, *The Army and Homeland Security: A Strategic Perspective* (Carlisle, Pa.: Strategic Studies Institute, US Army War College, March 2001).

16. This summary is based on David T. Twining, "The Weinberger Doctrine and the Use of Force in the Contemporary Era," in Alan Ned Sabrosky and Robert L. Sloane, eds., *The Recourse to War: An Appraisal of the "Weinberger Doctrine"* (Carlisle, Pa.: Strategic Studies Institute, US Army War College, 1988), pp. 11–12. The Weinberger Doctrine appears in Department of Defense, *Report of the SECRETARY of Defense to the Congress for Fiscal Year 1987* (Washington, DC: US Government Printing Office, February 5, 1986). A critique of the Weinberger rules appears in Eliot A. Cohen, *The Big Stick: The Limits of Soft Power and the Necessity of Military Force* (New York: Basic, 2018), pp. 214–216.

17. Colin Powell with Joseph E. Persico, *My American Journey* (New York: Random, 1995), p. 434.

18. Sam C. Sarkesian, "Who Serves?" *Social Science and Modern Society* 18, no. 3 (March/April 1981), pp. 57–60; and Sam C. Sarkesian, John Allen Williams, and Fred B. Bryant, *Soldiers, Society, and National Security* (Boulder, Colo.: Lynne Rienner, 1995), pp. 13–17.

19. Sarkesian and Connor, *US Military Profession.* See also Clark, *Waging Modern War.*

20. Sam C. Sarkesian, "The US Military Must Find Its Voice," *Orbis* 42, no. 3 (Summer 1998), pp. 423–437.

21. Sarkesian and Connor, *US Military Profession.*

22. Sarkesian, Williams, and Bryant, *Soldiers,* p. 147.

23. John Allen Williams, "The New Military Professionals," *US Naval Institute Proceedings* 122, no. 5 (May 1996), p. 42.

9

The Intelligence Community

The attacks on the World Trade Center and the Pentagon on September 11, 2001, were a chilling reminder of the importance of an effective intelligence establishment. This lesson was reinforced by intelligence failures prior to the 2003 invasion of Iraq and the difficulty of assessing the status of nuclear weapons programs in Iran and North Korea since then. Accurate and timely intelligence, analyzed realistically and used properly, is an essential ingredient for strong national security. Indeed, such intelligence is the nation's outer line of defense. Even exit strategies, as in the case of the US departure from Afghanistan in 2021, depend on accurate and timely intelligence for their success, and invite failure if intelligence is incomplete or misleading.

The Intelligence Cycle

The term *intelligence* refers to both raw information and the final product that comes from collecting and analyzing all available information on foreign nations and their operations, as well as information on group activities (such as terrorism) that are important for national security planning. To be useful, intelligence is dependent upon the intelligence services fulfilling their responsibilities in what is called "the intelligence cycle" (see Figure 9.1), described as "the process of developing raw information into finished intelligence for policymakers to use in decisionmaking and action."[1]

The intelligence cycle is a useful way to conceptualize the intelligence process. It can be viewed as a process of five steps: (1) planning

Figure 9.1 The Intelligence Cycle

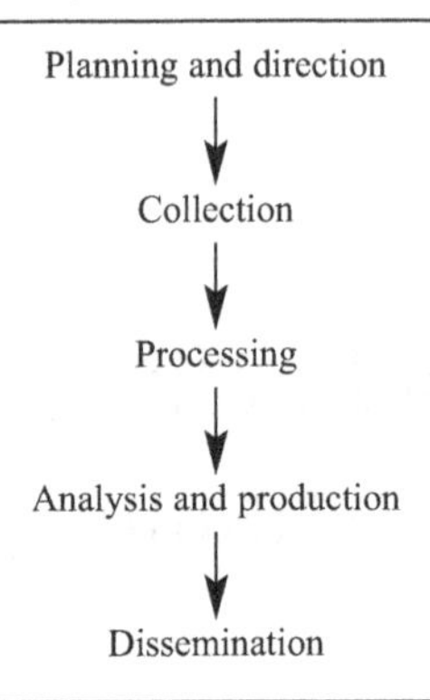

Source: Adapted from Central Intelligence Agency, *Factbook on Intelligence,* January 2001, p. 14.

and direction, (2) collection, (3) processing, (4) analysis and production, and (5) dissemination. One intelligence publication discusses evaluation as the sixth and final phase of intelligence cycle, noting that constant feedback helps members of the intelligence communities "adjust and refine their activities and analysis to better meet consumers' . . . information needs."[2]

The planning and direction stage involves "the management of the entire effort, from identifying the need for data to delivering an intelligence product to a consumer."[3] The process usually begins with a request from the National Security Council (or another department or agency) to collect intelligence on particular subjects. Some of these requests may be for onetime intelligence, or they may be a standing request for continuing intelligence on a particular subject, such as the development of Chinese strategic missiles.

The next stage, collection, "is the gathering of the raw information needed to produce finished intelligence."[4] Intelligence collection depends on a variety of operations and activities, including private sources and media accounts as well as clandestine sources. The methods used for collection are technical, including satellite and electronic means, as well as human.

The large amount of collected raw intelligence must now undergo the processing stage to make it manageable. This step "involves converting the vast amount of information collected to a form usable by analysts through decryption, language translations, and data reduction."[5]

The analysis and production stage "is the conversion of basic information into finished intelligence."[6] It turns the processed intelligence into an understandable and usable form for authorized consumers. All collected intelligence is not of equal quality and so must be evaluated for reliability and credibility (including the source of the information) as well as accuracy. The evaluation also includes the examination of other intelligence and sources that corroborate or contradict the original information.

Intelligence must be analyzed with respect to its relationship to US national security, that is, its importance with respect to enemy intentions, their strength, and their policy. This may be one of the most difficult parts of the process and relies heavily upon the experience, sophistication, and capability of the analysts. It is especially difficult to reconcile contradictory raw intelligence to give it meaning and make it useful for policymakers.

Given the vast amount of data collected, however, it is not possible to analyze all of it, especially in a short time frame. Indeed, this is the most convincing explanation for the failure to notice information that would have tipped the Japanese hand before their attack on Pearl Harbor in 1941. The relevant information got lost in the weeds until after the attack, when researchers knew what they were looking for and had time to find it.[7] The same is true for indicators of the September 11 attack that were missed or dismissed at the time as not important. It is sobering to wonder if there are similar indications available now that are also being missed owing to bureaucratic inefficiency, organizational politics, or otherwise beneficial restrictions on the intelligence community. The dramatic suddenness with which the pandemic of 2020 burst upon world governments and societies provides only one example of "black swans" or unexpected disasters that challenge intelligence resources and estimates.

Production is an inherent part of analysis and refers to the shape and form of the final product. This is followed by the final step: dissemination of the final product to the end users and receiving feedback from them as to its utility.

Although this process seems to follow logically from one step to the next, in reality the steps are blurred. Furthermore, it is difficult to manage and coordinate the agencies and services involved in the cycle and therefore to provide accurate and timely information to the consumer. Because the finished product is based on the judgment of analysts, the human equation enters into the picture, and with it the risk of mistaken judgments, human error, and the possibility of nonanalytical factors influencing the final product.

In the final analysis, regardless of how good the intelligence is, how analytically precise, and how timely, the utility of the finished product depends upon the consumer. Although the use (or nonuse) of intelligence is beyond the control of those in the intelligence system, it is an inherent part of the cycle. This creates dilemmas for the intelligence system. On the one hand, there is professional pride in producing worthy finished products and critical intelligence estimates for national security purposes. On the other hand, if these products are not used or are ignored, as President Trump was prone to do, the system is more likely to become self-serving, enmeshed in meaningless operations, and committed more to bureaucratic efficiency than to the purposes of intelligence. It is not difficult to move from this to leaks, policy advocacy, and even operations in the belief that the intelligence generated is correct and that policymakers should act on it in a timely fashion.

The Director of National Intelligence

In the wake of September 11, Congress established a bipartisan National Commission on Terrorist Attacks on the United States, more popularly known as the 9/11 Commission. It issued a public report on July 22, 2004, followed by two staff monographs on August 21 of that year.[8] Among the recommendations of the commission was the creation of a new post, the director of national intelligence (DNI), to replace the Central Intelligence Agency (CIA) director as the coordinator of the far-flung intelligence community. This recommendation was implemented by the passage of the Intelligence Reform and Terrorism Prevention Act of 2004, and a DNI was appointed.[9] Upon the passage of this act, "the DNI assumed responsibility as manager of the [intelligence community] and principal intelligence advisor to the President, leaving leadership of the CIA to the Director of the CIA."[10]

Since the office of the DNI was established, the position has evolved and grown in importance. In particular, the DNI's role in integrating the components of the intelligence community to produce overall strategic assessments has increased and the DNI has become more involved in acquisition and budgetary issues.[11] While the CIA director retains an important role in the intelligence community, it is the DNI, not the CIA director, who briefs the president and is his or her principal national security intelligence advisor.[12] The DNI relates as well to the National Security Council and the congressional intelligence oversight committees: the Senate Select Committee on Intelligence and the House Permanent Select Committee on Intelligence.

The Intelligence Community

The term *intelligence community* is a general reference and includes seventeen agencies (or components of such agencies) within the executive branch that deal with intelligence on some level. These include two independent agencies (the Office of the Director of National Intelligence and the Central Intelligence Agency); eight intelligence agencies organized under the Department of Defense, including the Defense Intelligence Agency (DIA), the National Security Agency (NSA), the National Geospatial-Intelligence Agency (NGA), the National Reconnaissance Office (NRO), and the intelligence elements within the Army, Navy, Air Force, Space Force, and Marine Corps; and seven components of other agencies, including the Federal Bureau of Investigation (FBI), the Department of Homeland Security, the Coast Guard, the Drug Enforcement Administration, and the Departments of State and Treasury.[13]

The Office of the Director of National Intelligence includes the National Counterterrorism Center (NCTC), whose director reports to the DNI and to the president. The NCTC "serves as the primary organization in the United States Government . . . for integrating and analyzing *all* intelligence pertaining to counterterrorism (except for information pertaining exclusively to domestic terrorism)."[14] The Department of Homeland Security (and the Coast Guard, which is part of that department) has a growing capability in counterterrorism. The Drug Enforcement Administration and the Departments of State, Energy, and Treasury also have intelligence elements.[15] In addition, states and larger cities have a significant capability to collect and process intelligence relating to their locales.

The Federal Bureau of Investigation has a particularly important role in counterterrorism and counterintelligence (that is, preventing the penetration of US institutions by hostile intelligence elements). The FBI has a long history of gathering information for successful criminal prosecutions on issues ranging from bank robbery to acts of terrorism. At the same time, the FBI has a record of success in counterintelligence operations and breaking up developing terrorist plots. It is the FBI, not the CIA, which is primarily responsible for intelligence gathering within the United States—indeed, there are strict legal limitations on what the CIA and other national intelligence agencies are permitted to do domestically.

The FBI's dual nature as both an intelligence collection and law enforcement organization posed a problem prior to September 11, when the understandable desire not to jeopardize criminal cases caused the FBI to limit information to flow across the bureau and between the bureau and other intelligence agencies. Despite strict guidelines in place for FBI intelligence gathering then and now, that type of information by its nature may not be admissible in court. Because this could contaminate a criminal case, a "wall of separation" was erected between the intelligence gathering and criminal investigation sides of the bureau and between the bureau and other intelligence agencies to prevent this from happening. After September 11, the rules for sharing information were relaxed to make intelligence information more widely available within the intelligence community so as not to hamper effective intelligence efforts. As then secretary of defense Donald Rumsfeld noted, "The Department of Justice and the CIA had to negotiate a delicate balance, devising new ways to cooperate and exchange intelligence while protecting our civil liberties."[16] In view of the many leaks of intelligence information since then, it may be that the pendulum swung too far in the direction of accessibility. On the other hand, some of the leaks indicate possible overreach by the intelligence agencies.

There are several important intelligence agencies within the Department of Defense. The largest of these is the Defense Intelligence Agency. The DIA coordinates many of the Defense intelligence elements and with them provides the secretary of defense the finished intelligence products required to carry out the responsibilities of that position. The National Reconnaissance Organization, National Geospatial-Intelligence Agency, and National Security Agency are especially secretive, given their advanced capabilities. The NRO plans, builds, and controls satellites (imagery and electronic) for use by other agencies—primarily, but not exclusively, the NGA and the NSA. NGA analyzes and exploits satellite imagery and geospatial information to provide intelligence and warning for national decisionmakers and the military. NGA is also a combat support agency of the US Department of Defense and produces strategic and tactical maps, charts, and other data necessary to support military weapons and navigational systems, based on open-source as well as highly classified data.[17]

The NSA focuses on cryptology to decode information gathered from foreign sources and to improve security of domestic cryptological systems.[18] It has been under scrutiny by our European allies owing to suspicions the NSA was collecting economic intelligence for the benefit of US corporations—an accusation the NSA denied. The George W. Bush administration tasked the NSA with analyzing intercepted phone conversations between persons in the United States and certain phone numbers abroad linked to terrorist networks. Although this seemed on its face to be a good idea, many expressed a concern over the lack of judicial supervision of the program. It was the source of great controversy in the wake of information from NSA contractor Edward Snowden that indicated the breadth of NSA operations that included collection against American citizens.

The military's intelligence elements—including Army, Navy, Air Force, Space Force, and Marine Corps intelligence—exist at another level. Each has its own community that includes a variety of subagencies and sources. The service intelligence centers focus on the battlefield intelligence necessary to support the individual services' tactical plans. This includes enemy order-of-battle information and analysis. The military intelligence elements are also intelligence collectors and coordinate with other agencies as appropriate. Also, the military maintains intelligence units directly subordinate to unified combatant commanders. Their purpose is to provide specific intelligence to support military plans and operations in a command's area of responsibility (European Command, Southern Command, and so on). In so doing, they serve their own particular regional command, but they also provide input into the total intelligence effort through the intelligence community.

The complicated intelligence relationships and responsibilities are difficult to coordinate, a problem magnified by the fact that each intelligence agency and service has its own parochial interests and organizational culture. Any distortions in the intelligence process are of concern, given the importance of accurate intelligence for national security and the associated civil liberties implications.

The DNI and other intelligence entities have relationships with the foreign intelligence services of US allies (as well as those of not-so-friendly countries, especially post–September 11). Close coordination is critical to counterterrorism operations and other clandestine activities. They also share intelligence information for less dramatic but mutually beneficial purposes in support of international efforts and treaties. This intelligence information is useful not only for national security purposes, but for humanitarian operations such as disaster relief.

In recent years, these relationships have been under considerable stress. The intelligence services of US allies were dismayed by the congressional investigations of the CIA in the 1970s, which revealed sensitive information and threatened the exposure of foreign sources. Contemporary revelations by investigative journalists, whether the reports are accurate or not, create embarrassing situations that do little to enhance the CIA's effectiveness. One result is that foreign sources become extremely cautious when dealing with US intelligence agencies for fear of exposure. This tends to chill relationships between the US intelligence system and its foreign counterparts. The massive exposure of secret communications by WikiLeaks was especially troubling and potentially damaging. Whatever the content of the leaked documents—even those that made the United States look good—the fact that such a massive security breach could occur has a continuing chilling effect on international intelligence cooperation.

Effective intelligence is necessary for defense, but can be problematic for civil liberties. Accordingly, supervision of the intelligence agencies is a continuing concern of interest in Congress. In an effort to minimize ill-advised operations in the future, Congress and the executive branch have attempted to reconcile secrecy and oversight. One example was the Intelligence Authorization Act of 1991. "[The act] represents the first significant remedial intelligence oversight legislation in more than a decade. The Act provides the first statutory definition of covert action, repeals the 1974 Hughes-Ryan Amendment governing notification to Congress of covert action, requires presidential 'findings' for covert action to be in writing, and prohibits the President from issuing retroactive findings."[19]

The act has been considered a "reasonable compromise between divisive political issues and competing interpretations of constitutional responsibilities."[20]

The key to developing a reasonably effective intelligence system is to create an environment of trust and confidence among national agencies, Defense Department intelligence services, and their subordinate services. The responsibility for that now rests primarily with the DNI. The structures are in place and the statutes spelling out the power and role of the CIA and other agencies are on the books, but there is a great deal of room for flexibility below the national level. How all of the pieces are brought together is not only a function of managerial efficiency but also of leadership, experience, and professional competence.

Equally important, the DNI and other intelligence officials need to nurture close relations with foreign intelligence agencies and services, despite the difficulties noted above. To be sure, Congress, interest groups, and journalists have a great deal of impact on images of the intelligence community that can degrade US intelligence credibility. This can be countered to some extent by a competent and skilled US intelligence service led by a DNI whose leadership skills are up to the task.

The Central Intelligence Agency

The key agency for intelligence collection and analysis is the Central Intelligence Agency, established by the National Security Act of 1947. Its charter reads in part:

> For the purpose of coordinating the intelligence activities of the several government departments and agencies in the interest of national security, it shall be the duty of the Agency, under the direction of the National Security Council . . . to correlate and evaluate intelligence relating to national security, and to provide for the appropriate dissemination of such intelligence within the Government using where appropriate existing agencies and facilities.[21]

The CIA has been both vilified and praised since its creation, and some of its more public and legendary activities have raised questions about the proper role of intelligence in a democracy. The CIA has been the subject of many congressional investigations, and there is continuing concern regarding its role and relationship to other instruments of government, especially in light of the war on terrorism. Unfortunately, the contemporary debate is often colored by political rhetoric and reveals much misunderstanding. As one scholar has written: "Much of the criticism of the CIA stems from the fact that its activities are secret. The public—and particularly the media—resent its being told that they cannot know something. Silence is interpreted as arrogance. Moreover, when people do not know what an agency is doing, they assume it is either doing nothing and not changing with the times, or that it is doing something wrong."[22]

In any case, the requirements of national security have created dilemmas for the CIA, the president, and Congress. Although this has been true for decades, it is especially so in the new strategic environment. On the one hand, national security requires a wide range of intelligence activities, many of them necessarily secret and covert; after all, the best information may well come from people who are not supposed to give it to you. On the other hand, some of these activities can stretch the notion of democracy and threaten individual rights and freedoms.

At the same time, questions have been raised as to what *strategic intelligence* means today. Should it include industrial espionage? How should the intelligence community respond to international terrorism? How to reconcile these demands yet still maintain an effective intelligence establishment is a persistent problem for political leaders, the intelligence community, and society at large. Issues with respect to covert operations and gathering information on US citizens are especially controversial.

Since 1947, many intelligence activities and covert operations—some successful, some not—have been attributed to the CIA. The publicized failures have caused a degree of embarrassment to the entire country, but the public is usually unaware of the successes. Even if such activities are successful, many citizens are uncomfortable with secret intelligence activities, for secrecy, covert operations, clandestine activities, and certain special operations do not easily fit into the moral framework of an open system.

The fact that such problems persist was confirmed by the 1987 Iran-contra hearings dealing with the Reagan-era scandal of diverting funds from arms sales to Iran to support antigovernment forces in Nicaragua called the "contras"—who were either rebels or freedom fighters depending on your point of view—in violation of congressional restrictions. This refocused the public spotlight on the activities of some CIA operatives and threatened to expose several covert operations. Critics in Congress, the media, and the public were quick to generalize from the problems that emerged. Others placed them in a more favorable perspective, noting that virtually all intelligence activities were conducted according to law and in full cooperation with Congress, Iran-contra notwithstanding. Nonetheless, the debate over the intelligence system continues, and it is not likely to be resolved anytime soon—especially in light of the intelligence failure to uncover the September 11 conspiracy beforehand or to note the absence of weapons of mass destruction in Iraq prior to the 2003 invasion or to have a better grasp on the nuclear weapons programs in Iran and North Korea. However, US and allied intelligence gave a clear picture of the buildup of Russian forces in 2022 prior to the Russian invasion of Ukraine.

More recent debates have centered on the propriety of interrogation techniques used on terror suspects and alleged mistreatment of prisoners at the US naval base in Guantanamo, Cuba, and abroad. Photographs of prisoner abuse at the Abu Ghraib prison in Iraq were especially shocking to a society that prides itself on moral behavior and resulted in much stricter and clearer standards for interrogation by the military intelligence community. A related controversy is the practice of "extraordinary rendition," in which such restrictions are avoided by transferring control of certain prisoners to third countries without such protections.

The improprieties identified during the Iran-contra hearings, including the role CIA director William Casey may have played in diverting funds to the contras, were reminiscent of the outcry against the CIA a decade earlier when Congress passed a series of laws limiting intelligence activities and establishing more rigid congressional oversight. This was a reaction to the CIA role in Watergate and other presumed domestic activities. The Church Committee hearings in the Senate (named for the committee chairman, Utah senator Frank Church) and its subsequent report in 1976 detailed many congressional concerns.[23]

The issue of CIA effectiveness was raised again in February 1994 with the revelation that career CIA officer Aldrich Ames and his wife were arrested for passing classified information to the Soviet Union and later to Russia. He was also charged with aiding the exposure and subsequent execution of several Russians working for the CIA.[24] Although his activities began in 1985, they were not exposed until 1994. This was a serious matter for the CIA, and it had repercussions for US-Russia relations, as members of Congress from both parties raised objections to US cooperation with Russia. Intelligence concerns about Russia were heightened by Russian penetration of computers of the Democratic National Committee and presidential candidate Hillary Clinton in 2016 and the subsequent release of damaging information via investigative websites. Russian penetration of social media platforms, especially Facebook, seemed designed to sow discord in the United States and damage the candidacy of Hillary Clinton. Concerns then turned to the security of the 2020 elections.

Constraints on intelligence activities expanded during the Jimmy Carter administration, especially with respect to covert operations. The overthrow of the shah of Iran, the abortive hostage rescue in that country, and Soviet activities in Afghanistan drew a great deal of criticism regarding the capability of the intelligence services. Criticism also surfaced as to the quality of CIA intelligence immediately prior to the Gulf War in 1991 as well as the failure to predict the attempt by Soviet hardliners to overthrow Mikhail Gorbachev and the collapse of the Soviet Union that same year. Later, a great deal of criticism was directed at the inability of technology to penetrate foreign political-social networks and provide on-the-ground analyses and judgments that can be done only by agents in the field.

Another review of the CIA was by the Commission on the Roles and Capabilities of the US Intelligence Community, led by former defense secretary Harold Brown and former senator Warren Rudman. This commission proposed in 1998 that the amount of funds used for secret intelligence be revealed and that the size of the intelligence agencies be reduced.[25] Given the nature of intelligence operations, however, it is doubtful that there will ever be a full public accounting of its funding—and in the present circumstances a reduction in funding to the intelligence community seems both unwise and unlikely.

The new administration under President George W. Bush quickly ordered its own review of the intelligence community, led by director of central intelligence (the DCI), CIA director George Tenet. Under National Security Presidential Directive 5, Bush called for the DCI (Tenet) to form a panel of internal and external members to consider the structure and operations of the members of the intelligence community.[26] This panel was headed by Brent Scowcroft, chairman of the president's Foreign Intelligence Advisory Board and a former national security advisor. Early reports suggested that the panel would recommend major changes to increase the control of the CIA director over all sources of intelligence, including photographic and electronic.[27] Evidence is emerging that indications of impending terrorist attacks were picked up prior to September 11, but their significance was not realized owing to lack of coordination among and within US intelligence agencies.

In a perceptive article written before the September 11 terrorist attacks, one scholar noted the importance of rethinking the intelligence establishment: "Despite the apparent consensus on the need for change, recent intelligence failures suggest that US intelligence has yet to leave its Cold War–era methods and structure behind."[28] Also, the United States launched a war on Iraq in 2003 based on, or at least justified by, the idea that Iraqi president Saddam Hussein had usable weapons of mass destruction. The failure to discover any such weapons after the invasion was a serious blow to the US rationale for the war and to US credibility. It also raised questions about the degree to which the CIA, in particular, could resist pressure to tell the administration what it wanted to hear. President Trump and his administration seemed especially unwilling to hear assessments from the intelligence community that ran counter to their own political or ideological positions, whether on the impact of Russian activities in the US political system, the degree to which Iran was actually complying with the nuclear agreement negotiated by President Obama, the lack of progress in denuclearization by North Korea, or even climate change.

Better intelligence and counterintelligence are crucial for effective policy and strategy. But better intelligence and its effective use depend on an understanding of the nature and purpose of intelligence and the knowledge of the intelligence community. Any study of national security policy must include the relationship between the president and the DNI, the structure and purpose of the CIA and the intelligence community, the role of intelligence, the intelligence cycle, and the system that tries to integrate all of these into a coherent whole.

The director of the Central Intelligence Agency is an advisor, a coordinator, and a leader-manager—all in different settings and with different political and professional relationships. His or her main function is to be directly responsible for the control and operations of the CIA. The CIA is no longer the statutory or regulatory coordinator of the intelligence system, but by virtue of its size and importance continues to play a central role an independent agency in the intelligence community.

The CIA's organization, resources, and operations cover a range of activities and require a vast managerial effort. Athough a high percentage of the human and other assets are controlled by the Department of Defense, the importance of the CIA in the intelligence community gives it a stature not enjoyed by other intelligence services.

The CIA is divided into five directorates that work together under mission centers that focus on "regional and high priority issues"—about a dozen as of this writing. There are also executive offices, such as personnel, to support the work of the directorates. The directorates are the Directorate of Intelligence and Analysis, the Directorate of Operations, the Directorate of Science and Technology, the Directorate of Digital Innovation, and the Directorate of Support.[29] The CIA describes these directorates as follows.

The Directorate of Intelligence and Analysis provides "timely, accurate, and objective intelligence analysis" on a wide range of national security and foreign policy issues. The Directorate of Operations (the National Clandestine Service) "handles the collection of intelligence acquired by human sources (human intelligence or HUMINT). When necessary, and under unique circumstances, they conduct covert action as directed by the president." The Directorate of Science and Technology "applies innovative, scientific, engineering, and technical solutions to support intelligence operations in support of our foreign intelligence mission. It develops and operates various intelligence collection and analytical systems using the most advanced technology." The Directorate of Digital Innovation "makes sure teams have the tools and techniques they need to operate in a modern, connected world and still be clandestine." Finally, the Directorate of Support is "responsible for key support functions, including security, supply chains, facilities, financial and medical services, business systems, human resources, and logistics."

These directorates are at the center of the politics and turf battles that occur within the CIA. For example, those in the Directorate of Operations considered themselves to be the cutting edge and looked upon others as paper pushers and administrators. "We got all the action. We make the world go around. Satellites can't tell you what people are doing."[30] The other directorates offer their own assessment of their importance to the overall intelligence effort. Even within the various directorates, separate elements tend to develop their own style.

The components of the intelligence community vary greatly in size and specialization, but all perform important functions. In particular, they are responsible for collecting and analyzing the many types of intelligence listed in Figure 9.2.

Figure 9.2 Types of Intelligence

- Human intelligence (HUMINT): Collected overtly or covertly by human agents
- Geospatial intelligence (GEOINT): Includes satellite and other imagery (IMINT) and geospatial information
- Signals intelligence (SIGINT): A general term that includes communications intelligence (COMINT), electronic intelligence (ELINT), and foreign instrumentation and signals intelligence (FISINT)
- Measurement and signature intelligence (MASINT): Encompasses seismic, electromagnetic, radio frequency, radar, acoustic, optical, and other data
- Technical intelligence (TECHINT): Information on foreign equipment to prevent surprise and devise countermeasures
- Open source intelligence (OSINT): Public domain information, such as in the traditional media, the Internet, and social media
- Counterintelligence (CI): Hardening the resistance of your own society to exploitation by foreign intelligence services

Source: Adapted from *Joint Publication 2-0: Joint Intelligence* (Washington, DC: Department of Defense, June 22, 2007), available at http://www.dtic.mil/doctrine/new_pubs/jp2_0.pdf.

Covert Operations

The most controversial aspect of the intelligence system is covert operations. The CIA has not been the only agency involved in such operations, although it is the primary one. Interestingly, covert action as it is now defined was not initially identified as a role for the CIA. That responsibility was assumed under the provision that the CIA was to perform such functions affecting US national security as the National Security Council directed.

The term *covert operations* is a convenient label used to identify a variety of clandestine activities, and in everyday parlance is the same as clandestine operations. They range from propaganda and psychological warfare to paramilitary operations and espionage. Indeed, published reports from post–September 11 actions in Afghanistan and Pakistan (including the location and elimination of groups and individuals deemed hostile) revealed that the CIA has a significant paramilitary capability. In the words of one authority: "Covert action is defined in the US as the attempt by a government to influence events in another state or territory without revealing its involvement."[31] The concept also extends to political action and various forms of intelligence gathering.

At the presidential level, however, the term *special activities* is used to identify a variety of covert operations. In Executive Order 12333, signed by President Ronald Reagan in 1981, special activities were defined as "propaganda, paramilitary and covert political operations. They specifically do not include the sensitive collection of foreign intelligence."[32] Thus the concept of covert operations can be assigned to virtually every part of intelligence activities as well as to some aspects of US military operations. Secret or concealed operations and activities are the stuff of covert actions.

The purpose of covert operations is to support the foreign policy of the state engaging in them. There are two major considerations: first, the state may be involved in operations that are best served by secrecy, that is, when the public and policymakers in the target state are not aware of such operations; second, the state may be involved in operations that are public yet wish to conceal or at least deny involvement. Paramilitary operations and political actions, for example, can be quite visible in the target state yet supported and encouraged by another state that does not want to be identified for any number of reasons, such as embarrassment or fear that the success of the operation would be jeopardized. The United States was involved in such operations against Cuba in the early 1960s during the John F. Kennedy administration, in particular the abortive Bay of Pigs invasion of Cuba in April 1961. There are credible reports of sabotage and multiple attempts to assassinate Cuban leader Fidel Castro around that time. (The success of such operations is never ensured; as of this writing Cuba is still undemocratic and until recently was governed by a Castro.) Other examples now known include US involvement in Chile beginning in 1970 and the activities revealed during the Iran-contra hearings in 1987.

As a result of the variety of covert operations in the 1980s, some of which were mismanaged, critics charged that such operations were contrary to democratic norms. In 1992, "a 20th Century Fund 'Task Force Report on Covert Action and American Democracy' blistered covert action as fundamentally at war with democratic norms."[33] Yet covert operations are an integral part of the intelligence system, often necessary to achieve foreign policy and national security goals. The nature of the United States—an open system committed to law and the norms of democracy and decency—places the US public in an awkward position with respect to certain secret operations. Moreover, officials and government representatives who must engage in them are also placed in difficult positions regarding what is proper behavior and moral conduct. A complicating factor is that even effectively managed and "successful" covert operations, such as the 1953 coup in Iran that returned the shah to the throne or the arming of mujahidin forces in Afghanistan in the

1980s with surface-to-air missiles to help defeat the Soviet forces there, may have been unwise from a long-term perspective.

According to a former CIA intelligence officer, one of the major aspects of this "third option"—the alternative to military and diplomatic actions—is "paramilitary operations or the furnishing of *covert* military assistance to unconventional and conventional foreign forces and organizations."[34] The reasoning is that the United States has been faced with a variety of "insurgencies" abroad since the 1970s that seemed to threaten several US national interests. Thus "the bottom line on the decision to use or ignore the third option will not be based solely on the quality of intelligence, analysis, or organizations. The decision will be made by those who have, or lack, the will to pursue policy goals through techniques that have preserved our interests in diverse areas."[35]

The nature of covert operations and the dilemmas they create highlight the fundamental uneasiness that any open system has with secret intelligence operations and the intelligence system in general. Although recognizing the importance of effective intelligence, critics feel that the requirements of democracy—to abide by the rule of law, adhere to democratic proprieties, and protect individual freedoms and rights—provide ample reason to oppose at least some intelligence activities and to demand oversight and accountability. These criticisms are not confined to the United States, of course; Great Britain's MI-6 intelligence agency and Israel's Mossad have also been subjected to considerable, and often highly critical, outside attention. Several actions apparently designed to stop or slow down the Iranian nuclear weapons program have been attributed to the Mossad, such as a computer worm in computers at Iranian nuclear facilities, unexplained explosions in sensitive facilities, and the assassinations of a number of Iranian nuclear scientists.

Democracy and the Intelligence Process

Many serious problems face a democracy that maintains an intelligence system. Aside from the relationship of intelligence operations to more transparent institutions in the political system, there are moral and ethical issues that compound the legal and philosophical problems inherent in such operations.

Intelligence successes (especially covert operations) are rarely revealed, but failures are often made public—to the embarrassment of the United States and to the detriment of foreign and national security policies. The media are quick to point out intelligence failures such as those revealed in the Iran-contra hearings, by September 11, by the justification for the war in Iraq, and by the length of time it took to track down Osama bin Laden. It is only a short step from this view to the support of constraints and limitations that preclude a wide range of intelligence activities, thereby reducing national security capability. In addition, revelations can become the focal point of media coverage and book-length exposés that are difficult to counter without revealing sensitive information and sources.[36]

The issue of successes and failures aside, the legality of certain intelligence operations is not always clear. As noted earlier, in the wake of Vietnam and Watergate, Congress and some interest groups attacked the CIA and the US intelligence system, creating an adversarial environment. Indeed, the second half of the 1970s may have represented the nadir of the US intelligence system and marked a significant decline in the US capability to pursue national interests. More than anything else, the Church Committee epitomized the adversarial approach to investigations of the CIA. According to one source, "consideration of the present mechanism of oversight and control of the US intelligence agencies should begin with the final report of the [Church Committee]."[37] The report identified the problems that arise between the executive and legislative branches in their efforts to determine the scope and purpose of intelligence activities.

A series of legislative acts evolved to regulate and control the CIA and other intelligence activities. To the earlier 1974 Hughes-Ryan Amendment were added a variety of procedures—from the expansion of congressional oversight to the Boland Amendments—that were at the root of the Iran-contra hearings.[38] These acts "created permanent oversight committees in both houses of Congress" with "principal budgetary authority over the intelligence agencies." In addition, statutory provisions were enacted to deal specifically "with the provision of information by the intelligence agencies to the two congressional committees"; new groups were added within the executive branch, including "the President's Intelligence Oversight Board, and . . . an Office of Intelligence Policy and Review within the Department of Justice."[39] Other procedures included the issuance of presidential executive orders, a process of judicial review, and the strengthening of the Office of General Counsel of the intelligence agencies with respect to intelligence activities.

The underlying assumption at this time was that Congress did not know about many intelligence activities, some of which did not comply with existing laws. In the majority of cases Congress was informed through its political leadership, yet many members of Congress would prefer to distance themselves from knowledge of such activities for fear that revelations, especially of failures, would harm them politically. Thus, denial of knowledge can be a useful political position.[40]

One authority had harsh words over the attacks on the intelligence community in 1975 in a comment that resonates today:

> No country had ever subjected its secret organs of government to such open and extensive review. . . . The Congressional Committees conducted their business openly and publicly, adopted an adversarial, accusatory, and investigative approach, and, perhaps, inevitably and irresistibly, dramatized its proceedings. . . . It rarely acknowledged any legitimate reasons for clandestine operations and operated under the assumptions that most clandestine or secret activities were indefensible.[41]

This statement is a sharp reminder of the persisting problems inherent in trying to reconcile intelligence needs with the norms and expectations of a democratic political system. Furthermore, it shows the divisiveness generated within the national security establishment over the role of an intelligence community in national security policy. The critical issue, however, is the proper role of an intelligence community that must adhere to the rule of law while remaining an effective instrument of US national security policy. Solutions to this dilemma are elusive and will continue to pose problems for the president and the national security establishment.

Another problem is the possibility of intelligence that is fabricated to favor a particular political-military posture or policy. This so-called cooked intelligence is closely related to policy advocacy: intelligence favorable to a particular stance is highlighted while contrary intelligence is ignored, downgraded, or allowed to slip through the cracks. Allegations of cooked intelligence and policy advocacy were leveled against CIA director William Casey with respect to US policy in Nicaragua.[42] The George W. Bush administration has been criticized for trying to shape the judgments of the intelligence community on Iraq to support the administration's rationale for the 2003 invasion. More recently, the Trump administration has been criticized for denigrating the importance of intelligence professionals in guiding policy decisions and discounting the intelligence community consensus that the Russians interfered in the 2016 election in order to assist Donald Trump. The Biden administration marks a return to the traditional relationship between the president and the intelligence community.

In any case, the credibility of the intelligence community rests primarily on its ability to avoid policy advocacy, retain institutional autonomy, and maintain professional competence. Any president who seeks intelligence as a basis for a preconceived policy position not only distorts the intelligence process but also erodes the credibility of the US intelligence community. Moreover, for the intelligence community to take any position—other than the most objective analysis of intelligence and the pursuit of the requirements in the intelligence cycle—will surely raise questions of competence and have a chilling effect on intelligence professionals whose horizons are not limited by agency protectionism or bureaucratic loyalties. An intelligence community in any country that is compromised by partisan political agendas is a ticking bomb of potential national security disaster.

Finally, the relationships among the DNI, the president, and Congress help determine the overall effectiveness of the intelligence community. It is important to note that how the DNI carries out his or her responsibilities to Congress, relationships with individual members of Congress, and relationships with the intelligence oversight committees has much to do with the ability of the president to play an effective role in the national security policy process. The trust and confidence between the DNI and the president remain critical, and an intelligence community without political motivations and purposes is the basis of this relationship. The DNI must therefore be responsive to the president's national security concerns while fostering objectivity within the intelligence community and maintaining good relationships with Congress—tasks that require an individual of strong character and personality, as well as competence and integrity.

All these relationships shape the public's image of the intelligence community. For example, Senator Church's 1975 comment that the CIA is a "rogue elephant" remains a pejorative used by many critics today. Although such a comment cannot stand close scrutiny, such an image still has an impact on public perceptions of CIA conduct. Continuing problems with and suspicions of the CIA have been reflected in various publications: "There are limits to human trust and gullibility; intelligence manipulation has more often than not threatened national security and the prospects of world peace. Therefore, the question of the CIA's standing invites close attention—and it cries out for redress."[43]

The twenty-first century brings with it several questions, not only about the proper role of the intelligence community in espionage and intelligence-gathering but also about its role in a democracy and the issue of secrecy and covert operations. This is complicated by the need to work in a world that is increasingly asymmetric, in which US opponents are not necessarily even nation-states, and the states of concern are not only peer competitors such as Russia and China, but smaller states previously of little intelligence interest.

Conclusion

The dilemma of how to maintain an effective intelligence institution in an open system is a continuing one. If the country is to be safe, it requires timely and accurate intelligence; at the same time, securing such intelligence is enormously intrusive, and there are no easy answers in the search for the proper balance between security and civil liberties. Developing an acceptable relationship, delineating proper boundaries and roles, and maintaining a dynamic and continuing assessment begin with the president and his or her leadership. He or she must be cognizant of these problems and recognize the need to maintain the moral and ethical credibility of the intelligence function. How this is translated and projected into the intelligence community rests with the leadership of the director of national intelligence and is affected by the president's trust and confidence in that person. Equally important, the most effective intelligence in an open system requires an enlightened and supportive Congress and public.

The difficulties involved in maintaining an effective intelligence establishment in an open system are complex. As one author remarked:

> To presume, however, that democracies must rigidly adhere to strict application of law, even to the point of self-destruction, is the height of immorality. Equally presumptuous is the view that democracy should take no action unless a clear and present danger exists. . . . To wait until there is a clear and present danger may be too late. Even if it is not too late, waiting for the outbreak of conflict may place the open system in an extremely disadvantageous position, considerably raising the costs of effective response.[44]

At the same time, it is imperative that the intelligence establishment function within existing laws and regulations. With proper oversight and skilled personnel, there is no reason to believe that the intelligence community cannot function in this way and be effective. This might require more effort, a more flexible system of regulations and procedures, and a more understanding Congress and public. It may also mean that the opinion makers, Congress, and the media must understand that the intelligence community belongs to the US public and serves it and the democratic system. But "the enduring irony of intelligence is its potential to destroy as well as to guard democracy."[45] As a colleague of one of us with knowledge of the intelligence community commented privately some years ago, perhaps in jest, "We can make you safe—but you will not be free."

Notes

1. Central Intelligence Agency, *Factbook on Intelligence* (Washington, DC: Central Intelligence Agency, 1987). Declassified March 13, 2013, p. 13.

2. Central Intelligence Agency, *U.S. National Intelligence—An Overview* (Washington, DC: Central Intelligence Agency, 2013), p. 6, https://www.dni.gov/files/documents/USNI%202013%20Overview_web.pdf.

3. Ibid.

4. Ibid.

5. Ibid.

6. Ibid., pp. 13–14.

7. Roberta Wohlstetter, *Pearl Harbor: Warning and Decision* (Stanford, Calif.: Stanford University Press, 1962).

8. National Commission on Terrorist Attacks on the United States, *9/11 Commission Report,* July 22, 2004, http://www.9-11commission.gov.

9. Office of the Director of National Intelligence, *The National Intelligence Strategy of the United States of America,* August 2009, http://www.odni.gov/reports/2009_NIS.pdf.

10. "The Director of National Intelligence (DNI)" (Washington, DC: Congressional Research Service, June 7, 2021), https://crsreports.congress.gov/product/pdf/IF/IF10470.

11. Ibid.

12. Information on the role of the DNI is from http://www.odni.gov.

13. *National Intelligence Strategy of the United States of America* (Washington, DC: Office of the Director of National Intelligence, 2019), p. 28, https://www.dni.gov/files/ODNI/documents/National_Intelligence_Strategy_2019.pdf. See also "Members of the IC" (Washington, DC: Director of National Intelligence, January 22, 2021), https://www.dni.gov/index.php/what-we-do/members-of-the-ic.

14. "About the National Counterterrorism Center," http://www.nctc.gov/about_us/about_nctc.html; emphasis in the original.

15. For a more complete discussion of member agencies of the intelligence community and their roles, see *U.S. National Intelligence 2013: An Overview* (Washington, DC: Office of the Director of National Intelligence, 2013), https://www.dni.gov/files/documents/USNI%202013%20Overview_web.pdf (this publication is updated continuously, but the original publication date of 2013 has not been changed). See also "Members of the IC."

16. Donald Rumsfeld, *Known and Unknown: A Memoir* (New York: Sentinel, 2011), p. 351.

17. *U.S. National Intelligence 2013,* p. 6.

18. Ibid.

19. William E. Conner, *Intelligence Oversight: The Controversy Behind the FY 1991 Intelligence Authorization Act* (McLean, Va.: Association of Former Intelligence Officers, 1993), p. 1.

20. Ibid., p. 39.

21. National Security Act of 1947, *United States Statutes at Large* 1947, vol. 61, pt. 1, 1948, pp. 496–505.

22. Ronald Kessler, *Inside the CIA: Revealing the Secrets of the World's Most Powerful Spy Agency* (New York: Pocket, 1992), p. 251.

23. For a detailed account of these matters, see Loch K. Johnson, *A Season of Inquiry: Congress and Intelligence* (Chicago: Dorsey, 1988). The volume is essential reading for those interested in the role and power of Congress in dealing with the US intelligence establishment.

24. See, for example, Sam Vincent Meddis, "CIA Officer Charged as Spy," *USA Today,* February 23, 1994, pp. 1, 3A. Virtually all US news networks on radio and television carried the story during the week.

25. Tim Weiner, "Commission Recommends Streamlined Spy Agencies," *New York Times,* March 1, 1996, p. A17. This report is sometimes referred to as the Brown-Aspin Report, as Defense Secretary Les Aspin was originally selected to head the commission.

26. Vernon Loeb, "US Intelligence to Get Major Review," *Washington Post,* May 12, 2001, p. A3.

27. Walter Pincus, "Intelligence Shakeup Would Boost CIA: Panel Urges Transfer of NSA, Satellites, Imagery from Pentagon," *Washington Post,* November 8, 2001, p. A1.

28. Bruce Berkowitz, "Better Ways to Fix US Intelligence," *Orbis* 45, no. 4 (Fall 2001), p. 609.

29. Information on the directorates and offices of the CIA are from "About CIA: Organization" https://www.cia.gov/about/organization.

30. Kessler, *Inside the CIA,* p. xxviii.

31. Roy Godson, *Intelligence Requirements for the 1980s: Covert Action* (New Brunswick, N.J.: Transaction, 1981), p. 1. For post–September 11 information on Afghanistan, see Andrew Feickert, *CRS Report for Congress, U.S. and Coalition Military Operations in Afghanistan: Issues for Congress* (Washington, DC: Congressional Research Service, December 11, 2006).

32. Standing Committee on Law and National Security, *Oversight and Accountability of the US Intelligence Agencies: An Evaluation* (Washington, DC: American Bar Association, 1985), p. 19.

33. Bruce Fein, "Official Secrecy and Deception Are Not Always Bad Things," *Insight,* June 8, 1992, p. 23.

34. Theodore Shackley, *The Third Option: An American View of Counter-Insurgency Operations* (New York: Reader's Digest, 1981), pp. 6–7.

35. Ibid.

36. See Gregory F. Treverton, *Covert Action: The Limits of Intervention in the Postwar World* (New York: Basic, 1987), p. 222: "If the United States remains in the business of covert actions, even under restrictive guidelines, it will continue to confront the paradox of secret operations in a democracy. That paradox is, if anything, sharper now because of the changes in the American body politic, particularly relations between Congress and the executive."

37. Scott D. Breckenridge, *The CIA and the US Intelligence System* (Boulder, CO: Westview, 1986), p. 230.

38. Standing Committee on Law and National Security, *Oversight and Accountability,* p. 7.

39. Ibid. See also Breckenridge, *The CIA;* and Loch K. Johnson, "Smart Intelligence," *Foreign Policy* (Winter 1992–1993), pp. 53–69.

40. Standing Committee on Law and National Security, *Oversight and Accountability,* p. 1.

41. Breckenridge, *The CIA,* p. 249.

42. Stafford T. Thomas, *The US Intelligence Community* (Lanham, Md.: University Press of America, 1983), p. 46.

43. Rhodri Jeffreys-Jones, *The CIA and American Democracy* (New Haven, Conn.: Yale University Press, 1989), p. 251.

44. Sam C. Sarkesian, "Open Society: Defensive Responses," in Uri Ra'anan et al., eds., *Hydra of Carnage: International Linkages of Terrorism* (Lexington, Mass.: Lexington Books, 1986), p. 219.

45. Johnson, "Smart Intelligence," p. 69.

PART 3

Making National Security Policy

10

The Policy Process

The policy process that exists within the US political system is extraordinarily complex. Examining it is like trying to find the beginning of a spider web. The process might begin as an idea from a member of Congress or a bureaucrat, be triggered by a special interest group, or be affected by political pressures or some change in the policy environment. The formal policymaking process can seem reasonably straightforward, but this is rarely the case. Whatever policy emerges at the end of this process may bear little resemblance to the original idea.

Scholars have studied the policy process using a variety of approaches. Despite the most earnest efforts of these scholars, the process remains somewhat of a mystery. As one authority observed some time ago, "anyone bold enough to undertake a serious analysis of how policy is made in the American political system must begin with the realization that he is examining one of the most complex structures ever conceived by man."[1] In the final analysis, whether a policy is approved and implemented has more to do with political forces and the ability and attitude of leaders than with any formal process. In an open system such as that in the United States, there are many opportunities for supporters and opponents of policies to affect the outcome.[2]

Understanding the US policy process is difficult for many reasons that relate to the characteristics of the system. These can complicate a seemingly straightforward process. This is not the case for all policymaking, however. Policies that respond to crises or that are grounded in a strong public consensus can pass quickly. This explains the rapid approval of security and antiterrorist initiatives in the wake of September

11. Nonetheless, some of the most important policies respond to ambiguous situations or those in which there is little agreement on the proper course and these are much more difficult to analyze.

A number of perspectives may be helpful in analyzing the policy process. We begin by discussing several approaches that serve as an introduction to our own.

Approaches and Models

There are many approaches and models to study how policy is made. The differing perspectives discussed here illustrate some of the problems inherent in this type of analysis.

Two approaches evolved from the distinct philosophical views of sociologist C. Wright Mills and political scientist Robert A. Dahl.[3] Mills's work describes the *elitist* approach and argues that policy is essentially in the hands of an identifiable elite (high-level bureaucrats, business interests, and the military) that is self-centered and does not necessarily reflect the public interest. Dahl's *pluralist* approach argues that even though elites have a disproportionate share of resources and therefore can have a strong effect on policies they care about, they are not all-powerful and the same actors are rarely involved with the same degree of intensity. Therefore, policy emerges as a result of compromise. Dahl's approach presumes that policy is affected by the people through a variety of means, such as public opinion polls, elections, constituent pressure on elected representatives, and interest group advocacy.

In the *statist* approach, the state is the primary actor with its own characteristics, objectives, and goals. Little attention is given to institutions and agencies within the state or, for that matter, to individual or group behaviors. The *bureaucratic* approach is based on the assumption that policy is driven primarily by the bureaucracies in the government, which create networks of like-minded interests. The *organizational* approach is similar in that it assumes that organizational perspectives and standard operating procedures drive policy. Thus, the Department of State has a particular worldview, which differs from those of the Department of Defense and the Central Intelligence Agency. It is often difficult to make clear distinctions between the bureaucratic and organizational approaches.

Another aspect of policy analysis relates to differing philosophies regarding policy procedures—the mechanics rather than substance. The *rationalist* view presumes that major policy alternatives are developed at the highest levels of government, based on the best possible information. Then, through a rational process, alternatives are studied, possible courses of action identified, the best course selected, and the policy passed through. In reality, however, policy is usually made *incrementally,* that is, in a piecemeal fashion.[4] Furthermore, policy initiatives occur at a variety of governmental levels or even outside the government structure. Policy is often a response to a particular situation rather than part of a grand design. Furthermore, important policies can also result from citizen action, bureaucratic activity, corporate lobbying, foreign nations, or presidential initiative. This is not to deny that there are grander schemes, but even they generally result in incremental programs. In times of crisis, policymaking and the policy process follow the rationalist model more closely if there is short time for decisionmaking and elite consensus and popular support for prompt action.

This study does not follow the elitist model; neither does it assume that the people can directly control policymaking and the policy process. Although a relatively small group of persons ultimately approves and implements policy, the legitimacy and credibility of policy are based on the perceptions of a broad spectrum of the populace. The pattern was well described in the following passage: "Few, if any, of the decisions of government are either decisive or final. Very often policy is the sum of a series of separate or only vaguely related actions. On other occasions it is an uneasy, even internally inconsistent, compromise among competing goals or an incompatible mixture of alternative means for achieving a single goal."[5] Therefore, policy often emerges by "halting small and usually tentative steps" full of "zigs and zags." Policy can also become the opposite of what was originally intended. Finally, it is possible that "issues continue to be debated with nothing being resolved until both the problem and the debaters disappear under the relentless pyramiding of events."[6]

Is there a more systematic approach? Policy phases and political dynamics, as well as dynamics emerging from the total process, help us perceive the big picture more clearly. Also, politics nurtures elusiveness, placing an opaque gloss over the entire process. Addressing the role of military officers in the national security policy process, two scholars concluded, "Because of the variety of purposes among subordinate national security professionals and especially among career military officers, the game of politics remains intense, marked always by the presence of vested interests, interorganizational conflict, intraorganizational rivalry, and the elusiveness of a 'best' policy."[7]

It is useful to consider policymaking as divided into several phases, each affected by political actors and resulting in an interconnected process. What occurs in one phase has an impact on the succeeding phases. Policymaking and the policy process are continuous, as established policies are constantly being revised and passed through the process once more.

Policy Phases

Scholars in this area often focus on the nature of the US political system, studying and identifying the various powers, how the public agenda is set, and the results of this process. Some scholars identify how policy is injected into the decisionmaking process, as well as the roles of agencies and groups in policy outcomes. These approaches have much in common regarding how policy flows through the process. Our framework for study is based on four phases: policy formation, approval, implementation, and feedback.

Policy formation refers to the shaping of a policy in response to a problem and its injection into the process. This includes the character of forces mobilized for and against it. The important elements are how policy is shaped and the source of the initiative—a bureaucracy, the Oval Office, interest groups, the federal court system, or Congress. This affects the environment and establishes the boundaries for struggle and compromise. The role of the media is especially important because of its ability to affect the policy agenda.

Approval is the process by which policy passes through formal executive and legislative procedures. Rarely can a policy be effective without congressional approval, whether direct or indirect. Major policy implementation usually requires the commitment of financial resources, meaning congressional action. Debates in Congress, congressional hearings, interest group mobilization, and corporate lobbying are all-important parts of the approval process.

Implementation refers to how policies are carried out. The bureaucracy has a special role in interpreting congressional and executive intent in this phase. It also translates intent into practical rules and regulations and applies them to the real world. Also included in this phase is how supporters and opponents affect how policy is interpreted and applied.

Feedback is the effect on the policy process of perceptions of success or failure of the policy in question. Policies may be revised, supported, or completely altered due to the evaluation of actors in the system as to their success. Especially for a large policy change, many of the outcomes will be unexpected surprises—and not always pleasant ones. Evaluation can involve a formal analysis, but it can also be informal and impressionistic. The role of the media in this stage is crucial, affecting as it does the impressions of various actors as to the effectiveness of the policy. These perceptions are injected back into the policy process, perhaps triggering new initiatives. The policy phases and their relationships are shown in Figure 10.1.

Figure 10.1 Policy Phases

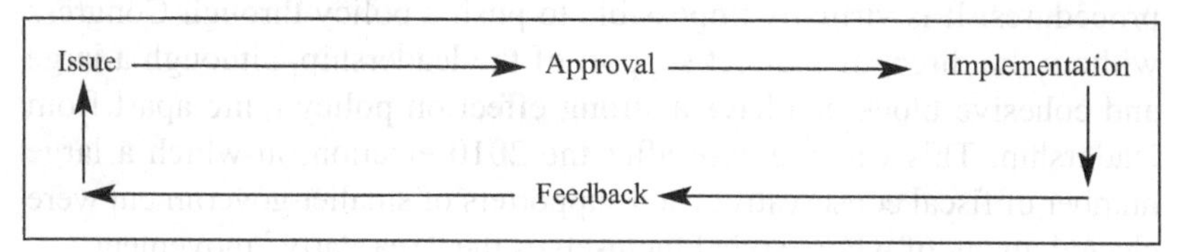

Congress and the Bureaucracy

Congress and the bureaucracy are critical political actors in the policy process. Congress has a constitutional role, broadened through oversight as well as constraints and reporting requirements it has imposed on the president and the national security establishment. Congress and the courts have traditionally allowed the president greater latitude with respect to foreign and defense policy, although congressional oversight will increase greatly if there is a perception that these policies are not going well. This was the case for the long conflicts in Vietnam, Iraq, and Afghanistan. Because domestic policies are generally of more immediate concern to the population, congressional initiatives and oversight of these policies have generally been more intense.

The bureaucracy grounds its power in the executive branch and the president's constitutional responsibility to ensure the laws are properly administered. As two bureaucratic policy scholars noted some time ago, "In the end, the President is heavily dependent upon the ability of bureaucratic organizations for his own success."[8] At the same time, Congress plays a vital role in the organization of the bureaucracy, its oversight, and especially its funding.

The Congressional Factor

The organization of Congress, congressional staffs, and constituent links are important elements of the policy process. Congress operates on the basis of standing committees. Those with the most continuous seniority in the majority party generally chair every committee, including the most important ones: Ways and Means, Appropriations, Budget, and Armed Services in the House; Armed Services, Foreign Relations, Budget, and Appropriations in the Senate. Furthermore, the party leaders are critical players: the Senate majority and minority leaders, the Speaker of the House, and the House majority and minority leaders. Together with the committee chairs, they form an inner sanctum with its own politics and procedures. It is virtually impossible to push a policy through Congress without the direct or indirect support of the leadership, although a large and cohesive block can have a strong effect on policy quite apart from leadership. This was the case after the 2010 election, in which a large number of fiscal conservatives and supporters of smaller government were elected, many of whom styled themselves the "Tea Party" movement.

The Democratic party is divided between those who consider themselves "progressives" and those who consider themselves "moderates," and this is reflected in their congressional delegations. Strong leadership by speaker of the house Nancy Pelosi after the 2018 and 2020 elections kept the Democratic caucus in the House of Representatives relatively unified. The Republican party has consolidated in a conservative direction, and has displayed a high degree of solidarity with former President Trump. Whatever their private feelings may be, Republican senators and representatives under the influence of Senate majority leader Mitch McConnell almost unanimously supported President Trump in the face of serious accusations of misconduct—including helping to foment a riot at the Capitol building on January 6, 2021—and did not seriously attempt to check his power or push back on unsupported allegations of voter fraud in the 2020 election. Despite animosity between Senator McConnell and the former president, this continued to be true as the Joseph Biden administration took office. This is a new phenomenon unanticipated by the Founders, and the implications of this for politics and policy are not yet clear.

This does not deny the importance of other members of Congress, especially when the partisan split in the Senate or House is close. In such situations individual senators or congress members have a very high level of influence, as they can prevent bills from passing. Individual members are elected from their districts and have an independent power base. They are, therefore, political entities unto themselves. Cooperation among members provides mutually beneficial political rewards, thereby strengthening individual power bases. The fact is, however, that much of the power of members of Congress resides in their staffs and constituencies. Staffs of twenty to forty spend considerable time strengthening and expanding the power base of each member. The bureaucracy and special interest groups are especially targeted. Congressional staffs have a significant input on policy, and efficient staffs increase the political power of members.

The congressional focus is generally domestic, although that changes in times of international crisis. Most members and their staffs develop their knowledge and political power in the domestic area.

Moreover, domestic politics is a direct link between policy and congressional behavior. Bread-and-butter issues determine whether members of Congress are doing well and are important for their chances for reelection. This links the political fortunes of members to constituents, whose concerns are employment, economics, and general well-being. This is not to suggest that foreign policy and national security are unimportant, but in normal times they generally give way to domestic matters in the voting booth.

It follows that most interest groups are concerned with domestic issues, and members must be sensitive to them. The result is a link among a member of Congress, special interest groups, and bureaucrats. Once this link is forged, it provides a basis for advocacy and opposition. Such "iron triangles," or broader and more inclusive "issue networks," can have a significant effect on the policy process.[9]

Thus, several important power clusters within Congress are critical components of the policy process. Although they have an important role in foreign and national security policy, they are especially prominent in making domestic policy.

The Bureaucratic Factor

The bureaucracy affects the way policy is carried out, and the result is sometimes quite different from the intent of Congress and the president. The bureaucracy also has an important impact on the kinds of policies that are injected into the higher levels of the process, whether civilian or military issues. The power of the bureaucracy rests primarily on its organizational character. The classic description is provided by Max Weber: "The decisive reason for the advance of bureaucratic organization has always been its purely technical superiority over any other form of organization. . . . Under normal conditions the power position of the fully developed bureaucracy is always overtowering."[10] Technical skills, administrative structures, and institutional loyalties are the bases for bureaucratic power.

The tendency for bureaucracies to protect their power base and responsibilities can distort policy goals. Bureaucracies often interpret policy according to organizational predispositions and bureaucratic advantage. In so doing they reinforce the existing state of affairs and resist major changes to the internal power structure. The dominance of the organization is reinforced by efforts to make the bureaucratic process a predictable routine. Individuality tends to become subsumed by the collective will of the organization. According to Carnes Lord, "A realistic approach to strategy at the national level must rest on two things, a careful distinction between the types or levels of policy-making and an appreciation of the bureaucratic faultlines that complicate and often defeat the development and implementation of national strategies."[11]

Also reinforcing the power of the bureaucracy is its influence over the emergence of policy. Power clusters within a bureaucracy, most identified with a particular organizational ideology, reflect a bureaucratic mind-set that advances only those policy goals and procedures reinforcing the organizational posture. Policy that is generated within the organization is vulnerable to the action and influence of gatekeepers, or the individuals with decisionmaking power over what is submitted to the higher policymaking levels. They control the flow of information, including recommendations, suggestions, and plans for policy and procedures. It is difficult for items that do not fit organizational preferences to find their way into the policy process.

In the final analysis, it is difficult to change the bureaucracy or persuade it to accept significant changes to the existing state of affairs. Indeed, the tendency is for the bureaucracy to justify the policy in the most virtuous terms and perpetuate as well as expand the commitment. In this way established policy becomes entrenched. President Trump came into office in 2017 intending to make difficult and rapid changes, and was repeatedly frustrated by various bureaucracies unwilling or unable to move quickly in his intended direction. President Biden came into office determined to undo many of the changes of his predecessor, but that could not be accomplished immediately. Over time, persistence on the part of the president and his main appointees can change bureaucratic direction, but it is a very slow process

National Security Policy and Process

Analysis of the national security policy process can apply the four-stage phase pattern described earlier, but the nature of national security and related issues shapes the policy process in several important respects. First, secrecy may be needed in order to respond to initiatives planned by an adversary. Second, in crisis or near-crisis situations, there is a need for speed. Third, in most cases national security policy must deal with external groups or foreign states outside the range of US laws. Fourth, the instruments for carrying out national security policy include the foreign service, the military, and intelligence agencies, many of which operate overseas. In general, national security policy issues evolve out of external sources and necessitate responses by instruments operating outside the United States.

The policy phases as they apply to the national security process are intermingled and collapsed into a relatively narrow time span. In addition, power clusters are usually limited to a few high-level political actors at some distance from domestic constituencies. Information is limited and often unavailable to the public, Congress, and bureaucrats outside the national security establishment. In this environment the media play an especially important, and sometimes adversarial, role.

The president confronts a difficult dilemma in dealing with national security policy. On the one hand, the sensitive nature of the policy may dictate quiet diplomacy and secrecy, including the undisclosed use of military, diplomatic, and intelligence instruments to affect foreign states and nonstate actors. Given the communication capabilities of the Internet and social media, maintaining secrets has become exponentially more difficult—especially given the willingness of some holding high clearances to leak them. On the other hand, the difficulties in conducting covert operations, maintaining secrecy, and explaining troop commitments only after the fact are challenging for an open system. The need to maintain the appearance of normalcy often places the president, other government officials, and bureaucrats in the position of telling half-truths or not informing the public of impending issues. Some domestic opponents of the administration and its policies are quick to suggest insidious intentions. A major challenge for future presidents will be to maintain and communicate a coherent foreign and security policy story amid the distraction of omnivorous social and other media driven by demands for instant gratification.

The most visible aspect of defense policymaking is the defense budget, which is approved under established legislative procedures in Congress. Although it is a tortuous process, it does provide opportunities for public discussion of national security matters. But the fact remains that the most serious issues are handled through a process that differs from the domestic process (see Figure 10.2).

Conclusion

In normal times, the national security policy process is primarily, but not exclusively, the preserve of the president and the national security establishment. Although this includes an inner circle of congressional leaders, the president generally has wide discretion in the use of military force, at least in the early days of its deployment. The primary instruments and agencies are under the direct control and supervision of the executive. Congress and the public depend on the president for information and for defining national security interests, but the president does not enjoy

Figure 10.2 Differences in Policy Phases

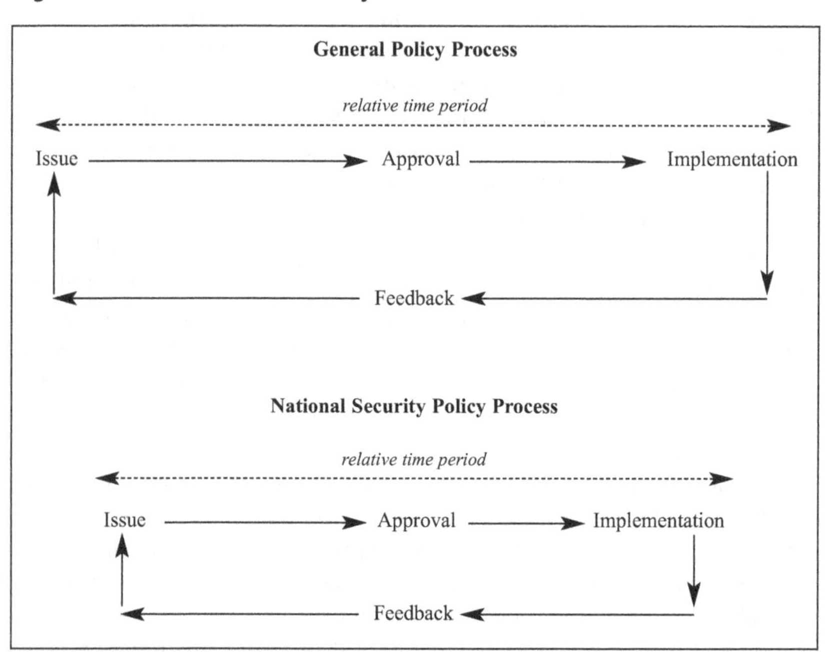

Notes: In addition to the differences in time elapsed for the two policy processes, the figure portrays the relative number of political actors involved. In the national security policy process, the circle of participants is considerably narrower than in the general policy process. It must be understood, however, that in a number of national security policy issues, the process follows a pattern similar to the general policy process. Much depends on whether the issue is a "crisis," whether it requires secrecy, and, in some cases, whether the president feels it is within his existing power to execute a certain policy without reference to Congress.

complete freedom. Indeed, he or she is considerably constrained in what he or she can do. In a time of diminished support for the president's policies, there will be increased challenges from Congress. This was clearly seen in greatly increased congressional opposition to the Vietnam War after 1967 and, more recently, the wars in Iraq and Afghanistan, with a complete withdrawal of US combat forces in the latter in August of 2021.

There are several problems associated with developing and implementing national security policy and strategy. First, national security policy and domestic policy can affect one another. In many cases, even the best national security policy can have a negative impact on domestic policy. For example, punishing an adversary through economic measures can have domestic repercussions, such as the 1980 grain embargo against the Soviet Union, which had a negative impact on US wheat sales, efforts to slow or stop Iranian nuclear weapons–related activities by imposing an economic embargo on Iran, and trade disputes with China beginning in 2019. This will be seen even more clearly if the harsh sanctions imposed against Russia in response to the Russian invasion of Ukraine have a strongly negative effect on US and allied economies. Similarly, national security policy and strategy can trigger domestic opposition, as with US involvement in Vietnam, Central America, the Middle East, and South Asia. Despite the usual prominence of domestic policies, a president focused only on those can be undermined if he or she is seen as ineffective on national security and foreign policy. Conversely, presidential effectiveness in national security and foreign policy can be eroded by ineffectiveness on the domestic front.

Second, national security failures are likely to become public and undermine the credibility of an administration. Yet many successes cannot be revealed, limiting the ability of the administration to generate support for its policies. Especially in the intelligence community, recognition for exceptional valor or sacrifice is often acknowledged only anonymously.

Third, inherent problems thwart the functioning of the national security establishment as well as policy formulation. The president must deal with the national security establishment, the bureaucracy, and Congress to develop and implement his or her policies. In this respect, Congress is an important factor in oversight and finances, as well as its ability to affect public opinion. Moreover, political action committees (PACs) and other nongovernmental actors, in particular the media, play important roles. All of this complicates the ability of an administration to develop and implement national security policy.

Fourth, the environment and constituencies of national security policy differ from those of domestic policy. Foreign states and groups outside US boundaries are major players in national security policy. Although there is considerable interdependence between domestic and national security policy, the focus of the latter is on external actors. This is also true for new missions such as humanitarian assistance, peacemaking, peacekeeping, and a variety of other peacetime engagements.[12] In domestic politics, the focus is on domestic constituencies and domestic political actors. Thus, there may be interdependence between domestic and national security policy, but there are also basic distinctions in the constituencies and in the strategies and instruments to implement policy. The character and power of international actors and the resulting politics differ in most respects from domestic politics and political actors.

Fifth, the strategic landscape is characterized by new challenges and potential threats, many of whose links to US national security and national interests are not immediately apparent. In such an environment it is difficult to design national security policy that can be understood by potential adversaries, allies, and the public.

In summary, the president is at the center of the policy process, yet national security is affected by institutional and interagency politics, individual mind-sets, domestic political and social forces, and foreign actors. Policy coherence, effective implementation, and credibility in commitment are presidential responsibilities. Ideally, these will evolve from a strategic vision articulated by a president whose leadership has a positive impact on national security policy.

Notes

1. John C. Donovan, *The Policy Makers* (New York: Pegasus, 1970), p. 16.
2. Given the checks and balances of the US system, it is easier to prevent change than to foster it. This contributes to the difficulty of making policy adjustments in response to changing international conditions.
3. C. Wright Mills, *The Power Elite* (New York: Oxford University Press, 1956); and Robert A. Dahl, *Who Governs?* (New Haven, Conn.: Yale University Press, 1961).
4. For more on the incremental nature of the US decisionmaking process, see David Braybrooke and Charles E. Lindblom, A *Strategy of Decision: Policy Evaluation as a Social Process* (New York: Free Press, 1970). On rationalist (or "rational actor"), bureaucratic, and organizational process models of decisionmaking and their impact on policy, see Graham T. Allison and Philip Zelikow, *Essence of Decision: Explaining the Cuban Missile Crisis,* 2nd ed. (New York: Longman, 1999).
5. Roger Hilsman with Laura Gaughran and Patricia A. Weitsman, *The Politics of Policy Making in Defense and Foreign Affairs: Conceptual Models and Bureaucratic Politics,* 3rd ed. (Englewood Cliffs, N.J.: Prentice-Hall, 1993), p. 347.
6. Ibid., pp. 67–69.
7. Richard Thomas Mattingly Jr. and Wallace Earl Walker, "The Military Professionals as Successful Politicians," *Parameters: US Army War College Quarterly* 18, no. 1 (March 1988), p. 43.
8. Robert T. Nakamura and Frank Smallwood, *The Politics of Policy Implementation* (New York: St. Martin's, 1980), p. 171.
9. For a clear discussion of iron triangles and issue networks, see Thomas E. Patterson, *We The People* (New York: McGraw-Hill, 2010), pp. 77–82.
10. H. H. Gerth and C. Wright Mills, eds., *From Max Weber: Essays in Sociology* (London: Routledge and Kegan Paul, 1984), pp. 214, 232.
11. Carnes Lord, "Strategy and Organization at the National Level," in James C. Gaston, ed., *Grand Strategy and the Decisionmaking Process* (Washington, DC: National Defense University Press, 1991), p. 143.
12. Military forces are exceptionally well suited for humanitarian response operations, as was seen in response to the Indian Ocean tsunami in 2004, Hurricane Katrina in New Orleans in 2005, and the Japanese tsunami in 2011. The military can provide a rapid augmentation in transportation, supplies, and medical facilities that are otherwise unavailable.

11

The President and Congress

Since the US Constitution was ratified, the evolution of power has placed the president in the dominant position in foreign affairs and national security. This remains the case even after passage of legislation designed to increase congressional power in these areas, such as the 1973 War Powers Resolution. Congress has an important role, but the nature of international security issues and the increasing complexity of international politics make it difficult for Congress to lead the nation or, for that matter, to check the president on policy initiatives. This is especially true during a national crisis. In addition, presidential power has grown in response to increasingly complex US economic and social systems, which have indirectly reinforced the president's power in national security policy.

In the new era, the traditional notion of national security is being questioned, and as a consequence of this presidential power in national security is undergoing change. Some prefer a stronger congressional role; others see a stronger presidential role. It is too early to predict the long-term effects of the so-called war on terrorism on this balance—let alone the less ambiguous national security problems from a resurgent Russia and an increasingly powerful China—but we expect the president's role to be significantly challenged, at least in the short term.

In any case, presidents are not monarchs and there are many checks in the US political system that prevent complete presidential domination in making and implementing national security policy. To understand the power that presidents wield, we must distinguish between foreign and domestic affairs, the specific issue involved, the particular circumstances, and likely domestic opponents.[1] Put simply, the exercise of presidential power is a function of the president's ability to understand the nature of the political process, the international climate, the power inherent in the office, and his or her own leadership and skills.

The Presidential Power Base

Several factors complicate the policymaking process, including the nature of the presidency, public expectations, and the demands of the international security environment. These all have philosophical, ideological, and political overtones. National security goes beyond a strong military and the ability to support it financially. It includes confidence in leadership, staying power, national will, political resolve, and agreement on national security goals. In today's strategic and political climate, it also includes some agreement on the meaning of national security.

The president faces potential opposition from several quarters. Although the president and Congress cooperate in many ways on national security matters, disagreements over policy often arise. Special interest groups, segments of the US public, and allies may also oppose the president. Add to this the disagreements within the administration and the national security establishment, and one can appreciate the extent of the problem, especially in an increasingly dangerous strategic landscape.

In addition to the built-in conflict between the president and Congress due to the checks and balances of the Constitution, the institutional characteristics of Congress and the power of individual members create conflict. Constituencies, terms of office, and mind-sets all play a role. Incumbents perpetuate their power to ensure reelection, and generally few are defeated. This has led some to view Congress as an institution of incumbents primarily intent on maintaining office. Over recent years, increasing partisanship in Congress has led to posturing for political advantage and near-gridlock on important issues.

The president can use several strategies to overcome opposition in Congress. Many are inherent to the office, but all depend on the effectiveness of the president using them. Every president beginning with Dwight Eisenhower has used a congressional liaison staff to establish and maintain relations. It targets key members in both houses, especially potential allies. The staff keeps the president informed of congressional power clusters and the general mood and recommends tactics to develop support for presidential initiatives. Similarly, the staff keeps members informed about presidential initiatives.[2] Although the liaison staff is usually concerned with domestic issues, national security policy and defense issues are also important.

Other tactics for developing support include bargaining, threats and intimidation, and rewards. A president who is popular with his or her base may threaten to support a challenge to an incumbent member of Congress in the next primary election in his or her district or state. Given the high percentage of safe seats, the only serious challenge to many incumbents is at the primary stage. This helps to explain the continuing strong support of congressional Republicans for President Trump, who remains popular in the Republican Party and could ensure that there is a primary challenger for any of his opponents. The president must be cautious in following certain tactics, however, because Congress can react negatively to extreme pressure from the White House and undermine the president's domestic and foreign agendas.

Presidents may decide that the best means to implement national security policy is to distance themselves from Congress and provide as little information as necessary. They can thereby maintain a degree of flexibility. The danger is that both Congress and the public may perceive the president as out of touch and isolated from major policy decisions.

In a direct confrontation with Congress, the presidents can take their case to the people. In 1987, President Ronald Reagan used this tactic to develop support for financial aid to the Contra rebels opposing the Sandinista government in Nicaragua. Earlier, he took the case for Vietnam to the people, labeling US involvement as a "noble cause" and honoring those who fought there. Other presidents have adopted this tactic when faced with congressional opposition to important presidential initiatives.

In 2007 President George W. Bush used public speeches, news conferences, and radio addresses to oppose congressional limitations on his freedom of action in Iraq. Indeed, with the 2006 election of a Congress controlled by the Democratic Party, congressional oversight of presidential actions in Iraq and Afghanistan began in earnest—with the Democrats determined to bring US troops back from Iraq on a strict timetable. By mid-January of 2007 the Congressional Research Service had prepared an exhaustive study of previous congressional attempts to restrict the presidential war-making power by restricting funding and by other means that could serve as a handbook for future attempts.[3] Generally these attempts ran into great difficulty.

The president must also provide a strategic vision for Congress and the American people. Critics of President Bill Clinton felt that there was no clear strategic vision articulated by his administration.[4] This was seen in the apparently muddled response and misjudgments associated with US involvement in Somalia, where eighteen US soldiers were killed in an engagement in Mogadishu in October 1993 (the subject of the popular film *Black Hawk Down*).

The president has an advantage in dealing with Congress in the national security arena, however, because the sources of intelligence, the basis of policy and strategy skills, and the operational instruments are centered in his or her office. The president's cabinet, presidential advisors, national security staff, and the national security advisor are reinforcing resources of power. Even though expanded congressional staffs are important, the presidential power base—the Departments of Defense and State, the military advisory system, the intelligence agencies, and the National Security Council and its staff—is dominant. Congress must rely on presidential sources for much of its information about national security.

President George W. Bush's efforts to prosecute the war on terror provided a dramatic episode in the ongoing struggle for power between the executive and legislative branches of government. The Bush administration put forth an ambitious and controversial agenda for expanded presidential power. This included warrantless surveillance of Americans' international telephone calls and e-mails, detention of suspected terrorists as enemy combatants without prompt access to legal counsel or the protection of the Geneva Convention, transfer of terrorist suspects to foreign countries ("extraordinary renditions") where barriers against torture were weak or nonexistent, and efforts to prevent the application of a congressional ban on torture during interrogations by US intelligence and military personnel. A number of these issues would ultimately have to be resolved by the courts. Activities of US intelligence agencies continued relatively unchanged into the Obama and Trump administrations, despite the great ideological differences between those presidents. It is not yet clear what effect the greatly increased transparency of intelligence operations due to leaks and publication of classified documents on the Internet will have on congressional oversight. In the wake of the trouble-plagued departure of US troops from Iraq in August, 2021, one would expect increased congressional oversight of national security issues in the Biden administration.

Congress has formidable weapons at its disposal if it chooses to use them, such as the investigatory power and the power of the purse, which were included in the Constitution to act as a check on the president. The president has different but similarly effective resources of power flow-ing from his many constitutional roles and his or her ability to take action while Congress deliberates.

Congress: The Legislative System

The relationship between the chief executive and Congress has always been contentious, just as the Founders intended. The US political system is based on the constitutional separation of powers among the branches of government and resulting checks and balances. The Founders expected that the members of each branch of government would defend their prerogatives against encroachment by the other branches. This has generally been true historically, but during crises presidential power is greatly increased and the assumption that members of Congress would put the interests of their institution ahead of party loyalty, personal political interests, or ideological considerations was severely tested during the Trump and Biden administrations.

To understand the congressional role in national security, we need to review the general features of the institution and the legislative process. The writers of the Constitution expected that Congress would be the most powerful branch of government. Although the president was given important powers in foreign affairs, those at the Constitutional Convention wanted to ensure that he would not dominate the policy-making process. In theory, the president would have power to react in emergencies, but Congress would determine war policy. Furthermore, the power of Congress in the legislative process and budget matters was to provide an effective counterbalance to the president.

The scope of congressional responsibilities has increased, yet Congress is finding it more difficult to respond because of the cumbersome legislative process and the characteristics of the institution. Nevertheless, effective national security policy depends on congressional support and public acceptance. Because of the representative role of Congress and its power over the purse, successful presidents cannot afford to disregard Congress, isolate themselves from the legislative process, or distance themselves from congressional leadership.

The organization and functioning of Congress rest primarily on the committee structure. In the normal course of the legislative process, bills first go to committees, with the chairs of committees and subcommittees exercising considerable power in determining their fate. Chairs are appointed by the majority party in each house, with seniority being critical to appointment. The internal power system of Congress does not rest solely with the committee structure, however. Congressional leadership positions, such as the Speaker of the House and the majority and minority leaders in both houses, carry power that exceeds that of committee chairs. Reforms in the 1970s placed final approval of leadership roles in the party caucuses and eroded the disciplined party system as well as the authority of the party leadership. Combined with the committee structure and the power of individual members, these reforms have fragmented power within Congress. Nevertheless, if party discipline is maintained, the leaders can be very powerful.

Power in Congress thus derives from a mixture of sources: power over the purse, the status of the membership, relationships with colleagues, the party, and the formal leadership offices. As long as constituent support remains high, members are powerful in their own right. Nonetheless, they can accomplish little by themselves and are dependent upon colleagues to get things done. Every bill needs supporters, and this leads to constant interplay among internal forces seeking accommodation and compromise (or leading to confrontation). Given the internal power fragmentation of the institution, effective leadership in Congress is essential for the legislative process to function effectively.

Congress and the Executive: The Invitation to Struggle

Congress has a critical role in national security. According to Frederick Kaiser, "National security is not a simple set of well-integrated subject matters neatly arranged along a single, consistent policy continuum. . . . It is a complex set of diverse subject matters that cross into many different policy lines; these in turn raise different issues and concerns, institutional interests, and costs that affect congressional roles."[5] At least two important distinctions need to be made with respect to the president and Congress. First, the institution of the presidency rests on one individual who heads a hierarchical branch of government. The center of power is clear, and the responsibility for executing the laws of the nation is focused on the president. There is little overt fragmentation of power or responsibility. In Congress a different picture emerges. Not only is there considerable fragmentation of power within the institution, but also it is often difficult to place responsibility in any single member. Responsibility falls on Congress as an institution, making it possible for individual members to shift blame to the institution as a corporate body. This affords members a great deal of flexibility in taking political positions, and they can disclaim responsibility for any institutional outcome that is unacceptable to their constituencies.

Second, the president is the only nationally elected official, aside from the vice president, whose power is dependent on his or her relationship with Congress.[6] Individual members of the House of Representatives represent districts within states, many of which reflect narrow segments of the population. Furthermore, such districts can be dominated by one or two special interest groups. In addition, a large number of districts are strongly Democratic or strongly Republican and winning the primary election generally ensures victory in the general election. (As noted earlier, a president's ability to affect primary elections is an important source of influence over many members of Congress.) Senators, representing states, also reflect a small part of the total population. Even in states, political power can rest with a handful of special interest groups.

If Congress is controlled by one party and the presidency by another party, the politics may result in partisan confrontation and gridlock, as was seen in the struggles of President Obama and President Trump, who faced Congresses in which one or both houses were controlled by the opposite party. The differing constituencies between the president and Congress not only reflect different power bases and interests, they create different policy mind-sets and the conditions of struggle over policy, programs, and budgets.

The War Powers Resolution of 1973 is an important reference point.[7] During the of Nixon administration, the Democratic Party held majorities in both houses of Congress. Congressional concern over continuing US involvement in Vietnam and the erosion of President Nixon's power as a result of the Watergate scandal prompted Congress to pass the War Powers Resolution over a presidential veto. (At the time, many Republicans saw the issue as one of congressional prerogative rather than party loyalty. As noted earlier, party loyalty and personal political considerations now weigh more heavily in such considerations than before.) The resolution required the president to consult Congress prior to committing US troops to hostile action and periodically thereafter. The War Powers Resolution provides that "the President in every possible instance shall consult with Congress before introducing United States Armed Forces into hostilities or into situations where imminent involvement in hostilities is clearly indicated by the circumstances, and after every such introduction shall consult regularly with the Congress until the US Armed Forces are no longer engaged in hostilities or have been removed from such situations."[8]

After sixty days, US forces would have to be withdrawn unless Congress declared war or passed an extension. The president has an additional thirty days to withdraw all US forces if he states in writing that "unavoidable military necessity respecting the safety of United States Armed Forces requires the continued use of such armed forces in the course of bringing about a prompt removal of such forces."[9] The resolution also provided Congress the option of passing a concurrent resolution ending US involvement in hostilities. Such a resolution could not be vetoed by the president. But the central features of the War Powers Resolution "have been deemed unconstitutional by every president since the law's enactment in 1973."[10]

Indeed, even when presidents consult Congress about military actions, they do not specifically invoke the War Powers Resolution, and Congress is often brought into the process long after it can affect policy.[11] Presidents may also ask Congress to approve on Authorization for the Use of Military Force (AUMF) in lieu of invoking the War Powers Resolution.

In their relationships with Congress and in trying to establish the necessary support for national security policies, presidents must deal with a variety of power clusters within the institution. In the past, given party discipline and effective leadership in Congress, the executive could focus attention on the Speaker of the House and the majority and minority leaders in both houses; today the president must also deal with other important members, especially key committee chairs and leaders of special interest caucuses. The increase in power clusters is especially pronounced in domestic policy, but it also affects national security policy.

If presidents lose popular support, or if their initiatives appear vacillating and ambiguous, Congress is more likely to take the lead. For example, following the killing of eighteen US Army soldiers in Somalia in late 1993, Congress set conditions for US involvement there. Some observers noted the marginalization of the presidency in the matter. President George W. Bush maintained high levels of public and congressional support for the war on terror and the war in Iraq during his first term. But second-term setbacks, including rising US casualties in Iraq, drove down his popularity ratings and emboldened congressional critics of his war policies, including some Republicans. Public opinion thus provides the base upon which assertive presidents who wish to expand presidential power can succeed—or from which presidents with low ratings can fall and lose control of their agendas.

Nonetheless, the public looks to the president for leadership in national security policy. This is true also for most members of Congress, even though they debate and criticize policy. Part of this acquiescence stems from a recognition that it is difficult for Congress to lead; it is better postured to react and engage in oversight. Another factor is the tendency for members of Congress to be cautious in initiating national security policy for fear of being associated with failures or controversies that might affect their popularity with constituents. The safest position is to keep some distance from national security policy until it becomes clear whether it is succeeding or failing.

This allows the president some latitude in initiating national security policy, although policy failures are easily attributable to him. Equally important, the complexity of national security issues, the changed external power relationships, and the difficulty the United States faces in trying to control external situations all mean success is never assured. Failure is no longer a remote possibility, but a continuing possibility—even for long time operations with significant loss of life, such as the attempt to democratize Afghanistan. One can understand, therefore, the reluctance of members of Congress to become too closely associated with presidential positions on national security, save for crises.

In summary, no president can ignore the congressional role in national security policy. Indeed, most successful policies depend on the bipartisan involvement of congressional leadership. Congress, when it is sensitive to its responsibilities and protective of its prerogatives, demands an equal, if different, role in national security. Ambiguity, shared power, institutional character, and the nature of national security policy thus create the basis for confrontation between the president and Congress.

The President, Congress, and the Policy Process

The president's ability to deal with Congress and to develop the support necessary for national security policy must be viewed from two dimensions: (1) the element of national security policy being considered, and (2) how the president's sources of strength can overcome the sources of conflict.

National security policy includes a range of sub-policies, from the defense budget and military manpower levels, to executive agreements and treaties, to covert operations. There is also a degree of overlap between national security and foreign policy. This overlap has become considerable in the new strategic landscape, in which national security increasingly encompasses nonmilitary matters.

The president has generally had a great deal of latitude in committing US military forces, especially in the early stages of a crisis. With President Truman's decision in 1950 not to ask for a declaration of war for the war against North Korea, such a declaration has not been deemed necessary for presidential war making.[12] Nevertheless, congressional oversight and budget power restrict the president in the long term. Congress, ever sensitive to negative reaction from constituents about US force commitments, will make its reaction known to the president. To be sure, in short-term commitments where success appears clear, and even in longer-term commitments where there is a clear threat to the nation, the president can enjoy popularity and support for his policies. But he or she will be blamed for any failure.

Support for the preceding points is evident in the US decision to invade Iraq in 2003, depose the regime of Saddam Hussein, and rebuild a stable and democratic Iraq. The US "shock and awe" military machine rapidly blew away the resistance of Iraqi conventional military forces, occupied Baghdad, and overthrew the government. The Bush administration proclaimed "mission accomplished" for Operation Iraqi Freedom on May 1, 2003, in a widely televised "photo op" for President Bush on a US aircraft carrier. At this point, Congress was in accord with presidential strategy. On the other hand, the postconflict reconstruction of Iraq was a mixed set of accomplishments and embarrassing setbacks. By the fall of 2005, increased public uncertainty about the stability of Iraq's new government in progress and about the clarity of the "endgame" for US completion of its political mission caused Congress to be much more assertive of its right to question the administration on war policy and military strategy. Congressional insistence on this point increased greatly with the 2007 takeover of both houses of Congress by the Democratic Party. Even the congressional Republican leadership demanded more clarity from President Bush about providing an exit strategy that would eventually replace departing US troops with Iraqi military and police forces.[13]

In Iraq, a withdrawal of US forces by President Obama that many viewed as precipitous permitted a newly formed "Islamic State"[14] to occupy much of that country with little resistance from Iraqi forces. US forces returned to help drive them out, but they remain as an insurgency threat. Neither President Obama nor President Trump found a military solution to the continuing fighting in Afghanistan, which became the longest lasting American war. In 2020 President Trump committed to removing all US troops from Afghanistan by May 2021, and President Biden amended the timeline to the end of October 2021 and removed the troops the end of August 2021. The chaotic nature of the withdrawal and the inability to save many Afghans who relied on US protection resulted in a great deal of criticism of the Biden administration. As the US departure approached, the Taliban forces quickly occupied the whole country as Afghan government forces collapsed due to a combination of lack of will and the withdrawal of critical US contractors and military advisors.

An essential part of the national security policy process is reflected in debates over the defense budget and the final shape of the national budget. The annual defense budget process focuses attention on general issues of national security. This usually does not involve serious discussion and debate over strategy, but there can be exceptions—especially in a severely constrained fiscal environment when not all projects can be funded and genuine strategic choices must be made.

Finally, some issues of US national security may be a continual source of debate, but there is continuity in important aspects of US policy. The fight against terrorism, the close relationship with Western Europe, the concern over weapons proliferation, protection of freedom and nurturing of democratic systems, and control and reduction of nuclear weapons stockpiles will continue to be important priorities. Despite this, there will be much discussion about the level of resources to be devoted to these issues.

Well-established and accepted components of national security generally do not create controversial or difficult issues for the presidents. It is when presidents want to change direction, add a new dimension to established policy, undertake new initiatives, or fail to clarify national security policy that they face more opposition in Congress and among the public. The end of the Cold War caused NATO to recast its missions and identity from the end of the Cold War through the end of President George W. Bush's first term in office. No longer a political-military bloc aimed primarily at containment of the Soviet Union, the North Atlantic Treaty Organization (NATO) expanded its membership among states in East Central and Southern Europe and accepted responsibility for crisis management, contingency operations, and "out of the area" peacekeeping and stability operations, as in Afghanistan following the defeat of the Taliban early in the war. NATO also provided essential military support for toppling the government of Muammar Gaddafi in Libya in 2011. Russian military activity, such as the forceful takeover of Crimea from Ukraine, continuing intervention in Russian-speaking eastern regions of Ukraine, and finally the Russian invasion, has led to revived concern with NATO's traditional mission of deterring a Soviet, then Russian attack. The continuing issue of burden sharing among NATO members has also become more salient due to pressure from the Trump administration and the realization by NATO members after the fall of Afghanistan that their reliance on the US military infrastructure (such as transport, communications, and intelligence) makes it very difficult for them to act alone in military operations if desired. Also, with the rise of Russia as a peer competitor to the United States, it's "back to the future" for NATO as concerns shift to its original mission of containing or, if necessary, fighting their giant neighbor to the east. Russia's mobilization of large contingents of combat ready forces on the Ukrainian border in the spring and autumn of 2021, as well as the 2022 invasion, were reminders of NATO's primary mission in Europe.

At the same time, pressure along the US southern border increased from large numbers of asylum seekers wishing to escape from conditions in Central America, seeking better economic opportunities, and trying to reunite with their families. That, combined with the draconian response from the Trump administration, quickly developed into a humanitarian crisis as the refugees overflowed the facilities available to house them, parents were separated from their children, and the issue was seen by many as not merely political, but moral. This put both authorized and unauthorized immigration at the center of a highly partisan political debate during the 2020 election. The Biden administration has attempted to roll back some of their predecessor's harsh methods, but the changes were not as great as would have been expected.

Presidential Leadership and Party Politics

Although developing consensus and support in Congress depends on presidential leadership, the direct involvement of the president in national security policy has an immediate bearing on his or her leadership success. In this respect, popular support and party politics are important factors underlying effective presidential leadership.

The relative strength of the political parties in Congress impacts the president's ability to shape national security policy through the legislative process. If the same party holds the White House and Congress, the president will likely have the advantage thanks to party loyalty, as in the first six years of the George W. Bush administration. The fact that presidents are also the leaders of their party reinforces this. Skilled presidents can use this to strengthen their position on national security policy and strategy, allowing them greater latitude for initiating changes.

President George W. Bush used his congressional majority to great advantage in moving his agenda in domestic and foreign policy forward until the Democrats won control of Congress in the 2006 election. President Obama also faced Congresses where the House or the Senate was controlled by the opposite party. President Trump's Republicans lost control of the House of Representatives in 2018, which increased Democrats' ability to thwart presidential policies and put him on the defensive by numerous investigations. The ability of presidents to use party support in Congress to their advantage is a function of their leadership skills.[15] But in general the advantage is with the president whose party holds both houses of Congress.

There is a close correlation between party support and popular support for the president. A perceived mandate from the electorate can translate into support in Congress. Even if members of Congress are opposed to presidential policies, they find it difficult to defy a popular president or party leadership publicly. At the same time, erosion of popular support has a similar impact on congressional support, regardless of party alignments. Lyndon Johnson and Richard Nixon discovered that the hard way, as did President George W. Bush after the shock of September 11 began to recede and concerns grew about his policies. By the fall of 2005, Bush's polling numbers were less favorable in the aftermath of a protracted war in Iraq, inept responses to Hurricane Katrina, and criminal investigations of key White House aides. The Democratic takeover of the House of Representatives after the 2018 election greatly increased President Trump's political difficulties, and Democratic control of both houses of Congress following the 2020 election (however tenuous) made it possible for Democratic priorities to get passed.

Sometimes presidents can make personal appeals to members of Congress. By approaching individual members, appealing to their sense of propriety, stressing the need to support the president and the nation in critical national security issues, and even promising support on future issues, presidents can overcome resistance to their policies. Lyndon Johnson was the acknowledged master of this tactic. A longtime member of Congress before he became president, Johnson personally knew most members and was keenly aware of congressional dynamics and politics. Making full use of this knowledge, he prevailed upon individual members to garner support for policies ranging from the Great Society social programs to US involvement in Vietnam. When popular support for the Vietnam War eroded, a similar erosion took place in the president's popular and congressional support.

If used prudently, personal appeals can be effective, but if the president resorts to such appeals too often, they lose their effectiveness. Only when individual members of Congress feel that personal appeals are focused on special issues that have a direct bearing on presidential performance and are essential to effective policies will they tend to respond positively. In other words, the president can rapidly deplete the power associated with personal appeals, but the possibility that the president will get involved against them in their next primary election helps keep members from straying too far from presidential positions.

Sometimes a president may decide to stand up for what he believes rather than make concessions. President George W. Bush indicated in 2006 his intent to "stay the course" in Iraq and refused to set a timetable for withdrawal of US forces despite a wavering Congress and plunging public opinion polls. The result was the "surge" of forces that contributed to the stabilization of the military situation there.[16] Ultimately it is the president who is held responsible for national security policy, regardless of the actions of Congress. As President Harry Truman's biographer put it some time ago, the president "is the only person in the government who represents the whole people."[17] That is still true, so long as the president chooses to do so.

Covert Operations and Secret Military Deployments

Covert operations are at the root of many controversial national security issues. Many times they reveal serious disagreements between the president and Congress and provide insights into the congressional role in national security issues. This was dramatically exposed during the so-called Iran-contra hearings in 1987, which provided a valuable education on certain aspects of national security policy.[18] The hearings revealed the character of covert operations and explained the role of Congress in the process. The Iran-contra episode highlighted the struggles between the president and Congress over covert actions that had been ongoing since the end of the Vietnam War.[19]

In the late 1970s, on the heels of investigations into alleged Central Intelligence Agency (CIA) abuses and secret operations, a series of legislative bills was passed strengthening the congressional role in intelligence matters. Two of the most important features of this legislation were (1) the creation of permanent oversight committees in the House and Senate and (2) provisions for dealing with the relationship of intelligence agencies and Congress. Although these focused specifically on intelligence agencies, information, and the relationship to Congress, they had a direct bearing on the relationship of the president with Congress on sensitive national security issues. As such, they help identify the direction and substance of the president's ability to deal successfully with Congress on national security policy and strategy.

In 1976, the Senate Select Committee on Intelligence (SSCI) was created, "composed of 15 members drawn from the Appropriations, Armed Services, Foreign Relations and Judiciary Committees, and from the Senate at large. . . . The SSCI [has] full authority to oversee the activities of US intelligence agencies and to authorize their funding."[20] Two years later the House passed similar legislation, creating the House Permanent Select Committee on Intelligence (HPSCI). "The HPSCI consists of 16 members, with membership drawn from Appropriations, Armed Services, Foreign Affairs and Judiciary Committees, as well as from the House at large,"[21] and has essentially the same authority as the SSCI.

The Intelligence Oversight Act of 1980 imposed several reporting requirements on the CIA director, as well as on "the heads of departments, agencies and other entities of the United States involved in intelligence activities to keep the committees fully and currently informed of all intelligence activities, including any significant anticipated intelligence activity."[22] This responsibility now lies with the office of the director of national intelligence, created by Congress in the wake of September 11 to improve intelligence coordination. The director of national intelligence (DNI) is now the titular head of the entire US intelligence community, including the CIA and other organs of intelligence collection and analysis scattered throughout the government. The congressional oversight provision does not require approval of intelligence activities; it is primarily consultative and informative.

The act requires reporting covert operations in a timely fashion, but it is not clear what "timely" means in this context. Moreover, according to some scholars, Congress does have important powers in reacting to or in limiting presidential initiatives: "It may cut off funding for foreign and defense policies. The Church Amendment to the appropriations bill for 1973 cut off US government support for the Republic of Vietnam . . . and the various Boland Amendments between 1982 and 1986 (named after Representative Edward Boland, D-Mass.) sought to prevent federal funds from being provided to the Contras in Nicaragua."[23] The success of this tactic has been mixed but more effective than use of the War Powers Resolution.[24]

Covert operations differ from secret military operations in important aspects. In the latter (e.g., the invasions of Grenada and Panama, as well as certain actions in Afghanistan and elsewhere) the initial phase of US involvement was concealed for reasons of security and safety. Secret military operations are difficult to keep under wraps for any length of time and eventually invite public debate and congressional involvement. They are often used to demonstrate US policy. Covert operations, however, are special activities cloaked in secrecy that are intended to conceal US government involvement, among other things.[25] Such activities range from propaganda and paramilitary operations to the use of small special operations forces for extended periods.

In response to September 11, and at the urging of the Bush administration, Congress passed the USA Patriot Act the following month. The act expedited the sharing of information and analysis across the "wall of separation" that had previously separated domestic intelligence gathered for the purpose of law enforcement and criminal conviction, on one hand, from foreign intelligence collected for the purpose of defeating foreign espionage and intercepting terrorist attacks, on the other. To some extent, this commingling of information from foreign and domestic surveillance had been foreshadowed by the traumatic cases of Soviet spies Aldrich Ames and Robert Hanssen. Ames of the CIA and Hanssen of the Federal Bureau of Investigation (FBI) passed vital secrets to Moscow for years until they were unmasked and arrested in the 1990s. Each did considerable damage to US security and likely caused the deaths of a number of US agents operating in the Soviet Union. Collaboration and teamwork between the CIA and FBI were required to indict and convict Ames, and Hanssen's case was closed by obtaining his KGB file from a retired Russian intelligence officer resettled in the West.[26] It's not publicly known the extent to which Russian intelligence was involved in more recent security breaches by National Security Agency (NSA) contractor Edward Snowden (now living in Moscow) or US Army clerk Chelsea Manning, both of whose information became widely available on the Internet.

Leadership and Policy

This study has stressed that the success of national security policy depends on the president's leadership and his relationship with Congress. The president has the key role, the constitutional authority, and much latitude in foreign and national security policy. This ability to build support in Congress, to control and direct the national security establishment, and to gain public acceptance of policies is a direct function of the president's leadership style.

No single model of leadership is sufficient. Yet certain principles of leadership are essential in dealing effectively with Congress. These principles, and the way they are applied, must lead to the development of trust and confidence. This in turn evolves from the perceptions of members of Congress that the president is in control of the national security establishment and that he clearly articulates a vision of US strength and commitment that they share. Furthermore, Congress must feel that the president's staff is knowledgeable, skilled, and supportive of his national security policy. Equally important, there must be mutual trust and confidence among the president and the national security staff, the military, and the intelligence establishment. Part of this evolves from the character, background, and experience of the commander in chief. In this respect, there is sometimes a decided gap between the president and the military.[27]

A great deal of presidential strength is a function of personality and character. According to Erwin C. Hargraves and Roy Hoopes, "The style and character of the president himself is every bit as important as the inherent power of the institution. And when we talk about powers of the presidency, we must consider three factors: a president's sense of purpose; his political skills; and his character."[28] The observation of James Q. Wilson (as mentioned in Chapter 6) is relevant here, as well: "The public will judge the president not only in terms of what he accomplishes but also in terms of its perception of his character."[29] For some, a negative view of the character of a president matters less than their support of his policies.

Trust and confidence between the executive and the legislature are strengthened if members of Congress, especially the leadership, feel that the president is sincere about consulting them and accepts the coequal status of Congress and the president. Furthermore, Congress must feel that the president is providing timely and useful information on matters of national security. This especially applies to covert operations and secret military movements, even though Congress initially is only a recipient of information and not an approving body.

Furthermore, presidents must make themselves reasonably accessible to members of Congress, especially to the leadership. Members become frustrated if they are ignored by the president and feel that such a situation damages their ability to deal with their own issues. In such an environment, confrontation and disagreement between Congress and the president are inevitable. The idea of consultation is ingrained in the two institutions. Consultation can pave the way for support, provide the perception of congressional power, and become a symbolic tool for fulfilling congressional responsibility.

Even if presidents do all these things, success will not be assured; but the environment will be the most favorable possible for pursuing such goals. Leadership is the key to relationships with Congress, and leadership must begin with an understanding of the important role played by Congress as well as an appreciation for the human motivations of individual members.

The performance of the Bush White House immediately after September 11 showed considerable skill in using the president's central position in the policymaking process for national security to advance his agenda. Bush chose a strategy of invading Afghanistan to dislodge al-Qaeda and the Taliban in 2001 and (more controversially) reprised his policy against Saddam Hussein in Iraq in 2003. Bush also signed a National Security Strategy in 2002 that highlighted preemption and unilateral actions as important options for dealing with terrorists or rogue states bent on attacking the United States. Congress, consistent with Bush proposals, reorganized the intelligence community by creating an interagency intelligence czar: the director of national intelligence. Congress also created the Department of Homeland Security as a cabinet-level agency incorporating more than twenty previously separate agencies and departments. President Bush's relations with Congress deteriorated as the war in Iraq dragged on and weapons of mass destruction, the ostensible cause of the war, were not found.

President Obama's relations with Congress were arguably better in the international sphere than the domestic, where sharp partisan differences made action difficult on social issues. In addition to the advantages presidents always have in leading foreign and national security policy, President Obama's policies benefited from not being much different from those of the preceding Bush administration. Former President Trump's support remains very strong among his base, but diminished for others by their perception that he may not have been suited by temperament or ability for the job. The perception of many that President Biden did not handle the removal of troops from Afghanistan in August, 2021 well could affect his public support and may have political effects.

Conclusion

The US system of government divides power and authority over national security policy among the various branches of government, especially between the president and Congress. The president must exploit the potential power of his office to drive the national security agenda toward his preferred goals. Congress must authorize wars and approve treaties, pay the bills for diplomatic and military actions, and hold the executive accountable by means of hearings, investigations, and reports. Presidents cannot simply respond to popular passions of the moment; neither can they simply ingratiate themselves with Congress. Such behavior can only lead to the erosion of executive credibility and project a picture of a weak leader. Perceptions of presidential weakness at home can have serious negative effects abroad. Foreign leaders may find it easier to resist US policy initiatives or, at worst, to attack vital US interests.

The president is ultimately responsible for the formulation and implementation of national security policy and strategy. For most elements of policy, he or she is in a position to receive the support of Congress and the public. But there are times in which the president stands alone, taking credit for its success but assuming full responsibility for failure. Unpopular wars or military interventions are the most controversial and lonely of these times.

Notes

1. Roger Hilsman with Laura Gaughran and Patricia A. Weitsman, *The Politics of Policymaking in Defense and Foreign Affairs: Conceptual Models and Bureaucratic Politics,* 3rd ed. (Englewood Cliffs, N.J.: Prentice-Hall, 1993), p. 145.

2. Richard A. Watson and Norman C. Thomas, *The Politics of the Presidency,* 2nd ed. (Washington, DC: Congressional Quarterly, 1988), p. 257.

3. Richard E. Grimmett, *Congressional Use of Funding Cutoffs Since 1970 Involving U.S. Military Forces and Overseas Deployments* (Washington, DC: Congressional Research Service, January 16, 2007). The Congressional Research Service, a branch of the Library of Congress that provides research for members of the House and Senate and congressional committees, has published other important papers relating to congressional-presidential struggles over the control of national security policy.

4. This was rectified somewhat later in his administration with the publishing of an annual National Security Strategy report. President Bush and President Obama also published National Security Strategy reports. The most contentious of these was the 2003 report of President Bush that sought to justify unilateral and preemptive military action by the United States under certain situations.

5. Frederick M. Kaiser, "Congress and National Security Policy: Evolving and Varied Roles for a Shared Responsibility," in James C. Gaston, ed., *Grand Strategy and the Decisionmaking Process* (Washington, DC: National Defense University Press, 1991), p. 217.

6. The vice president's constitutional power as president of the Senate can be important if the Senate is evenly divided, as it was for a short time after the 2000 election.

7. For a detailed analysis of the War Powers Resolution, including the degree of presidential compliance, see Matthew C. Weed, *The War Powers Resolution: Concepts and Practice* (Washington, DC: Congressional Research Service, March 8, 2019), https://crsreports.congress.gov/product/pdf/R/R42699.

8. The War Powers Act of 1973, Public Law 93-148, sec. 3.

9. Ibid., sec. 5(b).

10. Richard E. Grimmett, *The War Powers Resolution After Thirty-six Years* (Washington, DC: Congressional Research Service, April 22, 2010), p. i.

11. For greater detail, see Richard E. Grimmett, *War Powers Resolution: Presidential Compliance* (Washington, DC: Congressional Research Service, April 12, 2011), and *The War Powers Resolution: Concepts and Practice* (Washington, DC: Congressional Research Service, March 8, 2019).

12. Jennifer K. Elsea and Richard E. Grimmett, *Declarations of War and Authorizations for the Use of Military Force: Historical Background and Legal Implications* (Washington, DC: Congressional Research Service, March 17, 2011).

13. On shortcomings in the US plan for postconflict stabilization following Operation Iraqi Freedom, see David C. Hendrickson and Robert W. Tucker, *Revisions in Need of Revising: What Went Wrong in the Iraq War* (Carlisle, Pa.: Strategic Studies Institute, US Army War College, December 2005).

14. Officially "The Islamic State of Iraq and the Levant"—variously abbreviated as "ISIS," "ISIL," or "Daesh" from its Arabic language acronym.

15. This is not to suggest that all such lack of support is political in nature. For many, it is a matter of principle and genuine differences in policy preferences.

16. For a well-sourced and written discussion on how the "surge" came about, see Thomas E. Ricks, *The Gamble: General David Petraeus and the American Military Adventure in Iraq, 2006–2008* (New York: Penguin, 2009). Ricks's praise of President Bush in this instance was a sharp contrast to his highly critical earlier book on the origins of the Iraq War, *Fiasco: The American Military Adventure in Iraq* (New York: Penguin, 2006).

17. Merle Miller, *Plain Speaking: An Oral Biography of Harry S. Truman* (New York: Berkley Medallion, 1974), p. 445.

18. For details on the Iran-contra affair, see *Report of the Congressional Committees Investigating the Iran-Contra Affair* (Washington, DC: US Government Printing Office, 1987), and *Report of the President's Special Review Board* (Washington, DC: US Government Printing Office, February 26, 1987).

19. For a useful study, see Gregory F. Treverton, *Covert Action: The Limits of Intervention in the Postwar World* (New York: Basic, 1987).

20. Standing Committee on Law and National Security, *Oversight and Accountability of the US Intelligence Agencies: An Evaluation* (Washington, DC: American Bar Association, 1985), pp. 7–8.

21. Ibid.

22. Ibid., pp. 11–12.

23. Donald M. Snow and Eugene Brown, *Puzzle Palace and Foggy Bottom: US Foreign and Defense Policy-Making in the 1990s* (New York: St. Martin's, 1994), p. 148.

24. Grimmett, *Congressional Use of Funding Cutoffs Since 1970,* p. 1.

25. Thomas Powers, *Intelligence Wars: American Secret History from Hitler to al-Qaeda* (New York: New York Review of Books, 2002), pp. 391–398.

26. Standing Committee on Law and National Security, *Oversight and Accountability,* p. 19.

27. David Silverberg, "Clinton and the Military: Can the Gap Be Bridged?" *Armed Forces Journal International* 129, no. 3 (October 1993), pp. 53–54, 57.

28. Erwin C. Hargraves and Roy Hoopes, *The Presidency: A Question of Power* (Boston: Little, Brown, 1975), p. 47.

29. James Q. Wilson, *American Government: Institutions and Policies,* 5th ed. (Lexington, Mass.: Heath, 1992), p. 338.

12

Empowering the People

Popular control of government is a fundamental principle of democracy. The channels to establish, nurture, and expand presidential links to the people, and from the people to the president, are the media, political parties, and interest groups. These are usually considered linkage institutions, linking people to government and the president. The media transmit images, information, opinions, and the attitudes of the public to the president and vice versa. Media professionals also advance their own agendas, even becoming active players in the policy process.

Political parties attempt to mobilize the people to win office and are the main instruments for organizing Congress and controlling the legislative agenda and process. Interest groups reflect and shape the attitudes, opinions, and policy positions of important segments of the public. Interest groups are a means for individuals to have a voice and a channel for expressing their preferences.

As a general rule, the public holds broad views on policy, rather than informed opinions oriented toward specific issues. Some scholars are quick to note that public views are usually inchoate and oversimplified. But public views can be transformed into specific policy preferences as a result of interest groups and political parties. Interest groups (i.e., every type of organization that tries to achieve its goals by affecting policy choices and policymakers) can be transformed into single-issue groups, mobilizing segments of the populace. In the course of this mobilization and policy advocacy, interest groups provide a means to pressure policymakers and elected officials.

Public attitudes and the degree of public support are factors in influencing the relative success of a president's policies. This is the case not only in domestic areas but also with respect to national security policy. Public support over the long run is necessary to the success of national security policy and strategy. This is emphatically the case when the president seeks to place US armed forces in harm's way or when the public feels directly threatened, as on September 11, 2001, or during the pandemic of 2020–2022. Additionally, how these relationships evolve into support for or opposition to the president and national security policy is an important factor in the study of the policy process.

In this chapter, we place into context the role of linkage institutions and processes in making national security policy. First, we consider the public and its relationship to national security and defense policy formulation. Second, we discuss the enormous importance of the media and other means of communication on public attitudes and policymakers. Third, we review some important fundamentals about the role of political parties as linkage institutions; some feel that they are in decline relative to other components of the policymaking process. Fourth, we examine the power and influence of interest groups in the US political system generally, and national security in particular.

The Public

The ability of the president to deal with Congress and develop support for his policies is contingent in large measure upon perceived and actual popular support. Much of this stems from the mandate the president receives upon election. For example, the 1980 Ronald Reagan landslide became the basis of the so-called Reagan Revolution. Reagan's decisive reelection in 1984 might have provided the basis for continuing and broadening the revolution, but in its second term the administration struggled to maintain momentum. In part, this was owing to a cabinet reshuffling that placed key White House advisor James Baker in the Treasury Department, breaking up the team that had so effectively advanced Reagan's agenda during his first term. The loss of the Senate to the Democratic Party in the 1986 elections also created difficulties for the Reagan program. But the president retained a high approval rating, even though many Americans did not fully support some of his policies.

Bill Clinton won the presidency in 1992 without a majority of the popular vote, which meant that the president and his administration had to work hard to build credibility in foreign and security policy.

This was reflected in several polls in the fall of 1993 in which less than a majority—and at times a bare majority—approved of his presidential performance. Public skepticism about Clinton as commander in chief was a result of ambiguous or vacillating policies in Somalia, Haiti, and Bosnia-Herzegovina ascribed to the president's lack of military experience and possible distrust of the military profession. The controversy about openly gay service members added to the impression that he was tone-deaf on matters affecting the armed services, and his "don't ask, don't tell" compromise pleased few.

George W. Bush was elected to the presidency in 2000 with an uncertain mandate, having lost the popular vote and won a disputed outcome in the electoral college that had to be resolved by the US Supreme Court. Until the terrorist attacks of September 2001, Bush appeared uncertain of his footing in national security and other policy issues. After September 11, however, Bush found his voice as a spokesman for national unity and resolve in the face of the attacks on the World Trade Center and the Pentagon. The Bush reaction to September 11 created a favorable impression of his leadership and ability to protect the nation from further attacks. His image as a guardian of US security held up during his reelection campaign of 2004, when he ran against Navy Vietnam veteran and then–Democratic Massachusetts senator John Kerry. Only in 2005, following a slow and ineffective administration response to Hurricane Katrina and a stalemated war in Iraq, did Bush's polling numbers on national security begin to drop significantly. This continued in 2006 as many in the US public disapproved of the continuing US involvement in Iraq and the Democratic Party won control of both houses of Congress following the 2006 congressional elections.

President Barack Obama won the election in 2008 as a result of his charismatic personal appeal, a superb campaign organization, and widespread feelings of "Bush fatigue" that followed the nosedive of the US and other major economies after a burst housing bubble and a related banking and investment debacle. Obama promised "change you can believe in" and a shakeup of politics as usual in Washington, but traditions of partisan gridlock and Congressional inertia quickly reasserted themselves. Obama made health-care reform his major domestic priority during his first year in office, obtaining from Congress a compromise bill that satisfied neither conservatives nor liberals. Obama also called for the worldwide abolition of nuclear weapons and for a "reset" in US-Russian relations toward the accomplishment of nuclear arms reductions and nonproliferation objectives. In the war against terrorism and in the military campaigns in Iraq and Afghanistan, Obama also compromised between the expectations of his most liberal supporters and the preferences of US military commanders, intelligence professionals, and cabinet advisors in defense and foreign policy. For example, Obama increased the US troop commitment to Afghanistan until 2012, upped the ante on global strikes against terrorists and Taliban insurgents with drone aircraft, and backed away from campaign promises to close the detention center at Guantanamo and try high-value terrorist captives in US civilian courts.

President Donald J. Trump was elected in 2016 due to a wave of popular discontent with the US political establishment, a lackluster campaign by his Democratic opponent Hillary Clinton, and Trump's highly energetic campaign style that attracted voters otherwise bored or disinterested in national politics. Trump's first year in office was characterized by chaos in the appointment process for cabinet officers and other key aides, in setting domestic and foreign policy priorities, in relations with Congress, and in establishing a coherent public "face" in the White House press office. By the time of his campaign for reelection in 2020, President Trump had retreated from some of his more extreme early foreign policy positions and became somewhat less impulsive in managing affairs with traditional American military allies (e.g., the North Atlantic Treaty Organization [NATO]). The president also determined that he could not immediately depart Afghanistan without a meaningful peace agreement, not only between the US and the Taliban, but also between the Taliban and the Afghan government. Trump also demonstrated continuity with respect to his China policy, insisting on trade deals that were more favorable to the United States, pushing back against Chinese efforts to steal American intellectual property and to commit espionage on US soil, and resisting Chinese military assertiveness in the South China sea and elsewhere in the Pacific. Trump also took the initiative to establish a US space force and Trump administration defense guidance was the first to officially acknowledge that space was now a potential domain for war fighting. On the other hand, Trump's summit diplomacy with North Korea led to a dead end, his immigration policies continued to be controversial in US domestic politics, and his management of the coronavirus epidemic in 2020 was widely criticized.[1] The Biden administration marked a return to a more traditional approach to foreign and defense policies, although his decision to remove troops from Afghanistan was consistent with the policy of the Trump administration.

The relationship between the people and the president, which is complicated by the media's role in transmitting information and setting the public agenda, is thus more intricate than many might suppose. As the nation's leading political figure, the president is expected to develop and implement policies that are binding on the entire populace. People respond favorably or unfavorably to a president's personality and political style and to the events that occur while he or she is in office. They also assess the president by the way he or she relates to particular groups, as well as social (religious, ethnic, racial), economic (business, labor), and geographical divisions of the population.[2]

Although most authorities agree that public support is an essential part of presidential effectiveness, they disagree as to the specific nature of this support and its relationship to congressional support. How support is translated into presidential effectiveness and performance is subject to dispute. On the one hand, the public may give a president high approval ratings, yet have less than majority support for specific policies. The inconsistency is more apparent than real, since the public's perception of the president's reputation for success and integrity colors its judgments about individual decisions.

In addition, over time the support for the president and specific policies is affected by shifts in the public mood. Attitudes can change in the face of major policy issues or failures and because of a perception that the president has failed to live up to his or her promises or seems incapable of leading the nation. At times, the public may seem to act on a whim, shifting attitudes or suddenly dropping support for a president for no clear reason. Because of these intricate relationships, the president needs to be sensitive to the people and be aware of the fragile nature of public opinion.

This complex arena requires that the president respect the limits of public acceptance for national security policy and strategy. In times of peace and prosperity, it is difficult to energize the public and gain support for any policy and strategy that departs from the mainstream. In times of crisis and perceived national peril, the president can undertake broader initiatives for purposes of national security, defense, and intelligence. Following September 11, President George W. Bush ordered a review of US counterterrorism policies. Some of the proposals, such as arming civilian airline pilots and federalizing airport security, would have been unthinkable prior to the attacks. Congress authorized a broad reorganization of the US intelligence community after September 11 that tore down much of the legal and political "wall" between domestic and foreign intelligence (and between intelligence and law enforcement within the Federal Bureau of Investigation [FBI] itself) and created a new superordinate position, director of national intelligence (DNI), atop the entire intelligence community of agencies, including the Central Intelligence Agency (CIA).[3] Whether this move improved national intelligence collection and analysis or simply added another layer of bureaucracy to the existing system remains a matter of dispute.

Presidents must be especially wary of public sensitivity to secret and covert operations. Many Americans can understand the need for and accept them (albeit reluctantly), but there are limits. Moral and ethical boundaries exist for any operation. And even though the president has flexibility in national security policy and strategy, Americans expect an accounting of his decisions and actions, whether they are successful or not.

The Limits of Public Opinion

National security policy cannot be conducted solely according to public opinion. Public attitudes and opinions, except in crisis, are not geared toward specific national security issues. Moreover, there is some degree of secrecy involved in several contingencies, a double-edged sword: going public might justify the policy and strategy in the eyes of the people, but it would also telegraph US policy and strategy to adversaries, which could lead to failure as well as place US forces in danger.

The president cannot neglect public opinion on national security issues, however. Reaction after the event is an important component of the national security equation, part of the final accounting the public expects. The history of US public opinion in foreign policy demonstrates that presidents have the advantage of initiative and command of detailed and timely information. But the public, the Congress, and the media may define success or failure according to a different standard. The president and the government are sometimes preoccupied with the nuances and subtleties of small policy differences; the Congress, the media, and the public react more sharply to the impact of policy decisions.

Presidential Challenges

The president and the national security establishment must be especially sensitive to the fact that interest groups and the media, as well as adversaries and allies, play an important role in influencing the US public. For example, in 1987 the Sandinista regime in Nicaragua made serious efforts to develop political networks in the United States to convince the public of the legitimacy of the Nicaraguan system. Nicaragua hired a US public relations firm to paint the best possible picture and to lobby

Congress for favorable legislation. This is not necessarily an unusual or illegal tactic, as several foreign states hire local firms. During the Gulf War of 1991, the government of Kuwait hired a high-profile Washington, DC, public relations firm to coordinate its public opinion campaign against Iraq in the United States. But this strategy is not always available to the United States: a US president or presidential advisors cannot gain similar access to media in a foreign country with a closed political system in order to publicize US views to the foreign peoples.

In the final analysis, the public usually follows the president's lead in national security policy and overwhelmingly supports any president who takes bold and responsible action, especially during a crisis. President Reagan's orders for the raid on Libya in 1986 and the earlier invasion of Grenada were favorably received by a majority of the public. This was also the case with President George H. W. Bush's initiative in the Gulf War in 1990–1991. With the conclusion of that war, Bush's approval ratings were above 80 percent, and for a time he was considered unbeatable in the 1992 presidential race—which he would eventually lose to Bill Clinton. A decade later, President George W. Bush's approval ratings also reached 80 percent following the September 11 attacks, only to begin an accelerating slide. President Donald J. Trump's approval ratings for much of his time in office were deviations from the norm: his favorability ratings hovered around the 40 percent mark most of the time, regardless of his decisions and despite most media coverage of them. Although this base of support appeared to remain solid, Trump was less successful in expanding this base to improve favorability ratings among some Democrats and independents.

But national security policy and strategy problems, such as the debacle at the Bay of Pigs in 1961, the failed attempt to rescue US hostages in Iran in 1979, the Iran-contra affair in the mid-1980s, and the chaotic end of the war in Afghanistan, can lead to domestic political problems. John F. Kennedy recovered from his naïveté during the Bay of Pigs fiasco and managed the Cuban missile crisis in 1962 with greater awareness of the fallibility of national security decisionmaking and the tendency toward groupthink among some agencies. Jimmy Carter never recovered politically from the failed Iranian hostage rescue, and Ronald Reagan survived politically, but the Iran-contra scandal nevertheless tarnished his legacy. After the US military exit from Afghanistan in August, 2021, there was a reassessment on the twenty year US effort there and whether it was worth the human and fiscal cost. Many felt the effort was doomed from the start, others felt it should have ended sooner. Historians will revisit the roles of previous presidents in starting and continuing the war there, but the nature of the US withdrawal resulted in severe criticism of the Biden administration.[4] President Biden's so-far firm but measured response to the Russian invasion of Ukraine may help his public support, although as of this writing the final outcome is uncertain.

Mistaken decisions in national security and defense do not necessarily lead to the demise of an administration. Much depends on the president's previously established rapport with the public and the popular image of the president as trustworthy or slippery. For example, former president Dwight Eisenhower's ability to symbolize military excellence and personal rectitude carried him through the U-2 crisis in his second term, despite the embarrassment of the spy plane shootdown over Soviet territory and several official US misstatements about it. Kennedy's cinematic personal appeal helped to buoy him over the Bay of Pigs disaster, as did his disarming candor in taking personal responsibility for the decision (even as he planned revenge against the CIA leaders who he felt had misled him). On the other hand, in the wake of the failed Iran hostage rescue in 1980 and domestic economic issues, the Carter administration was voted out of office the following November. Unlike Kennedy or Reagan, Carter lacked the personal charisma that presidents sometimes need to get over the inevitable bumps in security (or other) policymaking. In this respect Carter resembled his Democratic predecessor, Woodrow Wilson. When Wilson felt he was right, he was righteously right, to the point of obstinacy—witness his failure to win congressional approval for US membership in the post–World War I League of Nations as part of the Versailles Treaty. In contrast, Democratic president Franklin Roosevelt was adept at playing "the lion and the fox" as he charmed his political supporters, opponents, and the public while winning four presidential elections during the Great Depression and World War II. Retired US Supreme Court justice Oliver Wendell Holmes, who had known Franklin Roosevelt for years, appraised the newly elected president in 1933 after Roosevelt paid him a courtesy call: "A second-class intellect. But a first-class temperament!"[5]

The George W. Bush presidency was impaired when the 2003 war in Iraq turned into a protracted and frustrating insurgency. When the public begins to lose confidence in a president's ability to respond effectively, or when it perceives that the president cannot respond positively to failure, credibility erodes, as does confidence in the president and US national security policy. This public attitude impacts the national security establishment and congressional attitudes.

In contrast to Clinton's indecisiveness in Bosnia during 1993 and 1994, George H. W. Bush's administration improved its performance by brokering the Dayton Peace Accords of December 1995, bringing an end to an exhausting civil war and ethnic cleansing in the Balkans. With NATO partners, the United States established the Implementation Force (IFOR), a large and highly capable military peacekeeping and peace enforcement operation initially involving some 60,000 NATO troops. IFOR was succeeded a year later by the Stabilization Force (SFOR). Although scaled down from its original size, SFOR remained in place at the dawn of the twenty-first century. The Clinton administration also claimed victory in NATO's US-led bombing campaign against Serbia in 1999, undertaken to end the ethnic cleansing of Kosovar Albanians by Serbs. Even though the air campaign was a one-sided affair, its political effects were uncertain and the US public was ambivalent.

The Media and National Security

An essential ingredient of any open system is the role of the media. Therefore, a free press is a fundamental principle of the US system. First Amendment freedoms often have priority, and individual freedom of speech and freedom of the press can have national security implications.

This gives the media a degree of power in an open system not enjoyed by other groups and institutions. This is not a new development. In the nineteenth century, the French aristocrat and political essayist Alexis de Tocqueville observed that even with some limitations "the power of the American press is still immense. . . . When many organs of the press do come to take the same line, their influence in the long run is almost irresistible, and public opinion continually struck in the same spot, ends by giving way under the blows."[6] A modern version of this view is described by one scholar as "pack journalism."[7]

There is an inherent dilemma with respect to the role of the media and national security policy. On the one hand, some security policy must be formulated and implemented in secret. On the other hand, the media's mission is a direct challenge to that secrecy. In addition, media technology has changed so drastically that it has become an unwieldy global network of electronic, print, and visual sources of information of varying accuracy and integrity. The rise of the Internet and social media such as Facebook and Twitter contribute to an even more complicated information sphere that challenges politicians on national security and other matters. For example, the unauthorized disclosure of sensitive US diplomatic and military activities, including after-action reports from the field, by an enlisted US soldier to the investigative organization WikiLeaks resulted in a global dump of information creating embarrassment for US policymakers and US allies. The demise of "mainstream media" in favor of niche outlets presenting extreme points of view has contributed to the polarization of US attitudes and the difficulty of reaching agreement even on basic facts. The most recent example of this is the continuing disagreement over the nature of the January 6, 2021, assault on the US Capitol, who and what might have contributed to it, and whether it represented one symptom of a larger disease afflicting the body politic. Quite aside from the regular media channels, the vast network of social media, such as Facebook, Instagram, and Twitter, can have a profound effect on popular opinion.

The quandaries posed to any administration are numerous. Some administrations have engaged in deception to avoid premature publicity of security strategy. In dealing with the Cuban missile crisis in 1962, for example, the Kennedy administration deceived the media and the public, at least initially, although most eventually accepted the need for secrecy in this case. The Iran-contra hearings revealed the half-truths and deceptions of administration officials in dealing with Iran and the Nicaraguan Contras.[8] Other administrations have openly tried to prevent publication of sensitive information. The Pentagon Papers, stolen government documents that revealed aspects of the Vietnam War, were published widely by the media, even though much of the content was classified.[9] When the government brought a lawsuit, the court ruled in favor of the media on the grounds that the First Amendment prohibited prior restraint of publications by government.

The media perform important functions which affect their role with respect to national security matters and the presidency. The media inform the public about what is going on in government, the country, and the world; they transmit information from political leaders both in and outside government to the public and political actors. The media are also used by political leaders and government officials to signal policy intentions and test reactions. In this sense, the media play a quasi-official role, knowingly or otherwise, and provide a channel to signal foreign adversaries and allies of government policy. For example, once the Cuban missile crisis became public knowledge in October 1962, President Kennedy and Soviet premier Nikita Khrushchev used the media as a means of negotiation: trial balloons were floated in the press, and at least one prominent US media personality was used as a go-between during a sensitive time in diplomatic negotiations.

The media business is highly competitive. Whichever reporter or corporate news structure breaks the news first has an important advantage. Too often television news anchors are rated on physical appearance and their impact on the viewing public, which appears to be more a function of symbols and gimmicks than of substantive news reporting skills. This tends to place television news reporting in the area of entertainment rather than information. Understandably, television ratings are key indicators of commercial success, which influences the way the news is selected and presented. It follows that news events and images of political leaders are shaped by a variety of factors that have more to do with sales than with content.

The role and function of the media in an open system are further complicated by the emergence of investigative reporting and adversarial journalism. Investigative reporting is the aggressive pursuit of news; it is an intense uncovering of facts and is associated with a presumption of wrongdoing. Watergate blasted into the public consciousness thanks to two reporters from the *Washington Post* who uncovered the political connection of the break in at the Democratic Party offices in Washington, DC—it was directed by high officials in the White House—ultimately implicating President Richard Nixon in its cover-up.[10] The unnamed source for many facts in the story, "Deep Throat," was revealed only decades later during the administration of President George W. Bush.

Adversarial journalism assumes that the best approach is to view government officials, individuals, and groups under investigation as the enemy. In one sense, the targets are presumed guilty until proven innocent. Presidential press conferences reflect one aspect of adversarial journalism: it is not uncommon for a reporter to ask the president questions beginning with a statement that presumes a blunder or lack of sincerity. Seeing themselves as adversaries, some reporters focus the spotlight on the president as well as on themselves. A similar dynamic appears in editorials, in the way the news is presented, and in the images of political leaders.

Investigative journalism and adversarial journalism serve a purpose in challenging government officials and policies, though. Indeed, according to some observers, the media emerge as the only visible counter to the government's national security policy and strategy—the only check on government excesses. Yet journalistic excesses occur as well owing to professional and commercial competition. These excesses can cause a well-conceived policy to fail and even endanger the lives of US officials and agents operating in foreign countries. As noted earlier, recent decades have also witnessed the rise of "niche" networks with political spin predominantly in one direction or another, commingling news, analysis, commentary, and opinion in a single package.

In addition, when investigative and adversarial journalism is combined with political agendas, there is the potential for distortion of the political process by interest groups promoting partisan interests. The seemingly unending debate about the personal conduct of President Clinton and his possible impeachment was distorted by many people working in or with partisan media, talk radio, and the Internet. As the president's domestic situation was perceived to have weakened, his ability to conduct foreign and defense policy was impaired. President George W. Bush's administration and supportive television networks complained regularly about a perceived liberal bias among the major news networks and their reporters. A poorly documented investigative report by CBS about President George W. Bush's military service led eventually to the early retirement of its stellar news anchor, Dan Rather. President Trump and his staff complained regularly about unfair coverage by the "mainstream media" and the president publicly and repeatedly referred to CNN as "fake news CNN." President Biden and his supporters likely feel the same way about Fox News.

Investigative journalism is a necessary part of media coverage. The government should never be trusted to reveal its own shortcomings, especially corruption and illegality. Yet investigative journalism can become a feeding frenzy, and the willingness to rely on unattributed or dubious sources, combined with a subtle antimilitary agenda, can lead to biased reports. Even history can be reenacted within the framework of a morality play critical of the US armed forces. The combination of the Internet and cable news creates a pressure for immediacy of news and against careful fact checking of the kind that media professionals ought to do.

A complicating factor is the alleged liberal political leaning of media elites. Although it is debatable, there is evidence to suggest the existence of "a media elite with a particular political and social predisposition that places it distinctly left of center of the American political spectrum."[11] One of the most authoritative studies of the media elite concluded, "Today's leading journalists are politically liberal and alienated from traditional norms and institutions. Most place themselves to the left of center and regularly vote the Democratic ticket."[12] On the other hand, comparable studies of the political views of those who own newspapers, television stations, and networks would be needed in order to evaluate the significance of working journalists' proclivities. British investigations of the Rupert Murdoch news empire in 2011 revealed some startling relationships between journalists and prominent United Kingdom public officials, including politicians and police.

Some journalists undoubtedly tend to see a world that is "peopled by brutal soldiers, corrupt businessmen, and struggling underdogs."[13] From this it is not unreasonable to conclude that such perceptions seep into news reporting, editorials, and the way political leaders are projected to the public. Of particular concern is that these perceptions can set the public agenda. Others have expressed concern that the media's most important potential bias is neither liberal nor conservative: instead, it lies in the media's control over the agenda of public policy debate:

> To control what people will see and hear means to control the public's view of political reality. By covering certain news events, by simply giving them space, the media signals the importance of these events to the citizenry. By not reporting other activities, the media hides portions of reality from everyone but the few people directly affected. . . . Events and problems placed on the national agenda by the media excite public interest and become objects of government action.[14]

Finally, some have questioned whether the allegedly liberal bias of reporters and commentators is offset by the conservative perspective of those who own media as opposed to those who work there. Wealthy entrepreneurs such as Rupert Murdoch and large media conglomerates are interested in maximizing ratings and audience sizes. This requires appealing to the centrist or conservative-centrist majority of Americans who channel surf or engage in Internet hopping for video and sound bites. Assessment of this issue is made more complicated by the alleged liberal bias of those who own and manage major American tech companies: including Google, Apple, and Twitter. Congressional hearings into the role of these tech giants in 2020 revealed diverse views as to whether these companies should be treated as "platforms" with no responsibility for regulating content or as "publishers" accountable for content selection on their platforms. The argument whether information technology companies were platforms or publishers reflected their enormous influence from business models that impacted directly on the "public square" and the behaviors of customers, including the companies' unprecedented access to, and use of, customers' personal data.

Reporting and a Political Agenda

The role of the media in a democratic society, the functions they perform, and their agenda-setting establish one set of important considerations. The political predispositions of the media elite establish another set. Combining the two creates a powerful profession, one that is able to set its own agenda and shape the image of reality according to its own views. Fortunately, there are some media professionals who place fairness and objectivity above personal or political agendas. In addition, the public can access a variety of news sources and has the opportunity to compare news reporting. For the concerned individual, analysis and comparison of news sources and the substance of news reporting can reveal misjudgments, errors, and political bias in reporters and editorial staffs. Of course, this requires that individuals are willing to make the effort to sort truth from falsehood.

A classic example of judgmental reporting was that of the Tet Offensive during the Vietnam War. In 1968, during the Buddhist New Year, Vietcong and North Vietnamese forces launched an offensive across South Vietnam in the hope of triggering a people's uprising. The US and South Vietnamese military reacted, with the enemy suffering a severe military defeat. Yet the media reported it as a military disaster for the United States. This version was accepted as true by many groups and was publicized by antiwar groups and others to such an extent that it became the common view. This perception played a role in prompting President Lyndon Johnson not to stand for reelection in 1968. As Peter Braestrup concluded: "The general effect of the news media's commentary coverage of Tet in February–March 1968 was a distortion of reality—through sins of omission and commission—on a scale that helped spur major repercussions in US domestic politics, if not in foreign policy."[15] There are other examples of distortions and predispositions coloring the news, but the coverage of Tet stands out for its massive impact on US politics and national security policy.[16]

In the 1991 Gulf War, the media, especially cable television, played an important role in informing an international audience and in shaping the images that impacted the political agenda and public perceptions of the war. "Daily, live coverage briefings from the headquarters in the Gulf and from the Pentagon via television and radio, reports from the 1,500 and then echoes—and there were lots of echoes from columnists, correspondents, consultants and assorted pundits in the United States and abroad—all served to keep the American public informed."[17]

Some critics contended that the US military's controlled press environment in the Kuwaiti theater of operations slanted coverage favorably toward administration policy. Yet there was serious criticism of television reporting from Baghdad, the enemy capital, by a CNN reporter who enjoyed exclusive access.[18] Many argued that his one-sided reporting undermined coalition efforts. Global television news also affected US and international perceptions, much of it based on television-driven strategy and policy. US forces exploited this for the purpose of perceptions management, or influence operations, by staging amphibious exercises off the Saudi Arabian coast in full view of CNN cameras.[19] Iraq's leadership was thus tricked into believing that an amphibious assault was imminent along the Kuwaiti border, and several Iraqi divisions were tied down awaiting an attack that never materialized.[20]

To say that the president must establish good working relations with the media is an understatement. But the president cannot control all of the news associated with national security policy and strategy. Neither can he or she control what members of the national security establishment say to the media. Confidential sources and leaks provide the media ample opportunity to gain access to classified material. Furthermore, partisan members of Congress can easily leak information to the media to thwart administration policy. To complicate matters, policy and strategy extend to the international arena, where a variety of sources and events can trigger exposure of US intentions and actions.

At times the president tends to court the media; although unseemly to traditionalists, media savvy is required for holding public office. Even before the advent of television, President Franklin Roosevelt masterfully exploited the media by means of carefully orchestrated fireside chats reaching nationwide radio audiences. Like it or not, the president must assuredly give the media their due, given the important role they play in an open system. It is best for the president to be confident in his national security posture, attuned to the dynamics of the national security establishment, and sensitive to the support of the public and Congress. These factors are not lost on the media. Yet it is also important that the president recognize the media's responsibility as a friendly adversary, remaining skeptical of government claims and actions until shown otherwise. President Trump was a contradiction in this respect. He showed incredible intuitive awareness of how to use the media to project his own image and views. On the other hand, his overreaction to media criticism gave it greater publicity than it otherwise would have received. President Biden's administration has been more savvy in handling the media, perhaps based on their previous government experience.

For their part, the media must be aware of the responsibility to provide objective and fair reporting as well as the risks of reporting classified information and prematurely disclosing policies and strategies that might jeopardize US national security interests. There is no clear line between the people's right to know and US national security interests. The media must therefore police themselves. Legal and political battles over the line between First Amendment freedoms and national security will surely continue.

One media responsibility is not to be a conduit for manipulation of public opinion by foreign adversaries, including state and nonstate actors such as terrorists. For similar reasons, they must not become the mouthpiece for domestic cranks and ideologues. Examples of manipulation of public opinion by foreign powers include atrocity accusations, demonizing the opponent, claiming divine sanction for one's global agenda, and hyperbole that inflates the stakes involved in a conflict ("the war to end all wars").[21] Examples of the unfiltered media transmission include accusations that the US Navy shot down TWA flight 800 in 1996 and the even broader coverage given to reports that the CIA sold crack cocaine to inner-city neighborhoods in California to raise money for the anti-Sandinista rebels in Nicaragua.[22]

After September 11, experts debated whether US television networks ought to publicize video- or audiotapes distributed by terrorists, especially by Osama bin Laden and other leaders of al-Qaeda. Media executives and reporters felt they were informing the public; critics feared that they were providing a global village for the distribution of Islamic extremism. This was one aspect of the information warfare contest in which the United States was engaged with state and nonstate actors who practiced what some theorists termed "Fourth Generation" war. As retired US Marine Colonel Thomas X. Hammes, explained:

> Fourth Generation war uses all available networks—political, economic, social and military—to convince the enemy's political decision makers that their strategic goals are either unachievable or too costly for the perceived benefit. It is an evolved form of insurgency. . . . Unlike previous generations of warfare, it does not attempt to win by defeating the enemy's military forces. Instead, via the networks, it directly attacks the minds of the enemy decision makers to destroy the enemy's political will.[23]

Hammes and proponents of the concept of Fourth Generation warfare emphasize that these new kinds of protracted wars or hybrid conflicts mixing insurgency with terrorism and possibly conventional war may go on a long time. Sometimes they are never really "won" but merely prolonged, and eventually one side outlasts the other, or the issues become moot. Examples of seemingly perpetual conflicts that nevertheless ended in the latter twentieth or early twenty-first century included the "troubles" in Northern Ireland and the civil war in Sri Lanka against Tamil Tigers. On the other hand, the off-and-on skirmishing between Israel and various Palestinian groups, including Hamas and al-Fatah, and between Israel and Hezbollah in Lebanon, continue without imminent prospect of political resolution. The United States was clearly outlasted by the North Vietnamese in the Vietnam War and by the Taliban in Afghanistan.

The Presidential Role

The president's leadership and personality shape the environment in which the media function, at least in terms of the national security establishment. Trust and confidence in the president and the perception of direction and initiative in US national security policy are key ingredients in shaping this environment. Even the best environment does not preclude political disasters and failures, but it does provide the president an opportunity to respond. The best presidents accept responsibility for failure while maintaining public confidence and trust—not an easy challenge but one that has been successfully met. The media's important role in this environment was recognized by Alexis de Tocqueville more than 150 years ago: "I admit that I do not feel toward the freedom of the press that complete and instantaneous love which one accords to things by nature supremely good. I love it more from considering the evils it prevents than on account of the good it does."[24]

All of this became more complex as cable television developed into a major information source, especially internationally, and the Internet and World Wide Web made "connectedness" across the boundaries of time, space, and territory a household word. A major characteristic of the strategic landscape is the role of global television.[25] Many conflicts and international problems become highly publicized, virtually dominating national security and foreign policy agendas. For example, the starvation and conflict in Somalia in 1993 became an international crisis thanks to the tragic images broadcast on cable television. Some in the United States bemoan the fact that national security and foreign policy have become video-driven in this way. (This used to be called the CNN effect, perhaps now more appropriately called the Internet-Web, Facebook, Instagram, and Twitter effect.)

Another phenomenon is the growing popularity of talk radio. Talk radio has expanded to include discussions on any variety of domestic and international issues. It provides an information source and allows listeners to express personal views over the airwaves. In 1993 and 1994, talk radio became a focus for attacks on the Clinton administration. Liberals complain that talk radio is overwhelmingly conservative in its political slant; conservatives respond that talk radio reflects the grassroots feelings of Americans whose views are not represented by the mainstream press and television media or Hollywood. In fact, there is no political view so extreme that it cannot find a voice on talk radio, cable news channels, or the Internet.

A global communications network is in place and continues to expand thanks to the Internet, fax machines, portable phones, computers, copy machines, and satellite television. In this information age, governments must rapidly respond to events as international audiences are exposed to issues almost instantaneously. With its variety of information channels and news sources, the information age has reaffirmed the case that the media shape the political agenda.[26] If policymakers are not careful, the media will drive, instead of merely influence, the agenda for security decisions. Involvement in major conflicts can be driven by media conglomerates thirsting for news.

It is also the case that "the public's evaluation of the incumbent president rises and falls in accordance with cues provided by the media. The more prominent the coverage accorded critics of the president, the lower the level of presidential popularity will be."[27] To reduce the amount of criticism, the administration can limit information provided to the media and make determined efforts to manage the news or at least shift its focus. Spin-doctoring is a proven White House strategy. In the Clinton administration there were many attempts to mute the criticism over Somalia and Haiti: One was by distraction, focusing on television events such as Hillary Clinton's visit to Chicago for health-care reform, conducting a trade fair on the lawn of the White House to drum up support for the North American Free Trade Agreement (NAFTA), and publicizing the need for free mammograms for women.

The George W. Bush administration during its first term succeeded in creating a media narrative that drove its Democratic opponents into defensive and reactive angst. This narrative positioned Bush as the chief warrior against terrorist and other threats to US security. Opponents of the president's way of waging this war on terror were depicted as lacking in patriotism or ignorant of the facts. The Bush White House also succeeded in commingling the war on terror with the invasion of Iraq to depose Saddam Hussein in 2003. Many Americans believed erroneously that Iraq had somehow been involved in the September 11 attacks. Others were convinced that Saddam Hussein would soon acquire a nuclear weapon and pass it to terrorists planning to attack the United States. The Bush military campaign against Iraq (Operation Iraqi Freedom) was accompanied by a public relations offensive that temporarily isolated Bush's critics and war skeptics as having marginal or fringe opinions. Later, however, as public support for continued occupation of Iraq plummeted in 2005–2006, more survey respondents separated their feelings about September 11 from their sentiments toward Iraq—to Bush's detriment.

The Obama administration in its public diplomacy was less about the manipulation of images and more concerned with the management of issues. President Obama was almost professorial in his interest in the details of policy and in his insistence on reaching consensus across diverse viewpoints within his national security bureaucracy. Some critics regarded this approach as dilatory and inconclusive. After some hesitation about policy toward Afghanistan, Obama announced a "surge" of about 30,000 additional US troops in support of NATO's counterinsurgency and stability operations in Afghanistan, with the expectation that improved political and military outcomes would permit the withdrawal of at least some US ground forces in 2012. Obama also presided over important changes in his national security team, including the replacement of General James Jones by Thomas Donilon as national security advisor. Among operational commanders, Obama appointed and then replaced General Stanley McChrystal as top US commander in Afghanistan (following a controversial story about McChrystal's staff's disrespect for the president in *Rolling Stone* magazine). General David Petraeus, then combatant commander in charge of US Central Command (CENTCOM) and former US commanding general in Iraq, was named to succeed General McChrystal in Afghanistan. In 2011, Obama announced that General Petraeus would replace CIA director Leon Panetta when Panetta changed hats to become secretary of defense in July, replacing Robert Gates. Most important for Obama's image as the president in charge of national security was the successful raid by US Navy SEAL Team Six into Pakistan in May 2011 that killed Osama bin Laden and captured many important documents without losing any US military personnel, all while concealing the raid and its timing from Pakistani officials who might otherwise have leaked it.

Public reaction to news good and bad reminds us that "the president is *the* big story in Washington."[28] By his presence and attention to certain events, he can (re)focus the media and the public. The president and the media have a competitive and cooperative relationship in national agenda-setting. The president requires media attention to explain and defend administration policy yet must remain wary of the media's ability to place that same policy under the microscope. The media need the president, a unique focal point for US national politics and a good source for boosting ratings. But the media must always remain alert to the possibility of White House spin-doctoring. This love/hate relationship was especially apparent during the second terms of Bill Clinton and George W. Bush.

Political Parties

The role of political parties needs little review here, but we will summarize the basic essentials. Parties are a vital connection between elected officials and the public; in theory they take positions on important issues of the day. Thus parties, more than politicians, can be held accountable for the passage of or failure to pass programs consistent with their views. Parties also relate to interest groups but perform a different function. Traditionally, US political parties, at least those that aspire to majority party status, have attempted to be large tents that accommodate a diversity of many intraparty factions and viewpoints. When the party's agenda is captured by one faction or wing, as with liberal Democrats in the 1970s, conservative Republicans in the 1990s, and by Donald Trump after 2016, presidential hopes dim. Parties seeking to win Congress and the White House must promote issues that appeal to the middle of the electorate. If either political party is captured by an intransigent and uncompromising group, it may be useful in the short run, but in the long run will be detrimental to the party and to democracy.

Some points need to be emphasized with respect to national security and parties. Often it is difficult to distinguish the role that political parties play in the national security policy process and in Congress because of the impact of interest groups. But the majority party in Congress has the power to control the legislative agenda and to select the leadership positions. At the same time, the party out of power is supposed to provide the "loyal opposition" essential to the nation's two-party system. It follows that a democratic system presumes that there is a viable two-party system at the national level and the periodic retransfer of power from one party to the other. This applies particularly to the president, who is expected to cooperate in a prompt and peaceful transfer of power to his or her successor. There were serious national security implications when that did not happen after the 2020 election, let alone the danger to democracy itself.

Thus in terms of national security, political parties are expected to offer viable alternatives when out of power and develop political resolve and strategic visions when in power. During the Cold War, it was generally presumed that partisan politics stopped at the water's edge—that both parties, regardless of their position, would support national security goals and, if necessary, mobilize party and public support for the president. Nonpartisanship did not always hold, of course, as partisan wrangling over national security issues has characterized every administration since Lyndon Johnson's.

In the post–Cold War period, however, with its multiplicity of challenges and uncertain "distributed" threats, the Republican and Democratic parties are challenged to fix on a particular national security policy or clearly articulate a strategic vision.[29] The parties do serve as instruments to debate and mobilize party members in focusing on the shape of the military, the defense budget, and responses to immediate national security issues. During President Clinton's two terms in office, immediate issues included humanitarian crises and civil wars in Somalia and Bosnia, the NATO air war against the former Yugoslavia, and relations with Russia, as well as the reduction of US military forces. President George W. Bush's agenda of security issues was driven by the attacks of September 11 and the wars in Afghanistan and Iraq. But most of the time, absent clear cases of threats to vital national interests, political parties tend to be driven by domestic issues.

Occasionally a candidate rises who takes on the party "establishment" or the Washington, DC, "swamp," but the path to the White House is difficult without support from one of the major parties. In 1992 Ross Perot emerged to energize voters disaffected by the major parties. With many voters alienated by the Republican and Democratic standard-bearers, the multimillionaire won 19 percent of the popular vote, precluding a majority vote for either Bush or Clinton. Bush supporters were quick to point out that Perot took more votes away from their candidate. Perot's use of the media and town meetings bypassed party structures. But Perot was unable to win votes in the Electoral College. In contrast, Donald J. Trump in 2016 upset his opponents to capture the Republican party nomination for president and went on to win the White House in November. In the razor-thin presidential election of 2000 between Republican George W. Bush and Democrat Al Gore, Gore supporters resented the Green Party candidacy of Ralph Nader, who siphoned votes from Gore. Nader probably tipped the balance in some swing states such as Florida. Nader, vilified by congressional Democrats, nevertheless committed himself to further pursuit of his environment-friendly and antiestablishment agenda in the twenty-first century. Ironically, the issue that Ross Perot rode to unusual popularity, budget deficits, returned after the hiatus of the 1990s to haunt policymakers of both parties during the George W. Bush and Barack Obama administrations.

Public opinion data demonstrate that the two major parties are losing support. The number of voters who decline to express any partisan preference or who declare themselves as independents indicates growing alienation toward the party organizations and their principal donors, whether corporate or other special interest groups. The rules of the game for electing presidents make it difficult for third-party candidates to compete with the two major parties, especially the electoral college that must be won on a state-by-state basis. The sporadic visibility of third parties reflected dissatisfaction with the partisan status quo at national and state levels. The early appeal of Senator John McCain in the 2000 Republican presidential primary and of former senator Bill Bradley in the Democratic primary suggested flagging enthusiasm for party establishments and their preferred candidates backed by prodigious amounts of so-called soft money.

At this time it is hard to tell how third parties and independent candidacies will affect national security policy. Third parties had negligible impact on the 2004 presidential race in which George W. Bush was reelected running against Democrat John Kerry. And in 2008, Barack Obama surprised favored Democratic presidential primary opponent Hillary Clinton by outlasting the former First Lady in a protracted contest for the nomination. Obama's innovative fundraising and magnetic personal style preempted the appeal of third-party or other novelty candidates. Obama's charisma helped to dissuade otherwise disaffected or distracted Democrats from bolting their party in favor of a different anti-establishment candidate. On the conservative side, Texas congressman Ron Paul has offered a libertarian alternative of small government against the tide of both Democratic and Republican party establishments, and Connecticut US senator Joseph Lieberman staked out an independent and more conservative foreign and security policy than was favored by most Democrats with whom he usually caucused. The experience of Donald J. Trump in 2016 seems to validate the argument that, at least for the time being, aspiring presidential candidates must first capture the nomination of one of the two major parties—however dissatisfied voters are with the established leaders and structures of those parties. The presidential election of 2020 was not much affected by third parties.

Interest Groups and Coalition Politics

Interest groups provide useful ways to mobilize the public and affect policy. Yet some groups can be more concerned with their own agendas than with serving the public good; some tend to serve their small group of leaders and do not necessarily seek out the public to mobilize votes and influence elected officials. In any case, interest groups serve an important purpose in an open system.

Interest groups have become increasingly important and powerful in the US political system. Part of this is in response to the vacuum left by the decline of political parties. The rise of single-issue politics (i.e., political activism based on one specific issue, such as abortion, gun control, or community policing) has reinforced the role of some interest groups. The performance of public officials tends to be judged according to that one issue—a litmus test—regardless of their record on other matters. Additionally, single-issue activism can be co-opted by broader-based interest groups to take on a particular policy or official. An example is the 1999–2000 controversy surrounding Elián González. He and his mother fled Cuba by raft in 1999; she drowned, but he was rescued off the coast of Florida. His relatives in Miami then fought a long battle against US immigration authorities and the Justice Department, going to court to prevent Elián's return to Cuba. Court decisions favorable to the Justice Department and a successful night raid to remove Elián from the control of his anti-Castro relatives were required to resolve the episode.

The emergence of political action committees (PACs) and soft money further complicated the relationships among the president, Congress, and interest groups. PACs have become major instruments used by corporations, business, and labor groups to influence elections and remain within the federal election laws. PACs donate to election campaigns and often become the vanguard in establishing political positions for larger groups. Recent federal court rulings (especially the US Supreme Court decision in *Citizens United v. Federal Election Commission* in 2010) have increased the difficulty of limiting campaign expenditures by companies, unions, and other entities not directly connected to the candidate's campaign organization. As expected, PACs have influence in Congress, partly because direct influence in the White House is difficult; also, the nature of congressional constituencies leaves members of Congress vulnerable to the influence of PACs. Soft money, that is, money not given directly to candidates but to issue development or under other auspices, indirectly, if not directly, supports the conduct of a political campaign. Both Republicans and Democrats have charged the other side with commingling soft and hard money, which is given explicitly and directly to the candidate. Only specialized lawyers fully understand the complexity of campaign finance law and of perennial "reform" efforts in rules affecting campaigns and lobbying. The public's eyes glaze over.

Think tanks of various political persuasions have established themselves not only as idea factories for political parties and candidates but also as political players in their own right. The reports and activities of think tanks can have political consequences and are used by interest groups to advocate their own policies. Yet certain think tanks are identified with partisan worldviews, whether liberal or conservative, such as the Brookings Institution (liberal) and the Heritage Foundation (conservative). The national security establishment as well as important political actors in the policy process can be, and has been, influenced by the research and publications produced by think tanks. In this sense, think tanks can serve, consciously or otherwise, as the basis for positions adopted by interest groups and political actors. Neoconservative pundits based in think tanks contributed much of the philosophical underpinning for the George W. Bush national security strategy, war on terror, and war in Iraq. With Bush in the ranks of former presidents after January 2009, neoconservative research institutes and opinion commentators made known their various displeasures with Obama policies in national security and other areas.

The fragmentation of power and the decline of central authority in Congress make it likely that the president will need to become involved in interest group politics. To develop support for legislation and promote consensus for national security policies, the president often must appeal to interest groups and form coalitions with them. Obviously, he seeks the support of groups most likely to favor his policies. For example, the president might turn to defense contractors and politicians in communities where military bases are located in order to build support for favored weapons programs and defense policies. Congress often spreads defense spending for a new weapons system around as many states and communities as possible to maximize the political reward, which also makes the program more difficult to oppose.

The underlying motivation for presidential involvement in interest group politics is that PACs, lobbyists, and other interest groups have significant influence on Congress, which is vulnerable to persuasion by external actors. In addition, there is a need for the president as well as Congress and the bureaucracy to respond to the growing power of organized interests to develop coalitions for supporting a policy and to thwart those opposed to a policy. The growing power and influence of interest groups was one major factor in establishing the Office of Public Liaison during the Gerald Ford administration. This office was created to help shape friendly relationships between the president and interest groups; its primary purpose is to mobilize support for the president.[30]

Interest groups also attempt to influence public opinion, hoping that public pressure will reach not only Congress but also the White House. For example, during the early 1990s, feminist groups and their political allies pressured the president and the Pentagon to prevent sexual harassment in the military. And throughout the 1990s, gay rights groups, with a great deal of support in the Clinton White House as well as Congress, led efforts to sensitize the military chain of command to the problems of discrimination against, and harassment of, gays in uniform. The same groups monitored Defense Department performance and contributed to the revisions of policies affecting women as well as gays in the late 1990s, especially the department's antiharassment directives of 2000. These and other aspects of broadening awareness during the George W. Bush and Barack Obama administrations on gay rights issues led to the Pentagon's decision under Secretary of Defense Robert Gates and chairman of the Joint Chiefs of Staff (JCS) Admiral Michael Mullen in 2010 to move gradually toward the acceptance of gays serving openly in the US armed forces. This example is one of many that shows how evolving social and cultural standards in the US body politic are eventually incorporated into the military and national security establishments, albeit with necessary adjustments for the uniqueness of professional military culture and traditions. Another example is provided by the issue of transgender service members and their status within the military. As he exited the office, Obama's last secretary of defense, Ashton Carter, issued a directive that transgendered persons could serve openly in the US armed forces, but this left undecided many procedural issues having to do with the conditions of entry and treatment of transgender personnel. The Trump administration reversed these permissive policies, which were restored early in the Biden presidency.

Domestic and National Security Policies

Some would contend that interest groups have less of an impact in the national security arena than in domestic politics. The intermingling of the two areas, however, blurs the distinction between interest group activity and national security policy. For example, interest groups and PACs involved in the domestic economy invariably become involved in defense spending. In addition, social-issues groups have an impact on the federal budget and defense outlays by urging spending on welfare and social programs and because of their belief that cuts in defense will help their programs.

In some cases the link between national security and domestic policy is straightforward. For example, it is difficult to boycott a country that imports US goods without causing some economic hardship at home. A case in point: during the Cold War, a decision not to sell agricultural products to the Soviet Union raised protests and triggered opposition by interest groups representing US farmers. Similarly, restricting imported goods to protect domestic industries can have an impact on US relationships with states that are important to US security policy. In February 1994, for example, President Clinton lifted the trade embargo on Vietnam that had been in place for nineteen years, primarily to stimulate the US economy. Prisoner of war/missing in action (POW/MIA) groups and several Vietnam veterans strongly opposed such a move. Many interest group activities were seen in the intense campaign by the Clinton administration to pass NAFTA in 1993. Labor unions mounted massive opposition campaigns, whereas some business groups as well as congressional Republicans spoke out in support of NAFTA. The agreement was passed by Congress. However, the Donald J. Trump administration regarded NAFTA as a swindle that benefited other countries to the detriment of American business and workers, but soon replaced it with a similar United States, Mexico, and Canada Agreement (USMCA). The Biden administration is unlikely to change this arrangement.

During President George W. Bush's administration, the connection between interest group politics and national security was evident in debates about control over illegal immigration. From the standpoint of national security, many advocated tighter border controls and more expedient deportation of illegals to their countries of origin. The governors of two southwestern states, Arizona and New Mexico, declared states of emergency with respect to loss of control over illegal border crossings from Mexico. On the other hand, immigration was also an issue that involved economic and social considerations and resonated with liberal and conservative interest groups. Some liberals supported President Bush's proposal for "guest worker" and amnesty programs that would legalize some previously illegal immigrants provided they met certain criteria. And some conservatives attacked Bush for not providing enough border security against possible infiltrators who could be terrorists or "coyotes" being paid to run terrorists and other illegals across the border. On the other hand, some business groups wanted cheap labor, and some humanitarian groups objected to immigration restrictions as denials of human rights. The issue of immigration propelled Donald Trump to the

White House in 2016 and draconian immigration crackdowns ensued. These included increased border controls and deportation of undocumented immigrants in this country when they were discovered. He also went to great efforts to extend the physical barrier separating the United States and Mexico to make unauthorized entry into the country more difficult, but not much was accomplished. President Biden does not support open borders, but has reversed some of the most draconian Trump administration policies—especially regarding the separation of families at the border.

Although most interest group activity is focused on domestic issues, some groups can have an important, albeit indirect, role in national security issues. These include church groups, friendship societies, cultural groups, and policy advocacy groups. Interest groups with concerns about national security can make alliances across ideological lines. An example is the coalition of liberal and conservative groups that warned against excessive zeal in prosecuting the war against terrorism. This ad hoc coalition defended US freedoms and included liberal groups such as the American Civil Liberties Union and conservative organizations such as the National Rifle Association. Others reacting to September 11 showed the diversity among US interest groups; some raised humanitarian aid, others rallied patriotic support for the president and nation, and still others clamored for improved security at airports and called for a housecleaning in the US intelligence community. During the Trump administration, in 2020 Black Lives Matter and other groups demanded greater attention be paid to racism in American culture and, more specifically, to racial bias in community policing. Opposite groups emphasized the extensive rioting in cities like Portland, Oregon, and Seattle, Washington, and the unwillingness or inability of mayors and city councils to take firm stands against arson, looting, and other illegal behavior. One side contended that there was no security without justice: the other, that there was no justice without security.

Other examples of interest group activity are aimed at the US role in the international arena. For example, human rights groups can affect US policy by publicizing human rights violations in other countries, which can lead to moral indignation at home, requiring some response. The events at Tiananmen Square in 1989 led to many demands for US sanctions against China. Years later, in 2020 Chinese treatment of its Uighur Muslim minority also led to US and other demands for retaliatory actions against Beijing. Amnesty International, which analyzes human rights activities, also raises public awareness of global abuses by governments and others. Also, veterans' organizations have a similar impact on the use of US military forces overseas. Human rights groups helped publicize widespread ethnic cleansing and humanitarian abuses in Somalia in 1992, in Bosnia from 1992 through 1995, and in the Serbian province of Kosovo in 1999, and presumed excesses by the troops fighting the growing insurgency in Iraq. During the George W. Bush and Barack Obama administrations, efforts by international human rights groups to bring legal actions against US policies on detention and interrogation of captured terrorists reflected the moral revulsion of these groups for US policies but had little impact on the American domestic policy debate.

Foreign countries also try to influence US public opinion and congressional views. "Large research and lobbying staffs are maintained by governments of the largest US trading partners, such as Japan, Korea, the Philippines, and the European Community. . . . Frequently, these foreign interests hire ex-representatives or ex-senators to promote their position on Capitol Hill."[31]

Bureaucratic Interest Groups

Informal groups and networks within the federal bureaucracy play a less visible but important role in influencing national security policy and strategy. Bureaucratic groups are active in defending parts of the defense budget, especially as part of the iron triangle that includes the bureaucracy, defense contractors, and supportive legislators. Bureaucracies are also capable of "slow-walking" or foot dragging against policies they prefer not to implement, or, in exceptional cases, in actively opposing the thrust of presidential or congressional policy. President Trump charged that high officials in the US Department of Justice, Federal Bureau of Investigation, and intelligence community conspired to subvert his candidacy in 2016 and to investigate his presidency with fabricated charges of Russia collusion.

Within the administration, power clusters may prefer one policy and strategy over another. During the Reagan administration, there was a considerable amount of infighting regarding the proposed Intermediate Nuclear Forces Treaty with the Soviet Union. On the one hand, some high-level military officers opposed the treaty, forming an implicit alliance with some members of Congress. On the other hand, at the highest levels of the State Department, there was considerable effort to reach an accord. During the Clinton administration, competing power centers doomed national health care. And pertinent to security issues, Joint Chiefs of Staff chairman Colin Powell publicly expressed disagreement with others in the administration on the issue of military intervention in Bosnia. And a 1992 Clinton campaign promise to permit gays to serve openly in the military was strongly opposed by the members of the JCS shortly after Clinton assumed office in 1993. Clinton and the JCS finally settled for the compromise "don't ask, don't tell" policy, which endured until 2011 when such service was no longer prohibited.

In the early and middle period of the George W. Bush administration, there were clear policy differences between the State Department under Secretary Colin Powell and the Defense Department headed by Donald Rumsfeld. Rumsfeld and Vice President Richard Cheney formed a political alliance that left Powell frustrated on account of the Rumsfeld-Cheney willingness to act unilaterally as opposed to emphasizing consultation with allies. Rumsfeld's dismissive reference to "old Europe" (France and Germany) compared to "new Europe" (former Soviet states now incorporated into NATO and more sympathetic to the US position on Iraq in 2003) sparked diplomatic fevers across the Atlantic. Similar interagency frictions were apparent in the relationship between the vice president's office and the intelligence community between September 11 and the onset of the US military campaign against Iraq in March 2003. Cheney aggressively pressed the view that CIA intelligence needed to be more responsive to policymakers' priorities and preferences, but some highly placed CIA analysts resented this interference with what they regarded as their professional assessments of Iraq's plans and military capabilities.

Interagency friction within the Obama administration was less obvious but apparent to close observers. These disagreements included the preferred strategy for fighting the Taliban and al-Qaeda in Afghanistan. A faction within the administration's national security team led by Vice President Joe Biden argued in favor of a more specifically focused "counterterror" approach to the war in Afghanistan, in contrast to the broader "counterinsurgency" approach favored by US military commanders and supported by Secretaries Gates and Clinton. On numerous occasions, the president expressed exasperation with the character of the policy debates and the tendency of agencies with competing views to box him in with a fixed choice of alternatives that really represented no choice at all.[32] President Biden soon resurrected his earlier opposition to "nation-building" efforts by US combat troops and ended US combat missions in Afghanistan in 2021, as discussed earlier.

It is interesting to note that in the early 1990s "the Defense Department [was] assisted by almost 350 lobbyists on Capitol Hill; it maintain[ed] some 2,850 public relations representatives in the United States and foreign countries."[33] During the Clinton administration, groups within the Department of Defense and the military services formed implicit alliances with members of Congress to expedite action on policy issues favored by the military (such as increased readiness spending) and to block unfavored proposals (such as permitting gays to serve openly in the armed forces). The point is that any number of interest groups evolve from the bureaucracy and become involved in advocating and supporting policy or strategy.

Iron Triangles

Domestic politics is characterized by interest groups' attempts to influence the public, Congress, the bureaucracy, and their own constituents. Equally important, interest groups become deeply involved in political campaigns and party politics. Although members of Congress can become captive to interest groups, especially if they dominate the member's district, members use interest groups to support their own legislative agendas. Furthermore, interest groups can become influential within the federal bureaucracy. Individuals with close links to interest groups are often appointed to administrative positions. Moreover, as bureaucrats implement a given policy over the years, they tend to acquire views similar to those of interest groups active in the field.

When the interests of members of Congress, interest groups, and bureaucrats come together, power clusters result, called "iron triangles." Such iron triangles are coalitions of interests that are very difficult to penetrate and influence from the outside.[34] Throughout the process, several iron triangles may be at work advocating or opposing policy; they can even form a network of triangles that can frustrate almost any policy. Iron triangles confront the theory of democracy with the actuality of elitism and inside-track policymaking.

Put simply, it is a fact of political life that the president faces power clusters and interest group activities while establishing national security policy and directing the policy process. Although far stronger in domestic policy, such influences are affecting national security policy and strategy. Thus the president needs to build a coalition of power clusters for policies that are likely to be debated in Congress or that require resources only Congress can approve. At the same time, if policies are undertaken in secrecy, the president must be sensitive to the fact that publicity, especially occasioned by failure, will require an explanation and taking responsibility. Support from various power clusters and interest groups eases the political damage and makes it less difficult to design and undertake even risky policies.

Conclusion

The impact of public opinion on security policy should not be underestimated. Although in normal times only small percentages of voters pay close attention to the details of security and defense policymaking, public awareness of broad trends is stimulated by the media, political parties, and interest groups. The distortion created by media reports, partisan politics, and self-motivated interest groups undermines public understanding of public policy, including security policy. Yet competition among the media, political parties, and interest groups within a pluralistic system ensures that voters are exposed to multiple and competing perspectives. At the ballot box, the voters will have the last word.

The president has more latitude, at least at the outset of a new initiative, in defense and foreign policy compared to domestic policy. Tradition and the advantages held by executive branch agencies vis-à-vis Congress ensure that a president who gets out in front of events and who seems to lead will have high approval ratings—at least for a while. Whether strong approval ratings can be sustained depends upon a president's skill in getting the message across to the government, Congress, and the media. In security policy, as in other aspects of the US political system, there is no single government policy but an array of competing policy preferences, each supported by powerful internal constituencies. If there is a threat of war or actual hostilities, the enemy and its perceptions and goals need to be taken into account as well.

In the final analysis, the president cannot simply assume that national security policy is self-executing. The president and the national security establishment, as well as other executive agencies, drive the instruments of policy and its supporting strategies. Orders must be passed down to apply the policy; then its implementation must be undertaken. When policies do not require appropriations from Congress or are hidden from view, the president and his staff must recognize that other political actors can shape the final result.

Even with an ethically dubious or controversial policy, the president must be prepared to accept responsibility and the political damage that results. Regardless of the nature and character of national security policy and its strategic implementation, presidents will be judged by their effectiveness in national security and how it furthered democracy and society. In the end, this final assessment is critical in the overall performance and credibility of the president. Even more important, it has a direct bearing on the capability of the United States to pursue its national security interests effectively. Presidential decisionmaking about peace, war, and crises leaves indelible legacies to successors in the Oval Office.

Notes

1. Stephen M. Walt, "How To Ruin a Superpower," *Foreign Policy,* July 24, 2020, https://www.realclearworld.com/2020/07/24/how_to_ruin_a_superpower_499932.html.

2. Richard A. Watson and Norman C. Thomas, *The Politics of the Presidency,* 2nd ed. (Washington, DC: Congressional Quarterly, 1988), p. 153.

3. Gregory F. Treverton, *Intelligence for an Age of Terror* (Cambridge, Mass.: Cambridge University Press, 2009), esp. pp. 7–11.

4. The military was credited for the massive airlift in the final days, but the security situation around the airport was nearly impossible with the limited number of troops available.

5. James MacGregor Burns, *Roosevelt: The Lion and the Fox* (New York: Harcourt, Brace, and World, 1956), pp. 156–157.

6. Alexis de Tocqueville, *Democracy in America,* edited by J. P. Mayer, translated by George Lawrence (Garden City, N.Y.: Anchor, 1969), p. 186.

7. Doris Graber, "Media Magic: Fashioning Characters for the 1983 Mayoral Race," in Melvin G. Holli and Paul M. Green, eds., *The Making of the Mayor, Chicago, 1983* (Grand Rapids, Mich.: Eerdmans, 1984), p. 68.

8. Bruce D. Berkowitz and Allan E. Goodman, *Best Truth: Intelligence in the Information Age* (New Haven: Yale University Press, 2000), pp. 133–135.

9. See, for example, *The Pentagon Papers,* edited by Senator Mike Gravel (Boston: Beacon, 1971).

10. See, for example, *Watergate Hearings: Break-In and Cover-Up: Proceedings* (New York: Viking, 1973).

11. Sam C. Sarkesian, "Soldiers, Scholars, and the Media," *Parameters* 17, no. 3 (September 1987), p. 77.

12. S. Robert Lichter, Stanley Rothman, and Linda S. Lichter, *The Media Elite* (Bethesda, Md.: Adler and Adler, 1986), p. 299.

13. Ibid., p. 95.

14. Thomas E. Patterson and Robert D. McClure, *The Unseeing Eye: The Myth of Television Power in National Elections* (New York: Putnam's, 1976), p. 75.

15. Peter Braestrup, *Big Story: How the American Press and Television Reported and Interpreted the Crisis of Tet 1968 in Vietnam and Washington* (Boulder, Colo.: Westview, 1977), p. 184.

16. See, for example, "The End of the 'Jimmy' Story," *Washington Post,* April 16, 1981, p. A18; Don Kowet, *Matter of Honor: General Westmoreland Versus CBS* (New York: Macmillan, 1984); and Edith Efron, *The News Twisters* (New York: Manor, 1972).

17. General Michael J. Dugan, USAF (ret.), "Perspectives from the War in the Gulf," in Peter R. Young, ed., *Defence and the Media in Time of Limited War, Small Wars, and Insurgencies* 2, no. 3 (special issue, December 1991), p. 179.

18. A balanced appraisal of this issue appears in McCormick Tribune Conference Series, *The Military-Media Relationship 2005: How the Armed Forces, Journalists, and the Public View Coverage of Military Conflict* (Chicago: McCormick Tribune Foundation, 2005). For a particularly scathing criticism of the media, see

William V. Kennedy, *The Military and the Media: Why the Press Cannot Be Trusted to Cover a War* (Westport, Conn.: Praeger, 1993).

19. "Perceptions management" is an older term that has now been superseded in military parlance by "influence operations," a generic term that includes psychological operations (psyops) and measures of strategic or operational-tactical deception. See John Arquilla, *Worst Enemy: The Reluctant Transformation of the American Military* (Chicago: Ivan R. Dee, 2008), pp. 132–155.

20. Dorothy E. Denning, *Information Warfare and Security* (Reading, Mass.: Addison-Wesley, 1999), p. 6.

21. Alvin Toffler and Heidi Toffler, *War and Anti-War: Survival at the Dawn of the 21st Century* (Boston: Little, Brown, 1993), pp. 167–168.

22. Denning, *Information Warfare,* p. 115.

23. Col. Thomas X. Hammes, USMC (Ret.), "Information Warfare," chap. 4 in G. J. David Jr. and T. R. McKeldin III, *Ideas as Weapons: Influence and Perception in Modern Warfare* (Washington, DC: Potomac, 2009), pp. 27–34, citation p. 28.

24. De Tocqueville, *Democracy in America,* p. 180.

25. Donald M. Snow and Eugene Brown, *Puzzle Palace and Foggy Bottom: US Foreign and Defense Policy-Making in the 1990s* (New York: St. Martin's, 1994), pp. 21–22.

26. See Bruce W. Jentleson, *American Foreign Policy: The Dynamics of Choice in the 21st Century* (New York: Norton, 2000), p. 215; and Stephen Ansolabehere, Roy Behr, and Shanto Iyengar, *The Media Game: American Politics in the Television Age* (New York: Macmillan, 1993), pp. 142–144.

27. Ibid., p. 201.

28. Ibid., p. 199.

29. Daniel Goure and Jeffrey M. Ranney, *Averting the Defense Train Wreck in the New Millennium* (Washington, DC: Center for Strategic and International Studies, 1999), p. 2.

30. Ansolabehere, Behr, and Iyengar, *The Media Game,* p. 199.

31. Steffen W. Schmidt, Mack C. Shelley II, and Barbara A. Bardes, *American Government and Politics Today, 1993–1994* (St. Paul, Minn.: West, 1993), p. 301.

32. Bob Woodward, *Obama's Wars* (New York: Simon and Schuster, 2010), provides important insights into that administration's war policy and national security decisionmaking process.

33. Ibid., p. 253.

34. Theodore White, *The Making of the President, 1972* (New York: Bantam, 1973).

13

Civil-Military Relations

The role of the military in a democratic society rests on absolute civilian control over it.[1] This is ingrained in the US system and in the US military itself. Although the fear of a politicized military has emerged in US history from time to time, the issue has become more challenging, especially in the wake of a presidential election in 2020 that tested the resilience of civilian institutions. Although the US military is now primarily concerned with the possibility of conflict with peer competitors such as Russia and China, its recent involvements have been in ethnic, religious, and nationalistic conflicts as in Iraq, Afghanistan, the former Yugoslavia, and Syria and it must be prepared to respond across the conflict spectrum.

Despite years of inconclusive combat operations, polls continue to show that the military is among the most admired of US institutions, respected more than many other (and far more democratic) institutions such as the Congress. This was true even before the terrorist attacks on September 11, 2001, and the subsequent response of the US military in Afghanistan and Iraq. Despite the popularity of the military, however, differences in experiences and culture ensure that US civil-military relations will remain troubled. This does not mean that the military is likely to disobey civilian directions or attempt to influence elections, other than by individual service members voting, of course, but the degree to which the military should try to influence government decisions (on personnel issues, force structure, and force employment, for example) remains controversial.

Hollywood films are a useful indicator of public opinion, as the media both shape the public's view of reality and react to it. This is especially true for their portrayals of the military. The post-Vietnam fall in the prestige of the military was reflected in such films as *Apocalypse Now, Full Metal Jacket,* and *Born on the Fourth of July.* Other films such as *An Officer and a Gentleman* and *Top Gun* show the military in a much more positive light. On the other hand, in *The Siege,* the military rounds up Arab residents of Brooklyn in response to a series of terrorist bombings. This suggests that US armed forces would be unresponsive to civilian control in an emergency, although it is more likely that a series of terrorist incidents in the United States would cause public demand for far more restrictive measures than the military would be comfortable carrying out. Other films have varied in their depiction of the military and military personnel. Some, such as *The Green Zone* and *The Hurt Locker,* were primarily negative, and others, such as *Taking Chance* and *Act of Valor* were strongly positive. Others were more nuanced, such as *Saving Private Ryan,* and reflected the views of society with empathy for soldiers if not always for their missions.

In the early 1990s, Air Force then–lieutenant colonel Charles J. Dunlap Jr. used a hypothetical military coup in 2012 as the backdrop for writing about his concern with the direction of civil-military relations. The fictional coup was sparked by the massive diversion of military forces to civilian uses, the monolithic unification of the armed forces, and the insularity of the military community.[2] Appearing as it did in *Parameters,* the journal of the US Army War College, the article received a great deal of attention.

The major questions for civil-military relations include these: How can society ensure that military authorities remain in their proper sphere and yet retain the capability to respond effectively across the conflict spectrum? Is there a gap between civilian and military society, and if so, is it serious enough to pose problems for civil-military relations and civilian control and military effectiveness? How can the military avoid even the appearance of political partisanship in a society where hyperpartisanship is so prevalent in the civilian population?

Theories

Two classic discussions of civil-military relations continue to frame the debate on how the military is controlled in a democratic society: Samuel P. Huntington's *The Soldier and the State* and Morris Janowitz's *The*

Professional Soldier.[3] For Huntington, civilian control is achieved through military professionalism. He argued that military officers exhibit three characteristics that define a profession: expertise (the management of violence), responsibility (for the defense of the state), and corporateness (institutional self-awareness and organization).[4] These properties distinguish the military from other professions, and their emphasis serves as the best basis for civilian control. The self-regulating norms of military professionalism would ensure that the military will remain obedient to civilian authorities.

In Huntington's view, the nature of the military makes it a poor match with liberal civilian society. Indeed, he suggested that "the tension between the demands of military security and the values of American liberalism can, in the long run, be relieved only by the weakening of the security threat or the weakening of liberalism."[5] As a result, too close an association between the military and society weakens, rather than strengthens, civilian control. The diminished security threat (at least for the moment) brought by the end of the Cold War has eased this dilemma but has not eliminated it, especially in view of the increased military challenges from resurgent peer and near-peer competitors.

For Janowitz, the founder of the field of military sociology, civilian control is achieved through the socialization process.[6] Put another way, the military comes from society and reflects its values in important ways. The military's sympathy with the values of society makes it a more willing servant. Although military members cannot enjoy all of society's privileges, they support the democratic system that makes these privileges available to the civilian population.[7] Even so, a distinct military culture is important: "In a market economy, the military establishment could not hold its most creative talents without the binding force of service traditions, professional identifications, and honor."[8]

The military as an institution prefers to stay far removed from partisan politics, although politicians like to use military personnel as a backdrop for speeches and photo opportunities.[9] A example of this occurred on June 11, 2020, when President Trump posed for a photo opportunity at a church across Lafayette Square from the White House after protesters were forcibly cleared from the park by civilian police. Included in his entourage were the chairman of the Joint Chiefs of Staff, US Army general Mark Milley and Secretary of Defense Mark Esper, both of whom later apologized for appearing in a partisan event. Milley remarked, "I should not have been there. My presence in that moment, and in that environment, created the perception of the military involved in domestic politics."[10]

There is also a hesitation among both military and civilian leaders to have too many current or retired admirals or generals in key civilian positions. Although they were eventually replaced when President Trump decided he no longer needed their advice, several served simultaneously in his administration in important positions.[11] President Biden turned to recently retired Army general Lloyd Austin to be his first secretary of defense.[12]

The extent to which society should impose its values and culture on the military is an unresolved question. Similarly, the degree to which the military should focus on military effectiveness as opposed to civil liberties issues is highly controversial. Historically, the military discriminated in the name of military effectiveness, making distinctions based on many grounds that would be illegal in a civilian context. Because of the special needs of the military and its role, these have been permitted by the Congress and the courts. Over time, the categories of people against whom the military could legally discriminate has narrowed greatly, however, as society's attitudes on gender, gender identity, and sexual orientation have evolved. For example, increasing public support for the open service of LBGTQ personnel made the eventual elimination of the ban on their service inevitable. Similarly, the sharply increased role of US women service members in Iraq and Afghanistan increased support for an expanded role for women in combat. President Trump's attempts to limit the service of transgender personnel continued, however, but was quickly reversed by the Biden administration.

When such changes occur, the military can build on its earlier success in ensuring that opportunities are not denied to service members based on their race. A focus on shared core values is an important way to emphasize commonalities among an increasingly diverse force. All of the services' core values are similar and useful in bridging sociological and attitudinal differences. The US Military Academy's "duty, honor, country," the Navy and Marine Corps' "honor, courage, and commitment," the Air Force's "integrity first, service before self, and excellence in all we do," and the Coast Guard's "honor, respect, and devotion to duty" all reinforce the idea of selfless service to the nation.[13]

These issues are not fully resolved and will need continuing attention if these core values are to be lived up to, while maintaining military effectiveness and cohesion. Although military actions to increase diversity and inclusion are sometimes portrayed as harmful to military effectiveness, US society is evolving and these trends will of necessity find their way into a military that draws its members from civilian society and needs its support. The issue is not an either/or choice; it is how to maximize both inclusion and effectiveness.

The Civil-Military Gap

Journalist and author Thomas E. Ricks noted some time ago what he saw as a widening gap between the military and society. Based on his personal observation of a US Marine platoon during and after basic training, Ricks saw a contradiction between the values inculcated in the military and those increasingly prevalent in civilian society, leading to feelings of estrangement by some in the military. He suggested three reasons for this: the end of the draft, the "politicization of the officer corps," and a more fragmented, less disciplined US society.[14] Trends in civil society since Ricks's book was published, including "cancel culture" and increasing political polarization, can only contribute to additional civil-military misunderstanding.

Political scientist Ole R. Holsti confirmed and extended Ricks's analysis. Among other findings, Holsti's extensive survey results documented what he called "a strong trend toward conservative Republicanism among military officers."[15] This is noteworthy in view of the previously nonpolitical nature of the military calling—indeed at one time many officers would even refuse on principle to vote. The military is voting these days, however. "Voting officers" in all units ensure that all who want absentee ballots receive them and have whatever help they need in meeting state requirements. In the 2000 presidential election, the votes of military personnel who were Florida residents stationed outside the country may well have been the deciding factor in the victory of George W. Bush.

There is a danger if society and the military that protects it become too disconnected from one another. Many in the military already feel estranged from civilians, whom they see as undisciplined, irresolute, and morally adrift. They view themselves as the true carriers of US values and tradition, swimming against the tide of a society gone morally soft. The degree and significance of this gap are open to debate. It may be simply a curious result of different socialization processes, but it could also mark a fundamental fault line that has implications for the nature of military service, military effectiveness, and the ability of civilian society to direct the military that defends it. The latter issue is of particular concern here.

There are also undercurrents of contempt for some civilian leaders. Public demonstrations of disrespect remain rare, but they do occur. An early visit by President Bill Clinton to an aircraft carrier was marred by discourtesy on the part of many crew members. Contemptuous public comments about the commander in chief by an Air Force general resulted in the rapid termination of the latter's service. An op-ed piece in the *Washington Times* written by a major in the Marine Corps Reserve calling for the president's impeachment effectively ended that officer's career. Although many service members felt the piece read well, most felt it went over the line of acceptable commentary by a serving military officer—even a reservist. Wiser commanders pointedly reminded their officers of the provisions of Article 88 of the Uniform Code of Military Justice: "Contempt toward officials: Any commissioned officer who uses contemptuous words against the President, the Vice President, Congress, the Secretary of Defense, the Secretary of a military department, the Secretary of Homeland Security, or the Governor or legislature of any State, Territory, Commonwealth, or possession in which he is on duty or present shall be punished as a court-martial may direct."

Note that Article 88 is directed at commissioned officers, not enlisted personnel, who can still be charged with disrespect to a "warrant officer, noncommissioned officer or petty officer while that officer is in the execution of his office."[16] Military personnel may not engage in partisan politics while in uniform, whether in support of or opposition to a political campaign. In that connection, many were disturbed when President Trump signed "Make America Great Again" campaign caps and a "Trump 2020" patch for service members in uniform during visits to Iraq and Germany in December of 2018. Defense Department guidance on such matters is clear: "Active duty personnel may not engage in partisan political activities and all military personnel should avoid the inference that their political activities imply or appear to imply Defense sponsorship, approval, or endorsement of a political candidate, campaign, or cause."[17]

For their part, most Americans have admiration and respect for the military but are not eager to put their own civilian pursuits aside to join it. Recruiting and retaining quality personnel, essential to a modern military, are continuing concerns for military leaders. For many civilians, military life is as unfathomable as life on another planet; military people are outsiders to them. This does not make them expendable—they are, after all, still Americans, and military casualties still strike a strong nerve—but it does not mean they are part of the mainstream. As noted above, military protection is appreciated, but there is little understanding of the individuals who have chosen the military as a career. The difficulty in understanding does not preclude respect for the sacrifices military people make that civilians do not, and perhaps could not, make, however.[18]

This is a great improvement from the Vietnam War era, when military service members endured great personal sacrifice, yet were tainted with the war's unpopularity and returned to a society contemptuous of those who served. After the successful 1991 Gulf War and especially after the September 11, 2001, attacks, positive feelings toward the US military as an institution and the people who serve in it reemerged, and growing public disaffection for the wars in Iraq and Afghanistan did not translate into lower support for the military as an institution or for military personnel. The United States has come a long way since Vietnam in that regard.

Civilian and Military Cultures

Many factors drive the civilian and military cultures away from one another. Perhaps the most serious is the diminishing number of civilians, especially elites, with personal exposure to the military, either through their own service or that of a family member. This may have been eased somewhat as large numbers of military reservists, with close ties to the civilian communities from which they came, served on extended combat tours in Iraq and Afghanistan.

Since the draft ended in 1973, the only remaining requirement is to register with the Selective Service System, an obligation that until now only applies to males ages 18–25. Compulsory national service plans (usually including a civilian service option) are occasionally discussed, but there is neither widespread political support nor military necessity for any such system. Sociologist Charles Moskos proposed an innovative plan for fifteen-month enlistments to attract college students into the military, but this plan did not receive widespread support. Yet as he noted, "If serving one's country became more common among privileged youth, future leaders in civilian society would have had a formative citizenship experience. This can only be to the advantage of the armed forces and the nation."[19]

Bill Clinton's avoidance of service in Vietnam was a lightning rod issue inside the US military. Despite Donald Trump's public embrace of the military, his avoidance of military service on a controversial medical waiver did not endear him to military personnel. Unfortunately, the lack of military service among US elites goes far beyond top leadership. Today, compared to years past, major presidential advisors, Defense Department appointees, members of Congress, senators, and their staffs are unlikely to have any military experience. Media and economic elites, not to mention the general public, are also less likely to have served.

Disassociation from the military on the part of large segments of society weakens society's ability to make informed judgments about military issues and influence military decisionmaking. It is not that uninformed civilians will necessarily distrust or dislike the military; on the contrary, there may be a tendency to like the military too much and to put unwarranted confidence in military solutions to international political problems. All this can have serious implications for establishing effective national security policies.

Although standards for the behavior of military personnel now include an expectation that basic civilian norms of propriety will be observed in their interpersonal relations, it remains critical for the military to fight effectively. As noted in a study of civil-military relations and the American way of war, "It is clear that the role of combat leader remains essential in the military that expects to win wars rather than deploy as armed social workers."[20] It may be that the improved popularity of the military after the 1991 Gulf War was due to the fact that it was a military victory, not simply that civilians developed feelings of guilt about the way soldiers were treated after Vietnam and decided to make it up to them decades later.

Integration in the Military

The military often finds itself ahead of society on issues of social change. Some of these policies have worked out better in the military than in civilian society, such as the generally successful attempts to achieve racial fairness.[21]

Perhaps the greatest sociological challenge for the military is the integration of women into the mainstream of military activities, including combat.[22] Of course, true equality would mean that women could not only *volunteer* for combat but also be *compelled* to serve in combat under the same conditions as men. Military guidance as early as 2011 increased the ability of women to operate forward in ground combat operations, although much of the guidance merely codified what had become common practice.[23] Military restrictions on women serving in close combat are falling as public opinion shifts on this issue.

The "don't ask, don't tell" policy of sociologist Charles Moskos, adopted by the Clinton administration to permit the service of gay and lesbian service members, withstood judicial scrutiny much longer than many had predicted.[24] The issue was not whether there would be gay and lesbian service members, because there have always been. It was how well the military cohesion so crucial in battle could withstand the stress of openly discussed homosexual orientations. Now that the "gay ban" has ended, it appears that fears about harm to unit cohesion and military effectiveness have not been borne out. Societal views in this regard have evolved and are reflected in the military—which is, after all, drawn from society. President Obama removed barriers to the service of transgender service members with similar results. Although President Trump rolled back regulations permitting their service, the Biden administration quickly restored the Obama rules, as noted earlier.

Civilian-Military Connections

Morris Janowitz's advice that it is best to increase the connections between the military and civil society in order to increase mutual understanding has serious implications for military recruiting and education. In that respect, it is good that there are a number of alternative sources for military officers, all of which are well connected to the civilian society. The highest percentage by far of officer accessions come from Reserve Officer Training Corps (ROTC) programs at civilian colleges and universities, followed by Officer Candidate Schools (a less expensive option because the services do not fund candidates' college degrees), service academies (with admissions from every congressional district), and direct commissions.[25]

The military services need a variety of sources for their officers, with some of them under their control. Successful efforts at several elite universities to remove ROTC programs from their campuses in the wake of the Vietnam War and disputes about military personnel policies are sufficient testimony to this. As an unfortunate result, it is less likely that the children of civilian elites who go to such universities would enter the military and perhaps continue to influence it through successful careers.

As officers advance in grade, a combination of training and education prepares them to assume higher levels of responsibility. This education needs to reinforce the lessons of proper civil-military relations, which it generally does. Whether academic standards at the war colleges are comparable to those at civilian universities is a matter of controversy. Acknowledging that the American military has some centers of intellectual excellence, Eliot A. Cohen nevertheless laments that "strategic thinking about the nature of war, and how to align military means with political ends, is a very different matter. There, arguably, it has done poorly. There are few major uniformed military thinkers. Writing about the nature of contemporary war is largely a civilian enterprise; and although the United States has designated strategists in the uniformed ranks, they are not always used as such."[26]

There is a strong argument for sending many of the best officers to top civilian universities. The education may well be even better, and military and civilian elites can interact and understand one another's perspectives.[27] In addition, getting a doctoral degree from a great civilian university does not make an officer less suited for military command positions. This is well demonstrated by Army general David Petraeus, former commander of the US Central Command, top US commander in Afghanistan, and later CIA director; Navy admiral James Stavridis, who rose to become the commander of the US European Command and Supreme Allied Commander Europe; and Army lieutenant general H. R. McMaster, who had a doctorate and a distinguished combat and noncombat Army career prior to his appointment as national security advisor for President Trump while still on active duty.[28]

Another civil-military connection is not as well understood as it should be: the military reserve and National Guard forces. These part-timers provide an invaluable connection between the active forces and society at large and will be increasingly important as budgets grow tighter and civilian technical and managerial expertise becomes more important for the military.[29]

The Military and the Media

The gap between the military and society benefits no one, and efforts by both the military and civilians will be required to close it. The military needs to ensure it is training people to understand and respect civilian values, institutions, and prerogatives. This includes inculcating respect for the role of the media and encouraging academic research bearing on military actions and the military as an institution.

In this connection, not all is well. It is too easy for the military to view criticism by outsiders as uninformed and mean-spirited and to see the free press as an enemy. Media coverage of the Iraq war was detailed, as it should be, and helped expose a number of improper actions, including the abuse of Iraqi inmates at Abu Ghraib prison, incidents of excessive or unwarranted use of force, inappropriate symbols, and desecration of bodies. Media coverage of military actions in Afghanistan and Syria has been similarly intense. Military educators are sometimes so sensitive to media criticism that they may inadvertently suggest to their students that such criticism is illegitimate. This is a bad lesson to teach future military leaders who will eventually be charged with defending a society in which a free press is of paramount importance.[30]

The evolution of the US media in recent decades has not necessarily been favorable to reliable coverage of political and military events. The entire concept of "broadcasting" to a mass audience has undergone a tectonic shift to "narrowcasting" to niche markets and to persons of similar political viewpoints. In addition, Americans are additionally turning to social media for political information at the same time that social media are being effectively "weaponized" by ideological political factions and even foreign governments. Thus, for example, special counsel Robert Mueller indicted a number of Russian nationals for attempts to interfere in the 2016 US presidential election and was pilloried for his effort by President Trump and his supporters. Experts warn that the weaponization of social media by domestic dissident groups, including dangerous ones, and foreign governments, should be expected as a new norm.[31] Within this climate, the risk of disinformation or misinformation about US uses of armed force is very high, especially among populations already predisposed to doubt US capabilities, resolve, or good intentions.

Military points of view are not as monolithic as many would expect. Moreover, the degree to which the military should express policy preferences in the media remains controversial. It has been forcefully argued that senior officers should voice their opinions in areas of their expertise. There has been a call for "constructive political engagement," which must steer clear of partisanship on such issues as "military democratization" (the imposition of civilian values and practices on the military institution) and the utility of military force in various contingencies.[32] Conversely, Huntington expressed many doubts about the degree to which the military can "participate in the good-politics of policy without also becoming embroiled in the bad-politics of partisanship."[33] In retrospect, the history of US intervention in Vietnam, in the Balkans, in Iraq, and in Afghanistan might have been far different had military leaders spoken out publicly.[34]

In the view of many military observers, the handling of the early stages of the 1999 Kosovo crisis helped turn a humanitarian tragedy into a catastrophe, as civilian leaders pushed for a military confrontation with Serbia yet ruled out the use of ground forces. In the minds of many, this showed how badly war can be waged by civilian policymakers who do not fully understand the uses and limits of military power. Perhaps worse, senior military leaders who knew better did not put their careers on the line by putting their doubts on the record in a timely fashion.[35] This is not a call to subvert civilian authority, but more candid and public advice would have better served the nation. More recently, many military leaders swallowed their professional reservations about plans for the Iraq invasion and subsequent occupation in the face of strong pressure from Secretary of Defense Donald Rumsfeld and other civilian leaders. As it turned out their reservations were well founded, and although civilian officeholders should listen carefully to military advice, they retain final authority in the US system. More recently, senior military leaders have been widely criticized for not expressing their longstanding doubts about the failed war in Afghanistan and how it was conducted.

In assessing the role of the US military in Vietnam, General Fred C. Weyand, US Army chief of staff, in 1976 stated, "As military professionals we must speak out, we must counsel our political leaders and alert the American public that there is no such thing as a 'splendid little war.' There is no such thing as a war fought on the cheap." As noted earlier, he went on to say that "the American Army is really a people's Army in the sense that it belongs to the American people who take a jealous and proprietary interest in its involvement. . . . The American Army is not so much an arm of the executive branch as it is an arm of the American people. The Army, therefore, cannot be committed lightly."[36]

Conclusion

The real danger in this country is not that military officers will defy civil authority or stage some sort of coup d'état.[37] For reasons laid out some fifty years ago by Huntington and Janowitz, that is quite literally unthinkable, despite the presence of some servicemembers with extreme political views. If military authorities ever do think in such terms, the future of US democracy will be in grave doubt. As Justice Robert Jackson noted in his dissent in *Korematsu v. United States,* the Supreme Court case that upheld the internment program for persons of Japanese ethnicity in World War II: "If the people ever let the war power fall into irresponsible and unscrupulous hands, the courts wield no power equal to its restraint. The chief restraint upon those who command the physical forces of the country, in the future as in the past, must be their responsibility to the political judgments of their contemporaries and to the moral judgments of history."[38]

To the extent that the estrangement of the military and society is real, it is not healthy. Fortunately, the remedies are straightforward. Mutual understanding between the military and society calls for increased linkages at all levels and an understanding that the military is a unique institution with standards that may not always be the same as those of the society it defends.

There is always a balance to be struck between civil rights and liberties within the military, on the one hand, and military effectiveness, on the other. Even within the military, there is a difference in the degree to which forces can or should be expected to mirror society. Ideologues at either end of the spectrum only make it harder to advance sensible policies.

The relationship of the military to society and the civilian-military culture gap is well stated by noted military author John Keegan:

> Soldiers are not as other men—that is the lesson that I have learned from a life cast among warriors. The lesson has taught me to view with extreme suspicion all theories and representations of war that equate it with any other activity in human affairs. War is . . . fought by men whose values and skills are not those of politicians or diplomats. They are those of a world apart, a very ancient world, which exists in parallel with the everyday world but does not belong to it. Both worlds change over time, and the warrior world adapts in steps to the civilian. It follows it, however, at a distance. The distance can never be closed, for the culture of a warrior can never be that of civilisation itself.[39]

Whatever other purposes are served by the military, it must remain a credible fighting force. If it cannot continue to fight and win a nation's wars, including state-on-state wars and the warlike operations inherent in peacekeeping operations and the war on terrorism, it is not fulfilling its mission.

Notes

1. Portions of this chapter first appeared in *The World & I* magazine and are reprinted with permission from the Washington Times Corporation.

2. Charles J. Dunlap Jr., "The Origins of the American Military Coup of 2012," *Parameters* 22, no. 4 (Winter 1992–1993), pp. 2–20. This piece is very interesting for a lay audience to read. It was followed by his "Melancholy Reunion: A Report from the Future on the Collapse of Civil-Military Relations in the United States," *Airpower Journal* (Winter 1996), pp. 93–109. A more detailed exposition of his argument, without the literary devices, is in his "Welcome to the Junta: The Erosion of Civilian Control of the US Military," *Wake Forest Law Review* 29, no. 2 (1994), pp. 341–392. Dunlap retired as a major general and deputy judge advocate general of the Air Force.

3. Samuel P. Huntington, *The Soldier and the State: The Theory and Practice of Civil-Military Relations* (New York: Vintage, 1957); and Morris Janowitz, *The Professional Soldier: A Social and Political Portrait* (New York: Free Press, 1971). For a post–Cold War perspective on these arguments, see Peter D. Feaver, "The

Civil-Military Problematique: Huntington, Janowitz, and the Question of Civilian Control," *Armed Forces & Society* 23, no. 2 (Winter 1996), pp. 149–178. Although the Huntington/Janowitz distinctions remain the most useful frames of reference for our purposes, those interested in a more detailed discussion of other perspectives should consult Giuseppe Caforio, *Social Sciences and the Military: An Interdisciplinary Overview* (New York: Routledge, 2007). With specific reference to political science, see John Allen Williams, "Political Science Perspectives on the Military and Civil-Military Relations," in Caforio, *Social Sciences and the Military,* pp. 89–104.

4. Huntington, *The Soldier and the State,* pp. 8–18. Although the focus here is on professional military officers, we are indebted to Robert B. Killebrew for pointing out that this traditional emphasis is becoming outdated. Senior enlisted personnel look more and more like junior officers in their talents and responsibilities and may be no less "professional" as Huntington uses the term.

5. Huntington, *The Soldier and the State,* p. 456.

6. Janowitz founded the Inter-University Seminar on Armed Forces and Society in 1960. Subsequently led by Sam C. Sarkesian, Charles C. Moskos, David R. Segal, John Allen Williams, James Burk, Patricia Shields, and Laura Miller, it continues to serve as an interdisciplinary "invisible college" of civilian and military scholars worldwide on issues relating to the interaction of armed forces and the societies they defend. Its journal, *Armed Forces & Society,* was also founded by Morris Janowitz and is a primary scholarly outlet for studies on civil-military relations.

7. This is reflected in a comment among those serving in the military, "We're here to protect democracy—not practice it."

8. Janowitz, *The Professional Soldier,* p. 422.

9. A dramatic example of this phenomenon is President Trump signing an executive order in the Pentagon's Hall of Heroes (where Medal of Honor recipients are honored) restricting the entry of persons from several Muslim states into the United States, claiming a national security rationale.

10. Dan Lamothe, "Pentagon's Top General Apologizes for Appearing Alongside Trump in Lafayette Square," *Washington Post,* June 11, 2020, https://www.washingtonpost.com/national-security/2020/06/11/pentagons-top-general-apologizes-appearing-alongside-trump-lafayette-square.

11. Retired Marine Corps general John Kelly was White House chief of staff, retired Marine Corps general James Mattis was secretary of defense, and active duty Army lieutenant general H. R. McMaster was national security advisor.

12. The National Security Act of 1947 requires a congressional waiver for appointing a recently retired flag officer (general or admiral) to be secretary of defense. Three secretaries have been appointed under that waiver: George Marshall, James Mattis, and Lloyd Austin.

13. The official motto of the US Army itself is "This We'll Defend."

14. Thomas E. Ricks, "The Widening Gap Between the Military and Society," *Atlantic Monthly,* July 1997, pp. 66–78; and Thomas E. Ricks, *Making the Corps* (New York: Scribner, 1997). See also Mackubin Thomas Owens, *US Civil-Military Relations after 9/11: Renegotiating the Civil-Military Bargain* (New York: Continuum, 2011); and Sam C. Sarkesian and Robert E. Connor Jr., *The US Military Profession into the Twenty-First Century: War, Peace, and Politics,* 2nd ed. (London: Routledge, 2006).

15. Ole R. Holsti, "A Widening Gap Between the US Military and Civilian Society: Some Evidence, 1976–1996," *International Security* 23, no. 3 (Winter 1998–1999), pp. 5–42. Another perspective is that the gap is a good thing but should be managed. See John Hillen, "The Civilian-Military Culture Gap: Keep It, Defend It, Manage It," *US Naval Institute Proceedings* 124, no. 10 (October 1998), pp. 2–4.

16. Articles 88–91, Uniform Code of Military Justice, https://www.law.cornell.edu/uscode/text/10/888, /-889, /-890, /-891.

17. Eli Watkins, "White House Says It Didn't Distrubute MAGA Hats Trump Signed in Iraq, Germany," December 27, 2018, https://www.cnn.com/2018/12/26/politics/iraq-trump-hats-us-troops-maga/index.html.

18. As a Navy flag officer remarked in a personal communication, "Most civilians equate military service with sacrifice—sacrifice of personal liberties, sacrifice of personal choices, and so on—and they are very much unsure that they could do the same thing, not to mention that the thought of possibly having to would be abhorrent to them." This idea is captured in a 2019 article in an irreverent and often profane satirical website popular in the military, *The Duffelblog:* "I Would've Joined The Military Too If I Didn't Think It Was Beneath Me." See https://www.duffelblog.com.

19. Charles C. Moskos, "Short Term Soldiers," *Washington Post,* March 8, 1999, p. A19.

20. John Allen Williams, "Civil-Military Relations and the American Way of War," in Stephen J. Cimbala, ed., *Civil-Military Relations in Perspective: Strategy, Structure, and Policy* (Surrey, UK: Ashgate, 2012), pp. 69–82.

21. See Charles C. Moskos with John Sibley Butler, *All That We Can Be: Black Leadership and Racial Integration the Army Way* (New York: Basic, 1996), for a positive interpretation of the US Army's attempt to create a "race-savvy" (not race-blind) force that maximizes combat readiness.

22. See Charles C. Moskos, John Allen Williams, and David R. Segal, eds., *The Postmodern Military: Armed Forces After the Cold War* (New York: Oxford University Press, 2000), for a discussion of the many changes in relations between the military and society in twelve democratic states after the end of the Cold War. The model was extended and updated in John Allen Williams, "The Military and Society Beyond the Postmodern Era," *Orbis* 52 (2008), pp. 199–216. Graphic reports of women soldiers who are maimed or abused as prisoners of war (POWs) would have an especially shocking effect on the US public, although the loss of two female sailors during a terrorist attack on the USS *Cole* on October 12, 2000, caused no greater outcry than did the loss of fifteen men in the same incident.

23. See Elisabeth Bumiller, "Pentagon Allows Women Closer to Combat, but Not Close Enough for Some," *New York Times,* February 9, 2012, http://www.nytimes.com/2012/02/10/us/pentagon-to-loosen-restrictions-on-women-in-combat.html.

24. The policy permitting gay and lesbian members to serve so long as they "don't tell" was an institutional version of the personal policy most service members had followed for years. Most could think of military associates whom they believed to be homosexual, but it did not become an issue so long as it did not affect job performance. It did not occur to them to "ask."

25. In fiscal year 2017, ROTC programs accounted for some 37 percent of officer accessions, officer candidate schools 20 percent, service academies 18 percent, direct commissions 15 percent, and other/unknown 10 percent. See Office of the Under Secretary of Defense, Personnel and Readiness, *Population Representation in the Military Services, Fiscal Year 2017,* tab. B-30, "Active Component Commissioned Officer Gains, FY 2017 by Source of Commission, Service, and Gender," https://www.cna.org/pop-rep/2017/appendixb/b_30.html.

26. Eliot A. Cohen, *The Big Stick: The Limits of Soft Power and the Necessity of Military Force* (New York: Basic, 2016), p. 83. Major institutions for military education include the National Defense University (Washington, DC), the Naval War College (Newport, R.I.), the Army War College (Carlisle, Pa.), Air University

(Maxwell Air Force Base, Ala.), and Marine Corps University (Quantico, Va.). Each of these institutions has an outreach program to civilians involving conferences, publications, and participation in college functions.

27. Such a case is made in Sam C. Sarkesian, John Allen Williams, and Fred B. Bryant, *Soldiers, Society, and National Security* (Boulder, Colo.: Lynne Rienner, 1995).

28. General Petraeus makes a strong case for the value of rigorous civilian graduate study in David H. Petraeus, "Beyond the Cloister," *The American Interest* (July/August 2007), http://www.the-american-interest.com/article.cfm?piece=290. For an amusingly presented contrary view, see Ralph Peters, "Learning to Lose" in the same issue, http://www.the-american-interest.com/article.cfm?piece=291.

29. As Army chief of staff after Vietnam, General Creighton Abrams integrated Army active and reserve forces so thoroughly that it would be impossible for the Army to fight a major war without calling up the reserves (as President Johnson refused to do in Vietnam, lest it reduce public support for the war). National Guard forces, normally under the control of state governors, are part of this equation as well.

30. See John Allen Williams, "The US Naval Academy: Stewardship and Direction," *US Naval Institute Proceedings* 123, no. 5 (May 1997), pp. 67–72.

31. David E. Sanger, *The Perfect Weapon: War, Sabotage, and Fear in the Cyber Age* (New York: Crown, 2018).

32. Sam C. Sarkesian, "The US Military Must Find Its Voice," *Orbis* 42, no. 3 (Summer 1998), pp. 423–424.

33. Huntington, *The Soldier and the State,* pp. 459–460.

34. For an extended discussion of this issue, see H. R. McMaster, *Dereliction of Duty: Lyndon Johnson, Robert McNamara, the Joint Chiefs of Staff, and the Lies That Led to Vietnam* (New York: HarperCollins, 1997). This book was widely read in Washington, DC, in the wake of US military operations in the Balkans.

35. Unidentified leaks from the Joint Chiefs of Staff suggesting they had reservations about the military strategy do not count here. If they had such reservations, they failed to convince the proper civilian leaders in a timely fashion. It is as if senior military officials began to speak on the record only after they feared that the failure of the original strategy would be apparent to all. See Robert Burns, "Reimer Reveals His Views on Kosovo Strategy," *European Stars and Stripes,* May 27, 1999, p. 1. On Iraq, see Thomas E. Ricks, *Fiasco: The American Military Adventure in Iraq* (New York: Penguin, 2006); and Michael R. Gordon and Bernard E. Trainor, *Cobra II: The Inside Story of the Invasion and Occupation of Iraq* (New York: Pantheon, 2006). It will be interesting to see how responsibility is apportioned for the futile twenty-year US attempt to transform Afghanistan into a Western democracy.

36. As quoted in Harry G. Summers, *On Strategy: The Vietnam War in Context* (Carlisle, Pa.: Strategic Studies Institute, US Army War College, 1981), pp. 7, 25. These originally appeared in *CDRS Call* (July/August 1976). Summers wrote, "General Weyand was the last commander of the Military Assistance Command Vietnam (MACV) and supervised the withdrawal of US Military forces in 1973" (p. 7).

37. There is widespread agreement on this among those who study issues of military and society. See, for example, Eliot A. Cohen, "Civil-Military Relations," *Orbis* 41, no. 2 (Spring 1997), pp. 177–186.

38. *Korematsu v. United States,* 323 US 214 (1944).

39. John Keegan, *History of Warfare* (New York: Knopf, 1993), p. xvi.

PART 4

Long-Range Perspectives

14

Emerging Issues

In the analysis of US national security, the policymaking process and the results of that process are both important. This book emphasizes the process of making US national security policy. Without understanding the process of policymaking, the results would make little sense. Yet a complete picture must also take into account the substantive issues of national security over which policymakers struggle. Such issues matter, for the process is not simply an end in itself. Nor is the relationship between "process" and "outcome" a simple or linear one. No template exists to guarantee that any particular mix of expert decisionmakers, fortuitous international circumstances, and clever military plans can guarantee success in war or preclude peacetime challenges to US national security.

We identify seven major considerations relating to national security that are important to understand long range issues: national security as a concept, strategic cultures, other ideologies and philosophical systems, geostrategy, homeland security, technology and economics, and the information and communications revolution. They raise long-range issues not amenable to short-range perspectives and ad hoc responses. Each issue requires constant evaluation, followed by adjustments and refinements to national security policy and strategy. The interrelationships among the seven considerations compound the difficulty in designing policy, since a response to one category of issues will affect issues within other categories.

National Security as a Concept

Our concept of national security is broad and inclusive.[1] Recognizing the problems of defining and conceptualizing national security, we offered the following statement in Chapter 1: US national security is the ability of national institutions to prevent adversaries from using force to harm Americans or their national interests and the confidence of Americans in this capability. Note that this definition implies that security is both "objective" and "subjective": people who do not feel secure within a political community are a potential nutrient for destabilizing or disintegrating forces in that community. From this, the definition was expanded to include a variety of components, ranging from economics and humanitarian concerns to refugee assistance and institution building. In fact, many of these components are not pure national security issues but are more closely related to foreign policy and finances.

Taking a page from Sun-tzu, if national security includes virtually everything, then it includes nothing. From time to time, policymakers have defined national security so expansively that it risked a loss of meaning and conceptual coherence. In this context, the primary function of the Department of Defense and the military got lost in a muddled maze of fragmented policy, clouded strategy, and questionable, politically driven issues that in the long term may erode US capability. For example, the George W. Bush administration had an expansive notion of presidential power and therefore favored a robust definition of US security purposes consistent with that. As a result, the US military—especially the US Army—found itself fighting two major conventional and unconventional wars in Iraq and Afghanistan and a global war on terror with intelligence, training, and advising missions in many countries. The conceptual bases of US national security in the current environment are also complicated by modern understandings of warfare and its relationship to politics, society, and culture. In the pursuit of all this complexity, the centrality of strategy and its relationship to policy and to the application or threat of force may be lost.[2]

We need to reexamine our definition with a view toward focusing the conceptual bases and clarifying the strategic purpose of national security policy. It may be useful to revise the definition as follows: the use of military force or the threat of force[3] to protect vital national interests based on core (first-order) priorities. Too often, when US policymakers argue about the "how" of military intervention, they have given insufficient prior deliberation to the "why" or "what" for that situation. The result of such thinking is that the country oozes or plunges into tactical engagements without a strategic compass. As Colin S. Gray has noted: "If strategy is conceived, understood and practiced as a bridge, at least the most relevant military commanders should have no difficulty understanding that theirs is, by its very nature, a bridging function. It is the task of higher military command to apply military force as threat and action that political interests are served as a consequence."[4]

Examples of the use of military power without the discipline of strategic connection abound in US history. For example, US military intervention in the Lebanese civil war in 1983 took place under the pressure of humanitarian impulses for conflict mediation by outside powers. But ambiguous policy guidance and inappropriate rules of engagement for US forces resulted in the tragic deaths of 243 US Marines, among other multinational forces, after a truck bombing of their barracks in October 1983.

Another example of plunging into military operations without clear strategic or policy goals was provided by United Nations (UN) and US decisions to intervene in Somalia in 1992–1993. The decision for intervention was taken by outgoing President George H. W. Bush during the final months of his administration in reaction to humanitarian disasters in Mogadishu and other cities in that troubled country. Over the course of the next nine months, UN and US commitments escalated into involvement in a civil war among Somali clans. In October 1993, a US military operation in Mogadishu by special forces and Army Rangers resulted in unexpected casualties and the loss of US helicopters in the infamous Black Hawk Down episode. Members of Congress were somehow surprised to learn that the United States was at war in Somalia, and shortly thereafter the Bill Clinton administration made a decision to withdraw US combat forces. The decisions by Bush and Clinton (with support from the UN) for military intervention and escalation were taken despite the fact that there was no functional central government in Somalia at the time; it was the personification of a "failed state." More recently, and far more costly, the rationale for US intervention in Iraq and Afghanistan changed over time over several presidencies and made continued military actions more difficult to sustain politically.

An important part of the concept of national security is strategic vision. The link between national security and national security strategy is vision—seeing how one proceeds from one to the other. This has yet to be fully developed by the United States in the new security landscape. For example, in the 1990s the Department of Defense and the military services outlined broad aspects of future warfare and how the United States should prepare to deal with twenty-first-century security challenges. Unfortunately, these studies (primarily *Joint Vision 2010* and *Joint Vision 2020*) placed little emphasis on asymmetrical warfare, the kind practiced by terrorists on September 11. According to General Rupert Smith, who commanded the British Armored Division in the Gulf War of 1991, UN peace operations in Bosnia in 1995, and British forces in Northern Ireland from 1996 to 1998, among other key assignments, an altogether new concept of war is necessary for strategists and policymakers: "It is now time to recognize that a paradigm shift in war has undoubtedly occurred: from armies with comparable forces doing battle on a field to strategic confrontation between a range of combatants, not all of which are armies, and using different types of weapons, often improvised. The old paradigm was that of interstate industrial war. The new one is the paradigm of warfare amongst the people."[5]

Since General Smith wrote these words, of course, elements of the old paradigm of conventional interstate warfare (and deterrence) have reasserted themselves and become an important focus for US defense planning.[6] The 2022 Russian invasion of Ukraine showed how difficult deterrence is when a powerful country is determined to act. General Smith's concept of "war amongst the people" goes beyond the ideas of military transformation, emphasizing technology, and asymmetrical warfare, which is somewhat trite because all warfare is asymmetrical in principle (each side seeks to win by doing the unexpected and unorthodox). "War amongst the people" emphasizes the difference between *deploying force* and *employing forces* to good effect. For example, over the years, various states have deployed conventional military forces for battle in a number of military engagements, but those deployments have "spectacularly failed to achieve the results intended, namely a decisive military victory which would in turn deliver a solution to the original problem, which is usually political."[7] Some military historians would argue that this is not an entirely new discovery; conventional militaries have often been bogged down or been defeated by unconventional or irregular warfare, from Napoleon's troubles in Spain to the US experience in Vietnam.[8] Smith and others, such as military historian Martin van Creveld, suggest, however, that what has been thought of as regular or conventional war (fought between professional militaries under state control according to standardized tactics) may be the exception, not the rule, in future conflicts.[9]

Van Creveld's prediction may not come about. The March 2021 interim strategic guidance of the incoming Biden administration dealt with a wide variety of challenges, including those from peer competitors. It notes: "In the face of strategic challenges from an increasingly assertive China and destabilizing Russia, we will assess the appropriate structure, capabilities, and sizing of the force, and, working with Congress, shift our emphasis from unneeded legacy platforms and weapons systems to free up resources for investments in the cutting-edge technologies and capabilities that will determine our military and national security advantage in the future."[10]

In sum, there is a compelling need to rethink the concept of national security to ensure its relevancy in the post–Cold War period and to design a strategic vision and operating principles for the US military. At present, there is a serious gap between the political rhetoric of national security and the capability and effectiveness of the US government to respond to security challenges in the twenty-first century. In addition, future concepts of national security must be multidimensional and capable of dealing with nonlinear and chaotic futures: the United States may be called upon to wage war and/or to practice coercive diplomacy and deterrence in the various domains of land, sea, air, space, and cyberspace in various combinations. So, for example, the United States may find itself employing counterterror and postconflict stability operations in one theater, deterring or fighting a sizable conventional war in another, and deterring space and cyber wars on a third front—all at the same time. Even in dealing with challenges from aspiring peer competitors and great powers, the United States will need to adapt to the new frameworks for strategy posed by the twenty-first century. As Anthony Cordesman has noted:

> Where possible, China and Russia use their military power in what might be called "wars of influence" and in ways that do not involve actual fighting. When they do use force, it generally takes the form of limited or demonstrative uses of their own forces; covert operations; or the support from forces of other states, non-state actors, or factions. If the United States is to deal effectively with such competition, then the U.S. must refocus its military strategy and forces to give gray area and hybrid conflicts at least the same or even more priority as it does to higher levels of warfare.[11]

Cordesman's prediction is generally correct, but Russia's 2022 war against Ukraine involved an initial use of coercive diplomacy, supported by large military deployments, followed by a large conventional invasion to bring about regime change.

Strategic Cultures

During the Cold War it was often said that the real struggle—with a direct impact on US national security—was between the tenets of Western democracy and those of Marxism-Leninism as practiced in the

Soviet Union and Eastern bloc. In the post–Cold War period, however, the tenets of democracy are confronting those of non-Western cultures such as Islam, Hinduism, Buddhism, and Confucianism. Religious fundamentalism, which can include anti-US virulence, is especially important in this context.

Philosophy establishes the moral tenets of a system. *Ideology* provides the political rationale and basis of legitimacy of that system. The struggles over these issues reach into virtually all parts of the world. The critical dimension is not so much the imposition of one culture over the other but whether the legitimacy and credibility of one type of political system prevail in the long run. This is not meant to imply that we are undergoing a war between civilizations, where one culture tries to overcome another. And this is not meant to imply that conflicts will necessarily have a military dimension. On the contrary, such struggles are primarily political-psychological. This is also not meant to argue that other states need to adopt one cultural system as their own. But there is an inherent and universal struggle between the open systems of the West and the closed systems of other cultures. As Samuel P. Huntington has written: "Western ideas of individualism, liberalism, constitutionalism, human rights, equality, liberty, the rule of law, democracy, free markets, the separation of Church and state, often have little resonance in Islamic, Confucian, Japanese, Hindu, Buddhist or Orthodox cultures."[12]

Philosophy and ideology have much to do with the degree of cultural affinity and commonality of purpose between the United States and a state. Obviously, US national security policy is best served by open political systems based on the Western Enlightenment-Renaissance heritage and democratic ideology. This is not to suggest, however, that commonality of purpose cannot be achieved with non-Western states such as India or those in the Middle East. If such states are striving for openness and a democratic ideology (not necessarily defined in Western terms), then it is important for US national security interests to develop friendly relationships with them. More problematic, of course, is the need to develop such relationships with states that are not striving for either.

The broader notion of strategic cultures can be better understood by reviewing Marxist-Leninist ideology and its attempt to establish a Communist cultural system. This background will help our understanding of what evolved from the Cold War and what remains in place in Russia, former Soviet republics, in China, and elsewhere. Although communism was a secular ideology rather than a sectarian one, its presumptions of universality, perpetual power struggles, and thesis-antithesis relationships between the existing and forthcoming political orders are shared with other antisystemic movements based on claims of injustice and promises of victory on earth or even heavenly vindication. Karl Marx's faith in the eventual triumph of a global proletariat, under enlightened Communist Party leadership, is replaced nowadays by the commitment of religiously politicized movements, like al-Qaeda and the Islamic State, to turn back the historical clock and restore a medieval Caliphate. Marx sought to move forward into an imagined future of material abundance and the disintegration of social classes; Osama bin Laden (and some other Islamists) preferred equally transformative, but retro, destruction of the existing order of economic globalization and cultural Westernization.

No single explanation does justice to the socio-ideological themes of Karl Marx and V. I. Lenin. The issue becomes even more difficult when the evolution of Marxist ideology is examined in the context of the Soviet state.[13] The Marxist-Leninist paradigm was based on several important concepts: the view that the party was the primary organ of power, the goal of establishing socialist systems throughout the world, and the notion that there was a basic contradiction between capitalism and Soviet communism that could lead only to the demise of the capitalist state. As Mikhail Gorbachev, the last head of the Soviet state, revealed in his address in Moscow on November 2, 1987, marking the seventieth anniversary of the Bolshevik Revolution: "The world communist movement grows and develops upon the soil of each of the countries concerned. . . . Born of the October Revolution, the movement has turned into a school of internationalism and revolutionary brotherhood. And more—it has made internationalism an effective instrument furthering the interests of the working people and promoting the social progress of big and small nations."[14]

Gorbachev's comments reflect Soviet culture and the Marxist-Leninist legacy. This culture was nurtured over a period of seventy years and was rooted in central control and disciplined social behavior with control of the state by a small elite, which perpetuated its own power and recruited its own. The party remained the real power in a system in which the party established policy and supervised the instruments of the state in carrying out that policy. This is seen most clearly today in China, where the Communist Party and the state maintain control with the help of increasingly sophisticated technical capabilities for monitoring the population.

There is a significant residue of Marxism-Leninism represented in the number of Communist Party representatives in the Russian Duma today, in the resentment by many elderly and poor Russians at their decline in status since the Soviet Union disintegrated, and in the nostalgia held by some elites and others in Russia for the imperial trappings and international respect that were formerly accorded the Soviet Union. As an ideology capable of capturing the government and being reestablished as an official public doctrine or faith, however, Marxism-Leninism lacks some of the attributes that would be required for restoration. It cannot offer a blueprint for reforming the Russian economy or making it competitive in international trade and finance. To the contrary, Marxism-Leninism would be a retreat into history. In the actual (as opposed to the idealized) Soviet Union, a few party and government elites lived well while the masses lacked adequate consumer goods and any real political influence. Thus, realists in contemporary Russia recognize that a historical flashback to communism in Russia is neither advisable nor possible, and President Putin is increasingly reminded that many middle-class and professional Russians feel the pace of democratization and government transparency in Russia is too slow or even moving in the wrong direction.[15] At the same time, Putin has long hoped to restore much of the old Soviet sphere of influence, as his 2022 invasion of Ukraine dramatically illustrated.

Other Ideologies and Philosophical Systems

The Chinese System

China presents a paradox to the United States in the twenty-first century. An economic partner and rival, it also presents rapidly increasing security challenges in the Asia-Pacific region and globally. China's embrace of Marxism-Leninism has been both nationalistic and opportunistic. With more than 1 billion Chinese on the mainland and a drive for modernity in the post-Mao period, the Chinese state is already an economic superpower and is already a military peer competitor for the United States.[16] By 2011, China's economy had passed that of Japan as the world's second largest, and by some measures, such as purchasing power parity, it equals its American rival. Yet serious questions remain regarding the future of China. Chinese influence throughout Asia and Southeast Asia increased considerably in the latter twentieth and early twenty-first centuries, especially in economics, and increasingly as a result of attempts to modernize the military system.[17] To be sure, the Confucian cultural system is part of China's overall development. Additionally, the Chinese mainland shares a long border with Russia, India, and states in Southeast Asia. Since the early 1980s, China has been involved in shooting confrontations with India, the Soviet Union, and the former North Vietnam, and skirmishes with other nations with conflicting territorial claims in the Western Pacific. China's ambitious "One Belt, One Road" approach to international primacy in supply chains for trade, manufacturing, and emerging technologies, including telecommunications, supports its growing web of transnational economic and political connections.

Confucianism, a somewhat pragmatic approach to external relationships, combined with an underpinning of historical power considerations of the Middle Kingdom and mandate of heaven, establishes an amalgam of Chinese ideology and the substance of Chinese culture. Chinese culture dates back centuries before the common era and remains deeply embedded in modern China, despite attacks by the ruling Chinese Communist Party. Yet it is clear that many changes are taking place that are transforming the country into an economic giant with a modern version of Chinese culture; such changes could also lead to civil strife or a balkanization of China, a modern version of the warlord system that predominated during the 1920s and 1930s.

The long history of Chinese influence throughout Asia and its potential as a hegemon in the twenty-first century make the People's Republic of China an important regional and global player, a pivot for US foreign and defense policy in the region. The Clinton administration took advantage of China's economic modernization in the post-Mao era by aggressively promoting trade and cultural exchange with Beijing. China was also asked to play a part in supporting a US-led solution to the problem of containing North Korea's nuclear ambitions by giving Beijing's blessing to the so-called Framework Agreement of 1994. Later the George W. Bush administration would also enlist China's aid on this issue, after North Korea declared itself a nuclear power and the United States sought to reverse that status. Later, repeated US, United Nations, and other states' efforts to leash or dismantle North Korea's nuclear program through multilateral negotiations or bilateral relationships came up short of expectations. The Donald Trump administration tried a new approach in 2017–2019, emphasizing bilateral summits between heads of state, but this Trump-Kim summitry achieved little beyond pageantry and headlines.

In this respect, the United States and China have also tangled over other nonproliferation issues. In the 1990s, the United States sought to persuade Beijing not to provide nuclear know-how and missile technology to states of concern in Asia and the Middle East, especially Iran, Pakistan, and North Korea. Issues of proliferation reminded US officials that China's relationships with Russia, Japan, and the two Koreas were important variables affecting the United States–China dialogue on arms control. China's relationships with North Korea are complicated because they intersect with China's interests and relations with South Korea, Japan, Russia, and the United States.

Although some US analysts have long argued that China seeks to become a global superpower, the Chinese political leadership, under President Xi Jinping, has only recently been open about their march to great power status. The phrase "China's peaceful rise" is belied by her aggressive posture in the Western Pacific. Part of China's economy remains underdeveloped by European, US, and Japanese standards. In addition, if China is to modernize its semi-marketized economy in order to compete with the leading financial powers, the degree of authoritarian political control over investors and entrepreneurs will have to loosen. For example, China is imposing controls over Internet access and content for political purposes and has had disputes with the Internet giant Google over alleged hacking condoned by the Chinese government. Even though China's attempts to modernize its armed forces by advances in computers, communications, and electronics are bearing fruit, its overall defense budget, large by international standards, still pales in comparison to that of the United States. However, given China's doctrinal emphasis on "unrestricted warfare" (China's comprehensive approach to the theory and practice of war including conventional, unconventional, and cyber conflict among other approaches), it can be assumed that China will seek niche capabilities that would offset any US or other state's comparative advantages, including an obvious Chinese interest and expertise in information warfare.[18]

After the Donald J. Trump administration took office, it promised to hold China accountable for what the US president regarded as decades of unfair trade practices. In addition, the Trump administration also confronted China over issues of intellectual property theft, cyber espionage, and construction and militarization of artificial islands in the South China Sea. In 2020, the Covid-19 pandemic led to charges by US officials that China had deliberately concealed the extent of the epidemic originating in China, permitting its global spread before other states were fully aware of the magnitude of the disaster. Further disputes arose over US charges of espionage by Chinese-American university researchers, exchange students, and others. The United States closed a Chinese consulate in Texas in 2020 and charged that it was a center for espionage. Both the Trump administration and its liberal critics condemned China's treatment of minorities, especially its oppression of the Muslim Uighur population in Xinjiang province. US critics charged that Uighurs and other prisoner populations were exploited under inhumane conditions and chided China for lack of transparency on this point. US officials also clashed with China over China's de facto repudiation of Hong Kong's special semi-autonomous status and its severe crackdown on political freedom there.

The Biden administration has continued to be tough on China, and is ramping up US military capabilities in the Western Pacific. Indications of a large planned buildup of Chinese intercontinental ballistic missiles (ICBMs) and other missile capabilities such as the development of a hypersonic missile that could evade US antiballistic missile defenses, of are of growing concern to the United States.

The United States must pay special attention to China's relationships with Japan, Russia, and India. India officially joined the nuclear club in 1998, partly owing to a perceived threat from nuclear-armed China. Russia, still powerful in North Asia, seeks rapprochement with Japan, including the settlement of territorial disputes outstanding from World War II. Japan is concerned with China's waxing military capability as well as any Chinese pretensions to superpower or hegemonial status. China, of course, has its own grievances against Japan stemming from Japanese actions there before and during World War II. Russia has taken steps to improve security relations with China, including Chinese participation in large Russian combined arms military exercises and joint combat air patrols over contested national air defense zones.[19] Both states suspect that US missile defenses will accelerate the nuclear arms race, and both Russia and China oppose a unipolar international order dominated by the United States. On the other hand, China's expansive Belt and Road Initiative threatens to compete for influence with Russia in Central Asia, and China and Russia have both convergent and divergent interests within and outside Asia. Russian declaratory policy, for example, favors nuclear disarmament of the Korean peninsula, but China's perspective on this issue is more opaque. As well, efforts to bring China into previously bilateral US-Russian nuclear arms reduction talks are unlikely to bear fruit any time soon. China's lack of transparency with respect to its nuclear arsenal, compared to the United States and Russia, makes a triangulation of nuclear reduction talks a challenging accomplishment.

The Middle East

The terrorist suicide attacks on the World Trade Center and Pentagon on September 11, 2001, stunned many US citizens and confounded many security experts. The sophisticated preparation required to carry out the attacks, combined with the high degree of fanaticism and self-sacrifice on the part of the perpetrators, offered a new variation on an old problem: terrorism and unconventional, or asymmetrical, warfare. The terrorists cleverly exploited US vulnerabilities (e.g., US civil society and democratic openness) in the classical manner of Sun-tzu. Many people asked whether religious hatred motivated the terrorist attacks, but it was more complicated than that. The problem was not religion per se, but rather politics with a religious undercurrent.

In the 1990s the Islamic and Arab worlds collided with globalization and its internationalization of finance, information, and technology. Traditional geographic borders and regional and local cultures were penetrated by outside influences, with billions of investment dollars moving at the speed of light. Governments began to feel a loss of control over their own destiny and citizens. This vulnerability to the forces of globalization divided those who were prepared to benefit ideologically or financially and those who were not. The nonbeneficiaries often included traditional religious leaders.

Traditional elites and politicians now feel that their values are under siege, and many blame the United States and the West. Symbols of Westernization often go hand in hand with globalization—particularly in the area of popular culture. For Muslims and Arabs with access to money and arms, their growing resentment expressed itself in revolts against secular regimes in their own states (e.g., the uprisings of the Arab Spring across North Africa and the Middle East in 2011) and sometimes through their support of global terrorist organizations. These include Osama bin Laden's al-Qaeda, responsible for the 2001 terrorist attacks in the United States, and the more recent Islamic State—now forced out of its original geographical caliphate in Syria and Iraq but a continuing danger in other states via regional affiliates. One of these, the Islamic State Khorasan Province (ISIS-K) was responsible for the death of thirteen American service personnel and hundreds of Afghani civilians in a suicide bomb attack at the Kabul airport in the final days of the US military withdrawal from Afghanistan.

In the immediate aftermath of September 11, some pundits foresaw a clash of civilizations between militant Islam and the West, between holy warriors and decadent capitalists. This pessimistic view mistakenly blames the entire Islamic culture for the actions of a few murderers using religion to justify terrorism. Islam, as in the case of Christianity, Judaism, Buddhism, and other belief systems, attracts followers with many different political agendas, and some of them use religious justifications for acts most would regard as sacrilegious. For example, many

Christians today would acknowledge with embarrassment the atrocities of the Crusades from the eleventh through the fourteenth centuries as well as the Spanish Inquisition. None of the world's great belief systems, properly understood, justifies mass murder; but many have called forth religious rationales for imperialism, persecution, and terror. We should expect this pattern to continue.

To his credit, President George W. Bush spoke forcefully in September 2001 about US determination to bring the terrorists to justice while emphasizing US unwillingness to wage a war against Islamic and Arab communities. Bush emphasized US multiculturalism and inclusiveness and deplored the attacks by bigots against Arab and Islamic US citizens. In addition, Bush sought to develop a broad-based coalition for the US war against terrorism and states that support terrorism, including Arab and Islamic states. His success in lining up support from Saudi Arabia, Egypt, and other Islamic and Arab states was a message to US citizens as well as to potential enemies of the United States.

Yet early victories against the Taliban and al-Qaeda did not end challenges to US security, especially in the Middle East. The Palestinian intifada continued to reflect secular and religious hatreds, and US support for Israel remained a sore spot throughout the Muslim and Arab communities. Absent a miraculous breakthrough, the United States had to strike a careful balance between its security commitments to Israel and its desire for better relations with Arab and Islamic communities. The Arab and Islamic sentiment toward Israel is also shaped by Israel's reputation for military effectiveness. The Abraham Accords negotiated by the Trump administration attempted with some success to normalize the relations of several Gulf states with Israel.

Another Middle Eastern challenge to US security is posed by Iran's drive to become a nuclear weapons state. The Obama administration in 2011–2012 combined diplomatic offensives among the five permanent members of the UN Security Council plus Germany, with UN and European Union (EU) sanctions against Iranian banks, oil exports, and other targets in order to coerce Iran into abandoning its uranium enrichment programs. The Iranian nuclear deal of 2015 (the Joint Comprehensive Program of Action) slowed at least temporarily Iran's march toward becoming a nuclear weapons state, and the Obama administration regarded the action program as a singular arms control success.[20] However, the Donald Trump administration abrogated the Iranian nuclear deal in 2018 and increased economic sanctions against Iran, arguing that the agreement was a bad deal for the United States and a policy of "maximum pressure" was required to end Iran's nuclear ambitions. US

European allies attempted to keep Iran in the deal by negotiating separate de facto agreements that would allow Iran to sell limited amounts of its oil without US secondary sanctions against allies and partners, but the US economic pressure on Iran was only increased. Ultimately in 2019 the Iranian leadership announced that, in its view, the agreement was no longer viable, and US relations with Iran deteriorated further. In mid-September 2019, cruise missile and drone attacks on Saudi oil facilities, presumably orchestrated by Iran, further ratcheted up tensions between Tehran and Washington,[21] as did the US drone strike that killed Iranian Revolutionary Guard general Qassim Soleimani on January 3, 2020. Biden administration attempts to reanimate the agreement have been hampered so far by continuing hostility and political pressures in both countries, but efforts continue.

On the other hand, the United States also supports some conservative and moderate Islamic and Arab regimes. Sometimes this US support for authoritarian governments in the Arab and Islamic worlds has resulted in unwelcome "blowback" as US aid and policies are, correctly or not, conflated with those of local rulers. For example, when uprisings in Egypt demanded the end of the Hosni Mubarak regime in the spring of 2011, the United States found its policy torn between the principle of promoting democracy and its prior reliance on Mubarak as a regional support for US policies and peace with Israel. US interests across the region during the Arab Spring protests of 2011 varied from one country to another: for slow movement toward democratic regime change with sporadic violence (Egypt for a time and Tunisia); for prompt overthrow of the existing regime by internal opposition supported by external military forces (Libya); for regime change over time provided it does not have drastic spillover effects into other US allies (Bahrain, without destabilizing Saudi Arabia); or for destabilizing or even revolutionary movements against the regime in power, although the likelihood of accomplishing this in the near term is slight (Iran, Syria). Meanwhile, the pot boiled in Yemen, where the Ali Abdullah Saleh government provided willing support against al-Qaeda in the Arabian Peninsula for US antiterrorist campaigns, but US officials remained uncertain of the regime's sustainability or enduring legitimacy. More recently, the US military and intelligence support for Saudi Arabia's war in Yemen against Iranian-backed Houthi rebels resulted in blowback from Iranian-sponsored attacks against Saudi oil fields and infrastructure. Elsewhere in the region, Russia's military intervention in Syria beginning in 2015 appeared several years later to have succeeded in rescuing the Assad regime from overthrow by various enemies, including some factions that were US-supported.

Other Systems

Other parts of the third world reflect combinations of geography, socialism, democracy, communism, authoritarianism, military rule, and personalism, often influenced and shaped by Islam, Hinduism, Buddhism, Christianity, and animism. The point is that US and Western forms of democracy and open systems compete with a variety of other philosophical and ideological systems.

South America and Central America are important in terms of geography, ideology, and philosophical principles. To be sure, South America has a cultural dimension influenced by a colonial legacy from the Old World, including Britain, France, Spain, the Netherlands, and Portugal—all states with a Christian heritage. This legacy reflects a range of political orientations, from monarchical to authoritarian to democratic. But many states in the region developed political cultures with distinct interpretations of Christianity and social structure, in which the concepts of openness and democracy take on a different connotation. There is a distinct Hispanic culture with roots in the Old World intermixed with indigenous cultures. Additionally, the increasing Hispanic population in the United States adds a cultural as well as emotional link to Central America and the Southern Hemisphere.

Sub-Saharan Africa, although far from representing a coherent philosophical or ideological entity, presents a unique dimension nevertheless. Its colonial history and legacy of the slave trade have put sub-Saharan Africa in a special relationship with the United States. African Americans account for about 13 percent of the total US population.[22] Many Americans of African ancestry feel a link with the continent of Africa, reinforced by the history of slavery in the United States and by the civil rights movement. Although in the short term there may not be a compelling national security issue in sub-Saharan Africa, in the long term this link is sure to be influential in shaping US national security interests, not least because of the many resources found there. Although much political turbulence and regime change took place in sub-Saharan Africa in the 1990s, there were some developments favorable to the spread of open systems, but even there governmental mismanagement caused a crisis of legitimacy.

The breakup of failed states and the resulting chaos that may lead into local or regional wars are among the major challenges to US foreign policy in sub-Saharan Africa in the twenty-first century. For example, protracted civil wars partly motivated by ethno-religious strife in Congo and in Sudan pushed themselves onto the foreign policy agendas of the

George W. Bush and Barack Obama administrations and the United Nations. In 2011, South Sudan separated from Sudan to form a new nation-state, although of uncertain durability and amid some border clashes with Sudanese troops. The problem of post-Qaddafi Libya, divided between a government recognized by the UN and rebel militias seeking to overthrow it, continued to confound US policymakers into the Trump administration. The US Department of Defense has established a unified command for Africa reflecting its growing significance in posing challenges of international conflict management, economic development, environmental despoliation, demographic change (including migration), and potentially failed states or tottering governments (for example, Libya and Somalia) creating havens for terrorists or other malefactors.[23]

In summary, US security interests must deal with a variety of cultures and ideological and philosophical tenets that perceive the world in distinctly different terms. US values, norms, and political system may be seen as a threat by some states, as imperialistic by others, and as interventionist by still others. Thus, US security policy and strategy, although appearing reasonable, rational, and moral from the US perspective, are often seen as arrogant and immoral by others.

The design of US national security policy and strategy must therefore be based on intellectual and policy horizons that are not strictly bound by Western culture, ideologies, and philosophies. Thus, to design national security policy in the context of our homegrown democracy with the expectation that it will find favor elsewhere is naive and dangerous to US interests. The concept of democracy is shaped by the distinct experience and cultures of particular states. In dealing with the variety of cultures that exist in the world today, the United States must understand them and pursue US national security policy with prudence and patience. This does not require that the United States and Western cultures be compromised. But it does require an appreciation of the limits imposed by the external environment and a recognition that the pursuit of US national security policy takes time. Indeed, it sometimes results in failure, as the Biden administration seems to have realized as it abandoned the democratization project in Afghanistan.

Geostrategy

Major powers usually place themselves at center stage in world politics. They see the world from their own physical location and with their own worldview. In the early twentieth century, this tendency became systematic, elevating geopolitics to a major dimension of the strategic maneuvering of major powers. Geopolitics stressed the importance of geography as a factor in determining the power of a state. In more current versions, geopolitics has spawned a geostrategic component, combining geography and strategy into a national security perspective. Differences in geostrategic perspective are among the variables that make it more difficult for the United States and other powers to see the world in similar ways. For example: the United States asserts a "responsibility to protect" that includes the right of military intervention against governments that commit violations of human rights against their own citizens. Russia, on the other hand, has rejected arguments for regime overthrow on this basis as a contradiction of state sovereignty and as a Western attempt to destabilize leaderships in Russia's "near abroad" and, perhaps, in Russia itself. Of course, these protests ring hollow in the face of the 2022 Russian invasion of Ukraine in order to achieve a change in regime to one favorable to Russia. In addition, Russia's unique geostrategic location astride both Europe and Asia creates not only an expansive territorial reach, but also a more inclusive set of borders to defend. For this reason, Russia is highly sensitive to perceived encroachments on her security, such as the possibility of North Atlantic Treaty Organization (NATO) membership for Ukraine to its west and Georgia to its south. This was an impetus for wars against Georgia in 2008 and Ukraine in 2022.

The location of states, their natural resources, and their power relative to other states are critical elements of geostrategy.[24] For example, oil resources make many states in the Middle East extremely important to the national security of other states dependent on foreign oil. Sitting astride important sea-lanes and trade routes (e.g., the Panama and Suez Canals, the Strait of Hormuz, the Caribbean passage, and the Straits of Malacca) also enhances a state's geostrategic importance. The location and policy of a state with respect to major power strategies can also make it geostrategically important. Nicaragua, Cuba, and the eastern coast of Africa are but a few examples. Although seemingly minor states such as Nicaragua or Somalia might not appear to be important in and of themselves, manipulation by other powers as conduits for power projection and the pursuit of their national security goals carries national security implications for the United States.

Geostrategic components as of US national security policy and strategy cannot be considered in isolation. Cultural issues influence geostrategic considerations and, in some instances, reinforce geostrategic importance. For example, the geostrategic importance of Israel is reinforced by the fact that it is a relatively open system whose values are generally compatible with those of the United States. In other instances, ideological and philosophical issues can be temporarily subordinated to geostrategic considerations, creating a condition in which the United States has few options other than to deal with a nondemocratic state. Past US support for governments in Saudi Arabia, Pakistan, and Egypt provide pertinent case studies.

In relating geostrategic elements to the specifics of national interests, the United States faces contradictory forces and contending ideologies. As a result, US policy and strategy can appear vacillating and unclear. Geostrategic considerations must be included in formulating policy and designing strategy, but they cannot simply be quick fixes and short-term responses. Rather, geostrategy must be included as part of a broader concept, one that provides a systematic application based upon a geostrategic vision. For example, the containment policy adopted by the Harry Truman administration just after World War II was rooted in geostrategic concepts that implied a principle of selection. The United States would not attempt to defend everything, everywhere: primary security interests lay in preventing any single power, especially the Soviet Union, from imposing hegemony over Western Europe, North Asia, and key parts of the Middle East. When the idea of containment was extended (through the domino theory) to include every trouble spot under threat of Communist revolution, as in Vietnam, it became muddled, provided less compelling policy guidance, and got the United States into trouble.

The George W. Bush administration stirred considerable controversy at home and abroad as a result of its ambitious geostrategy in response to global terrorism and threats of weapons of mass destruction (WMD). Bush declared the days were numbered for regimes making up the "axis of evil" (Iraq, Iran, and North Korea) and for other regimes that supported transnational terror or spread WMD to terrorists. Bush objectives included a realignment of Middle East politics and security through a transformed military landscape and a hoped for but as yet unrealized spread of democracy in its wake. The Middle East and Central Asia were no longer peripheral theaters of operations for US regional military commanders but central pivots for US policy and force operations planning.

The Obama administration that followed Bush had to deal with the management of this enlarged canvas of security commitments and force deployments while, at the same time, accepting resource constraints imposed by overstretched US ground forces (including reserves and National Guard forces) and a querulous anti-spending Congress elected in 2010. In this respect, the Trump administration followed an uncertain trumpet with respect to foreign policy, military strategy and overseas commitments. On the one hand, Trump campaigned in 2016 against additional US military involvement in the Middle East and called for termination of the wars in Iraq and Afghanistan. On the other hand, US commitments to allies and unexpected challenges from abroad required a more nuanced approach to the Middle East and to foreign policy in general. Thus, for example, when Iran attacked Saudi oil fields with cruise missiles and drones in mid-September, 2019, President Trump and his advisors sought proportionate military responses and diplomatic support from US allies, while leaving open the door for future negotiations with Tehran.[25]

As an integral part of national security policy and strategy, therefore, the United States must develop and articulate a broad geostrategic vision. This can be the guiding framework for those in the national security establishment to design strategy on a global scale. Equally important, there must evolve a sense of where the United States stands with respect to adversaries and potential adversaries, the international security environment, and its national security policy. It is in geostrategic terms that global vision, policy, and strategy are given specific meaning (i.e., commitment to a specific state or region in the context of US national security).

The Contemporary Period

The modern age of weapons technology appeared to reduce the relevance of geopolitical and geostrategic approaches to national security issues. Advanced aircraft, surface vessels, and submarines, combined with space technology and long-range, reconnaissance-strike systems, have tended to replace the views of Halford Mackinder ("heartland" and land power), Nicholas Spykman ("rimland" and sea power), Alfred Thayer Mahan (sea power and bases), and Giulio Douhet (air power). The new theorists had more advanced strategies of forward presence and crisis response as well as new concepts of land and sea warfare.[26] But the new approaches fail to grasp the fundamental global perspectives inherent in the earlier geopolitical-geostrategic views (although some elements have been borrowed and assimilated). New viewpoints focus more on war-fighting, operations other than war, and regional issues than on global visions of security relationships.

Critics of geopolitical models have long argued that geographical determinism is a dangerous concept because it assumes geography alone determines national power and policy. It is as great an error, according to critics, to translate geopolitical notions into geostrategic ones—that is, to make geopolitical factors into strategic ones by rationalizing their importance in national security. According to Hans Morgenthau, "geopolitics is a pseudoscience erecting the factor of geography into an absolute that is supposed to determine the power, and hence the fate, of nations."[27] Nonetheless, geostrategic theory has taken on prominence in the contemporary period primarily because of global industrialization and economic interdependence. Ironically, the demise of the superpower era created situations in which some states adopted policies precipitating less than major wars and engaging in unconventional conflicts to achieve political goals. All of these issues include geostrategic elements and clusters that have important consequences for US national security. Morgenthau countered, "There are areas that are crucial to the functioning of the world economy and our interests."[28] Put simply, this means that some geostrategic and power clusters are critical to US national security policy, but are not absolute. An example of rising power centers that leapfrogs traditional geostrategy is provided by the growing influence of multinational information technology companies, especially in the realm of social media, and their impacts on communications and governments. Companies like Facebook, Google, Twitter, and others have resources and power comparable to that of many state actors and have demonstrated considerable power, within and among states, to influence the political debates in the public square. Some members of the US Congress continue to demand increased government regulation of the tech giants in the public interest, and other commentators raised the possibility of breaking up one or more of these mega-information platforms.

A Global View: Competing Power Clusters

The global view of US national security, based on competing power clusters, recognizes that the world has moved away from the bipolarity of past decades. In addition, it acknowledges the emergence of a variety of states whose power and security postures preclude, or considerably reduce, the major powers' control and influence. With multiple centers forming power clusters, the focus of US policy and strategy, in geostrategic terms, should be to stabilize balances or create equilibrium among competing ideologies and systems in order to establish a basis for resolving conflicts through alliances. This approach assumes that conflict between major powers must be avoided, partly by recognizing legitimate interests, partly by removing conflicts from the battlefield, and finally by effectively resolving conflicts between competing power clusters. (This view incorporates elements of Spykman, Mackinder, and the balance-of-power theories.) For US national security, the global view of competing power clusters has some essential premises.

First, it does not reduce the US need to maintain a deterrent capability in its nuclear strategy or the US need to pursue arms reductions, arms control, and conflict limitation. It does require a concentrated effort to develop realistic alliances and form mutually supporting power clusters within a modern vision of geopolitics and geostrategic perspectives. In such alliances, the United States does not always need to take a dominant role; indeed, a supportive role with lower visibility could often better serve US security interests. Even advisory or midwife roles in regime changes can lead to unexpected complications and occasional blowback. For example, the US "lead from behind" strategy that supported the overthrow of Muammar Qaddafi in Libya, on the heels of the Arab Spring, beginning in 2011, initially appeared as a low risk, high payoff intervention. On the other hand, the post-Qaddafi turbulence in Libya left a security vacuum occupied by insurgents and terrorists fighting against the UN-recognized regime in Tripoli. Another example of unintended or unexpected consequences was provided in Ukraine in 2014, when the United States supported the overthrow of the pro-Russian government there. Russia responded by invading and annexing Crimea and destabilizing eastern Ukraine in 2014 and invading the country in 2022. These examples notwithstanding, a promising approach to international influence would imply a more consistent use of US intelligence and special operations forces as trainers and supports for other countries' indigenous counterinsurgency and counterterror forces, so-called foreign internal defense, as opposed to having the United States take the lead by conducting major combat operations and maintaining a large and protracted military footprint.

Second, the overarching purpose is to create a directed balance of power against the most likely aggressor in a particular region of importance. The Gulf War coalition mobilized by the United States against Iraq in 1991 is an excellent case in point. But this directed balance should not be so rigid as to limit US flexibility in responding to unforeseen security challenges. Moreover, it cannot preclude the United States from developing political-psychological and economic support networks of mutual benefit among a variety of states, even those that are not democratic.

Third, an important part of US security policy and strategy is to ensure that centralized systems—totalitarian and authoritarian—and other groups with expansionist ideologies face a network of alliances shaped by geostrategic considerations. Alliances can serve as roadblocks as well as containment, deterrent, and defensive forces. These alliances, although primarily a security response to potential and real aggressive behavior and thus concentrating on security issues, also need to provide a framework for other policy issues. Finally, alliances can become part of an effective strategy only if states feel that entering such alliances will enhance their own national interests. For example, some wonder whether NATO, arguably the most successful military alliance of modern times, can continue to expand its membership along with growth in its responsibilities across the entire spectrum of conflict. Since the United States is the strongest military power in NATO, any gap between NATO promise and performance reflects adversely on the reputation of US as well as alliance leaders.

In summary, competing power clusters are an important characteristic of the world security setting. An effective response, including a security posture that can provide for sound policy initiatives, rests on the earlier global visions of Mackinder and Spykman, refined to accommodate characteristics of the contemporary period. These initiatives in turn are driven by the broader policy issues for a stable world order, peaceful conflict resolution, and democratic values. The global strategy evolving from the competing power-cluster view can serve as the context within which other supporting strategies are designed and implemented.

The Limits of Geostrategy

Political, psychological, and economic considerations are beyond simple geostrategic influences. For example, the effectiveness of governments, ideologies, and nationalistic movements is not bound or necessarily influenced by geostrategic considerations. Islamic fundamentalism, for example, cannot be explained by geostrategic factors.

Furthermore, technological advances can reduce the impact of geostrategic factors. For example, airpower reduced the effectiveness of the physical containment based on land power inherent in the Mackinder thesis. Similarly, "system of systems" or information-driven military transformations, including enhanced long-range strike weapons, reconnaissance, and precision aiming, have reduced US requirements for massive numbers of troops with which to fight wars of attrition. In an important study of international relations, Quincy Wright examined political geography and concluded:

> It has been the hope of some geographers that because of the apparent permanence of geographic conditions, geography might become the master science of international relations. This hope seems vain.

> Geography is primarily a descriptive discipline . . . [and] does not determine international relations. . . . The apparent permanence of geographical conditions is illusory. Civilized man uses his environment to serve ends which come from other sources. . . . Geography cannot develop concepts and conceptual systems applicable beyond a limited time and area in which a given state of the arts, of population, and of society can be assumed.[29]

Geostrategy and National Security

A geostrategic vision nevertheless remains an important component in designing global strategy. The competing power-cluster approach to geostrategy, although shaped in somewhat new terms and requiring concentrated effort in developing alliances, is not so different from established US policy and strategy as to cause massive restructuring. It does, however, require some strategic rethinking. The approach does not demand the presence of US military forces in quantity to conduct effective military operations; rather it requires a specialization within alliances and various networks so that each state contributes its fair share (i.e., what it is good at and what it can afford). Conversely, the states within the alliance or network best equipped and positioned to implement a particular alliance strategy can be called upon to do so. Finally, Wright's criticism of political geography is equally appropriate for geostrategy. Geostrategic determinism distorts the importance of the dynamics of the security environment and ignores the changing nature of the physical environment. Furthermore, geostrategy cannot account for the nonphysical dimensions of security policy and strategy, such as national will, political resolve, and staying power, which in the long run are more important than geostrategic considerations. Yet geostrategic considerations have some influence on the posturing of forces and strategic alternatives and must be included as part of the security equation.

Homeland Security

September 11, 2001, changed the US national security agenda overnight. The terrorist attackers did not use weapons of mass destruction—nuclear, biological, or chemical weapons launched from offshore or released on US soil. That phase came later, in October, when either domestic or foreign terrorists mailed anthrax spores to various US government offices and prominent news media. The resulting publicity set off a national anthrax scare and numerous warnings from the US government that more attacks might be imminent. President George W. Bush established an Office of

Homeland Security and, eventually, a cabinet-level Department of Homeland Security to coordinate domestic policy efforts against terrorism.

The phrase "weapons of mass destruction" conceals more than it reveals, however. Nuclear weapons remained the ultimate deterrent in the hands of five permanent members of the UN Security Council (the United States, Russia, China, Britain, and France). Other states aspired to acquire nuclear weapons as symbols of great power status: it is generally assumed that Israel has nuclear weapons, India and Pakistan went publicly nuclear in 1998, and North Korea declared its capability to produce and deploy bombs during the George W. Bush administration. Since then, North Korea has continued its missile testing and nuclear buildup in the face of UN and US opposition under the George W. Bush, Obama, and Trump administrations, despite the different approaches to North Korea taken by them. Iran has increased its nuclear capabilities greatly and is likely to become a nuclear power absent a renegotiated treaty. Also, the spread of nuclear weapons and long-range delivery systems in Asia create the possibility of dangerous regional rivalries and attacks on the US homeland.

Biological and chemical weapons can also be used in long-range missile attacks. But chemical strikes would be destructive on a scale well below even small nuclear attacks. And the weaponization of biological pathogens for long-range missile attacks is a tricky business; atmospheric conditions and other uncertainties make the impacts very unpredictable. On the other hand, the pandemic continuing as these words are written illustrated the potential of biotoxins to create international havoc, whether deliberately contrived as bioweapons or resulting from accidental or natural causes. Nuclear weapons were road-tested in actual detonations and in simulations throughout the Cold War. Societies struck by even limited nuclear wars of the kind imagined by Cold War planners would be returned to prehistory.

The US national security and intelligence communities were unprepared for the October 2001 anthrax attacks, which used the US mail system and hit various US government and media offices. Some people suspected foreign terrorists, especially al-Qaeda. But US law enforcement did not rule out the possibility of domestic terrorists seeking to disguise their activities under the cloak of suspicion about foreign adversaries, and at least one US research scientist was investigated extensively for involvement with anthrax. Regardless of the identity of the perpetrators, the anthrax attacks led to demands for new powers of investigation for the Department of Justice against suspected terrorists, especially immigrants holding temporary visas that had expired.

By the end of October 2001, about a thousand persons were being held in US detention facilities based on alleged ties to terrorist organizations, related charges, or immigration violations. Repeated warnings from Attorney General John Ashcroft and Homeland Security director Tom Ridge to the effect that more attacks at home or abroad could be expected raised public and media awareness of the lethal potential of biological weapons. Despite the media frenzy and public anxiety, military professionals cautioned that biological weapons were probably not as useful in combat operations as in creating mass fear among civilians. Even the world's largest militaries had little experience in testing biological warfare under realistic conditions. Biological warheads and chemical munitions on ballistic missiles might be used for strikes deep into an enemy's rear to disrupt command and control and logistics. Such attacks could force the defenders to don protective suits and shield equipment, thereby inhibiting personnel and vehicular movement, which might slow down forces at the front. Yet forward-deploying US forces might adopt several responses, not excluding their own use of WMD, against enemy forces, depending on the exigent circumstances. The policies of successive US presidencies deliberately left open the possibility of using US nuclear weapons for missions in addition to the obvious response to a direct nuclear attack on the US homeland.

Cruise missiles provide another weapon for attacking the US homeland and US forces with biological and chemical weapons. Cruise missiles are pilotless projectiles designed to exploit aerodynamic lift and precision guidance to attack surface targets on land or at sea. Cruise missiles designed or modified for land attack can deliver biological and chemical munitions over thousands of kilometers against shore-based targets, including cities, military installations, and economic targets such as the Iranian attacks against Saudi oil fields in September 2019. Cruise missiles can be launched from land installations, from vessels at sea, and from aircraft. Cruise missiles can be guided during flight to attain precision targeting; they can also fly at high and low altitudes to confuse air defenses.[30] It would be easier for third world countries and terrorists to acquire, conceal, and modify cruise missiles, as opposed to ballistic missiles, for land attacks.

Strategic realism suggests that enemy states might be cautious about attacking the US homeland with long-range ballistic or cruise missiles and WMD. States have geographical locations and vulnerable populations that would surely be subjected to severe US retaliation. Saddam Hussein arguably did not employ chemical or biological weapons against US forces during Desert Storm because he feared possible US nuclear retaliation. Undisguised attacks on US soil that caused mass casualties would be suicide for heads of state and many of their citizens. Therefore, states that wanted to use biological or other WMD against the American homeland would probably disguise their efforts by attributing them to third parties (e.g., terrorists) or accidents. The coronavirus epidemic of 2020 that originated accidentally in China and caused millions of deaths globally shows the destructive potential for deliberately planned, mass biological attacks with viruses that are previously unknown, or covertly manufactured and purpose built for mass destruction. Few analysts suggest that China deliberately unleashed the Covid-19 virus, but the pandemic demonstrated the detrimental potential of such an attack.

Terrorists are another matter, however. Terrorists have no permanent address and usually lack the assets, such as infrastructure, that can be threatened by retaliatory strikes. The best deterrent is preemptive—anticipating attacks and capturing or killing the planners and perpetrators. But preemption requires extensive human intelligence from inside the terrorist organization, something very difficult to acquire in a reliable and timely fashion. The effectiveness of deterrence through intelligence is still an open question, but if the absence of further attacks is the definition of success, then it has been successful, at least in the short term.

Terrorists' uses of WMD and other mass-casualty weapons (such as fuel-loaded airliners) are examples of asymmetrical warfare, something US adversaries may use more frequently in the future. US superiority in technology, conventional warfare, and military information systems would make it suicidal for state or nonstate actors to confront the US military one-on-one. Future non-state US opponents will likely pursue the strategies established by Osama bin Laden and al-Qaeda. WMD in the hands of terrorists (even state actors) provide a potential force multiplier, an asymmetrical counterweight to US superiority. To defeat WMD and asymmetrical strategies, the United States must emphasize brains instead of brawn. Improved intelligence overall, better exploitation of human and technical sources, and intelligence support for theater commanders are necessary parts of any US response to asymmetrical threats.

WMD thus pose unique problems for US policymakers today. During the Cold War, nuclear weapons were symbols of great power status, the objects of envy by dissatisfied nonnuclear states. Today, nuclear weapons and other WMD are potential instruments of the weaker against the stronger. US and allied superiority in advanced technology conventional warfare will be tested by asymmetrical strategies that include the use of WMD as threats or actual methods of attack. The US homeland is no longer a sanctuary of safety, and homeland security is now a strategic mission of first importance. The use of WMD, including nuclear weapons, against US citizens en masse remains a frightening possibility.[31]

Technology and Economics

The US public has become accustomed to the daily conveniences of technology. Modern appliances; electronic gadgets at work, home, and school; and sophisticated medical treatments have become the rule rather than the exception. The enhanced quality of life has tended to personalize the impact of technology. As a result, the US public too often overlooks the impact of technology on the power of the state and its effect on the economy. In addition, many have little appreciation for the relationship between technological advances and the course of human history.

Technology reaches virtually all levels of society and the economy. Furthermore, technological developments tend to precipitate other technologies, creating a technological impetus that in turn can drive technology for technology's sake. Technology spawned the Industrial Revolution in the United States and Europe. Because industrial growth and technological change feed off one another, technology ultimately became part of the culture of industrial states. Indeed, competitiveness in the world economic system relies upon technology. In turn, technological capability depends upon a society that is postured to accept and demand technological change and growth. Similarly, this is dependent upon research and development within society and a commitment to scientific education.

Put simply, continuing economic growth requires a society that is attuned to technology and can adapt to it while incorporating technology into the political-social culture of the system. In this respect, the US education and military systems will be challenged by artificial intelligence that pushes the boundaries of "man-machine" interface further in the direction of reliance on the latter. Another obvious example of challenging new technology is the appeal of cyber war to state and nonstate actors.[32] Indeed, hypersonic missiles, stealthy cruise missiles, and weaponized artificial intelligence might so reduce the amount of time US leaders have to respond to warnings of nuclear attack that some deterrence experts have proposed giving an artificial intelligence (AI) system control over the launch decision and retaliatory options.[33] Further challenges from "technology creep" will be provided by newer systems for space deterrence and conflict, including anti-satellite weapons and specialized defensive satellites to protect against those weapons.[34]

Two Dimensions of Technology

The two major national security dimensions of technology are the power of a state and its military capability. The initial period of the Industrial Revolution brought the rise of Great Britain as a world power and the era of Pax Britannica. Technological breakthroughs in steam power, metallurgy, and weapons development provided Great Britain a technological edge in developing its economy and military power, which led to the development of a superior navy and also protected British interests. Mechanization—the substitution of mechanical for human power—allowed the more efficient use of resources and nurtured the design of new industrial organizations to manage the production of goods. Improved means of communication brought societies closer together and allowed states to consolidate power and develop more effective control over the governing instruments. The combination of mass mobilization of populations for warfare, bequeathed to Europe by the French Revolution and the Napoleonic era, combined with advances in manufacturing and industrial power to create the potential for unprecedented destruction that was realized in World War I.

The state possessing advanced technology could develop a dominant power position, allowing it to create conditions favorable to policy goals. Its quick reaction to events, its ability to gather information, its control over foreign policy, and its military were, to a large degree, dependent on technological developments in communications, transportation, economic capacity, and weapons development. The shift to mechanization, as well as efficient industrial organization, was the springboard to effectiveness in the international field.

The second national security dimension is military technology, which has shaped the conduct and characteristics of war. Thus the state possessing advanced military technology had a partial advantage on the battlefield. The invention of gunpowder and the refinement of rifling in guns changed battlefield tactics. The introduction of the Gatling gun, followed by the machine gun, had a dramatic impact on infantry tactics. More accurate and longer-range artillery engulfed civilians, threatening towns and cities well behind the defined battle area. The introduction of the battle tank late in World War I opened the battlefield to further mechanization and tactics that stressed mobility and rapid movement. The rapid development of aircraft for military use changed the face of war. Similar advances in shipbuilding made naval warfare more destructive and far-reaching. These dramatic technological innovations of the nineteenth century and first part of the twentieth century have their contemporary counterparts: the nuclear era, the electronic revolution, and space-age capability have again altered the dynamics of the battlefield and pushed the qualitative as well as quantitative dimensions of war almost beyond comprehension. States possess a destructiveness well beyond anything envisioned in the past.

But technological advances do not guarantee a dominant power position. Historians are quick to point out that Great Britain lost its technological edge to Germany in the latter part of the nineteenth century, as German innovations in chemicals, electronics, and the steel industry gave rise to a powerful state. In the twentieth century, the growing industrial capacity of the United States provided the economic base for it to develop into a superpower, militarily and economically. This technological advantage helped to vault the United States into the "postmodern" era of high technology in computers, electronics, and communications ahead of other powers.

This brings us to an important phenomenon of the technological dimension: the diffusion of technology from one state to other states. According to one authority, "The diffusion of military and economic technology from more advanced societies to less advanced societies is a key element in the international redistribution of power. Although technology is expensive and not easily created, once it is created it usually diffuses relatively easily."[35] Diffusion of technology allows new powers to emerge while reducing the power of existing technologically advanced states. It is clear that industrialization and a variety of "economic factors in particular have become an important source of national power and advantage." This in turn is contingent upon technology. But "in time . . . this technological advantage disappears."[36] In short, there is no "last move" in competitive military technology. Next generation military competition will privilege technology developments in computers, networks, and communications, but also areas with less obvious military applications including nanotechnology, robotics, and bioengineering.[37]

The diffusion of technology, with the resulting decline of one state and the rise of others, has led many to develop a sense of historical determinism about technology. Once a state reaches the limits of its expansion, then it begins a period of decline; "It has great difficulty in maintaining its position and arresting its eventual decline."[38] If one accepts this view, then retrenchment and consolidation follow, and an increasingly inward-looking culture evolves. Some point to US technological diffusion, a view reinforced by the limits of US power. If this is true, it follows that the United States must substantively change its national security posture, reducing its commitments and developing a better match between capabilities and national security policy. On the other hand, the United States must also defend its vital interests against more sophisticated threats from both state and non-state actors: from future cyber stalkers to terrorists with tribal mind sets. Technology can help the United States to deal with both high-tech and low-tech enemies of this sort, but it will never be a panacea. The quality of US military-related research and development, together with the flexibility of the political process and the "learning curve" of politicians with their hands on national security and defense portfolios, will determine whether America remains a great power or just one state among many. Impressively, US military history shows an adaptive capability for recognizing and correcting mistakes in good time. But strategic competency is in short supply even in the best administered and funded militaries—and governments.

Although technology is not the sole determinant of national power, there is no question that it is a very important one. It has a direct impact on industrial and postindustrial military capacity and has a pervasive effect on all aspects of life. The penetration of technological culture into society is a sure sign of its secular character. As one authority concluded, "taking mankind as a whole, there has been an irreversible movement of technological progress. Techniques once invented have seldom been wholly lost although they may have been lost in local areas."[39]

Technological Aspects

It is difficult to develop and maintain high technology without a strong economic base, although the ability of North Korea to develop nuclear weapons stands out as a dangerous exception. Furthermore, an industrialized or postindustrial society needs to develop and nurture a scientific commitment within its educational systems and research communities. With a broad-based scientific culture and an industrial base that can translate scientific advances into technological developments, society can improve quality of life.

Such advances are contingent upon a diverse and modern economy that is stimulated by consumer buying as well as continuous and growing investments in new plants and products. Government fiscal and monetary policies are important factors in shaping economic growth, as are market forces and international markets and trade. Thus a strong industrial economy, scientific culture, and technological advances are inextricable and essential in maintaining a state's ability to pursue national security goals.

In this environment, technology benefits military development, which in turn benefits the economy. But these mutual benefits are dependent upon the nature of the political system and its ideological orientation. The link between the economy, technology, and military development can be considerably distorted by government action and ideological and cultural aspects of the system.

In the United States, a key indicator of technological effort is the resources committed to defense and military research and development. According to some experts, the current US "arsenal of democracy" is subject to stress from various twenty-first century forces, including the erosion of the US manufacturing base, a growing scarcity of vital engineering skills, shrinking budgets that reduce cash flows, and systemic flaws in the military procurement process.[40] On the other hand, both economic and political forecasting are bounded universes, since history is not deterministic and leaders, going forward, have options with respect to national security related ends and means. Crises like September 11 also have a way of generating resources and new ways of thinking not previously admissible within the highest circles of power. American entrepreneurial spirit and "Yankee ingenuity" should never be underestimated. In 2020 a private US space company (cooperating with the US government) launched astronauts to the International Space Station and recovered them in an oceanic landing for the first time in more than forty years. The US example of institutionalized innovation in research and development, the Defense Advanced Research Projects Agency (DARPA), continues to set the pace for embedded innovative thinking in defense related technology. Each of the US military services also has its own specialized centers for advanced thinking about technology as it relates to military missions and strategy.

Economic and military challenges are not mutually exclusive. In fact, a state's rising economic power can lead to improvement in its military capabilities and, therefore, its aspirations on the global stage. The United States in the nineteenth and twentieth centuries and China in the twenty-first century offer examples of economic growth supporting increased military power and international ambition.

The issue is not simply military technology, but civilian technology from which the military can benefit. Thus technology, military developments, a strong military establishment, and the need for a strong economic base are sure to be persistent political issues within the US administration and political and military policymaking circles. To this date, there is little to suggest that anyone has designed a magic formula to balance these competing interests. There seems to be little question, however, that in the United States a strong industrial and post-industrial base, scientific culture, educational system, and technological commitment are essential in providing for an effective national security posture.

One fear is that technology, whether civilian or military, increasingly tends to drive itself: that is, a technological culture may develop to a point that technology will be seen as a good in its own right, with little reference to its impact on the quality of life, values, and morals of society. In its extreme form, the fear is that technology will advance without recourse to its social utility, and society will blindly adapt to it rather than technology's adapting to society. Biotechnology is a good example; some scientists claim that human cloning is inevitable and will someday become routine.

Several important issues emerge from these observations on technology. Technological developments are critical in nurturing economic progress, and military modernization and capability are contingent upon military technology. The links between technological progress in the economy and military modernity seem fairly obvious, as both factors are extremely important for US national security policy and capability. It is also clear that in an open system the strength of the economy is a basic priority in sustaining an effective national security posture.

Another dimension is related to the economy as a whole. In the 1980s and the 1990s, there was serious concern about the increasing involvement of foreign firms in the US economy, even though most were from friendly countries. British, West German, French, and Japanese enterprises gained control over several US manufacturing firms—including defense-related industries—as well as supermarkets, banks, and other businesses. Large tracts of real estate in major US cities were bought by foreign companies, and brokerage houses on Wall Street acquired several foreign partners. As a result of foreign economic penetration, some feared economic control by foreign companies as well as political control evolving from economic leverage. These fears raised more questions about allowing transfer of technology and access to defense products, even among friendly states.

The paradox of contemporary international economics is that states are unavoidably interdependent with respect to trade, finance, and transnational "supply chains," but, at the same time, sovereign governments are reluctant to yield control over vital national assets and interests. These vital interests include control of immigration, protection of corporate and government secrets, and defense against terrorism.

Thus the fundamental problem facing the president and those in the US national security establishment is how to balance resources within the defense budget, ensure economic growth and technological advances on a broad scale, and minimize the unauthorized transfer of high-tech material and information to foreign states. At the same time, the United States must share certain technologies with allies to strengthen alliances and provide important national security benefits to the United States.

The problem is compounded by the fact that the president and members of his national security staff do not have the ability to direct or even shape all the instruments of national security, nor can they put policies into place without the cooperation of other political actors, especially Congress. Even when there is general agreement between the president and Congress about national security issues with respect to economics and technology, there is likely to be disagreement over military technology and the means to pursue national security goals—except in the most extreme cases when there is a direct threat to US national security interests. The controversy during the George W. Bush administration over letting a Dubai company manage a number of ports illustrated how economics, technology, and national security can intersect over a single controversy. Thus, except during crises, the US economy is driven by market forces that are outside the purview of government. The amorphous, free-market system—despite its efficiency—is not well suited to central direction and control.

Even in an economy as vast and diverse as that in the United States, government resources are not infinite. Therefore presidents, defense secretaries, and members of Congress must decide among competing needs for the military. The US defense budget must provide for force structure, readiness and sustainability, modernization (investment for the next generation of weapons and infrastructure), and connectivity (linkages across the electronic spectrum and cyberspace that make joint operations involving more than one arm of service in a single military action or campaign possible). Important compromises must be negotiated among these various requirements and among the military arms of service (Army, Navy, Air Force, Marines, and Space Force). A trade-off of particular importance is that between readiness (how well we can fight now) and modernization (investment for tomorrow's wars). Finally, the various budget decisions relative to forces and missions are supposed to be made consistent with the president's overall national security strategy and defense policies.

Times of growth in US defense budgets can be as contentious politically as periods of decline. During the George W. Bush administration, US defense budgets were increased to support the war on terror and the wars in Afghanistan and Iraq. Defense Secretary Donald Rumsfeld's vision of "transformation" anticipated a future in which "platform-centered" warfare (aircraft, ships, and tanks) would be supplanted by "network-centric" warfare based on information principles and technologies. This vision was highly controversial within the military services. In addition, as costs rose for the US postwar occupation of Iraq and counterinsurgent warfare, members of Congress began to question how long the United States would stay and to demand an explicit "exit strategy."

During the Cold War a symbiotic relationship grew among the US military establishment, defense contractors, and members of Congress seeking defense spending in their districts. This triangle of influence will not disappear entirely, but it will be subjected to considerable turbulence as technology from the civilian sector begins to drive military innovation toward a revolution in military affairs. The computer, communications, and electronics revolutions were not driven by government laboratories but by entrepreneurial spirit and inventiveness in the private sector. The Internet and the globalization of information and finance now make it possible for medium and small powers to acquire advanced technology, and even terrorists now fly unmanned aerial vehicles or drones. A more diffuse international security power structure will almost certainly be one ramification of the information revolution. Hierarchies will be replaced by networks as the organizational frameworks of choice for problem solving in military and government circles. Traditional approaches to intelligence collection and analysis based on compartmentation and specialization may not work well in this new era.[41]

Information and the Communications Revolution

Given the development surrounding the information superhighway since 2002, there is little question that the information age has matured. Microchip and laser technology, computer advances, and satellite networks have all made access to information and entertainment global in scope. In addition to the Internet, the ability of viewers to interact with television messages (interactive television) will offer a variety of options for making political, commercial, and buying decisions through the television set. Furthermore, numerous political, social, educational, and advocacy channels will be available to viewers. Add the ability of the print media as well as social media to extend their reach globally and one can begin to appreciate the impact of the communications revolution. The individual has been put into the driver's seat of communications with access to Twitter, Facebook, and other social media, and to other means of self-directed, digitally based communication; as a result, large organizations such as governments and armed forces are hard pressed to keep abreast of new technologies.

The impact on national security will be no less dramatic. Continual monitoring of the conflict arena will provide commanders in the field with up-to-the-second and comprehensive views of the battle arena. Penetration of the adversaries' military formations, political and social infrastructures, and target areas will be commonplace. Furthermore, access to the same information will be available to policymakers in Washington, DC, global viewers, and the media. And in the case of terrorist attacks intended to cause mass casualties for the purpose of visible defiance and exposing US vulnerabilities, the media are perceived by terrorists as instruments of strategic effect in a protracted war. Vivid images of the destruction would be transmitted globally in an instant.

Moreover, the communications revolution will have a major impact on military command and control, and technology and communications will revolutionize weapons systems and the characteristics of conflicts—at least conventional conflicts. The electronic battlefield will be enhanced by sophisticated communications and standoff capabilities (e.g., an exchange of missiles using highly sophisticated targeting).[42] Military training and education will need to incorporate the lessons of the communications revolution into the battle arena. In summary, it is conceivable that conflicts can be more closely controlled from the safety of the operations room in Washington, with civilian leaders being able to see and talk directly to commanders on the battlefield down the chain of command. Radio centralized control will need to be used sparingly lest military effectiveness be compromised.

One danger is that perceptions of war can take on the feel of a video game, depersonalizing it to the point that only those in harm's way will really understand the personal nature of ground combat. Those outside the immediate combat area could well make decisions that have little to do with realities on the ground.

In the broader sense, it will be difficult to maintain secrecy in the national security policy process. The media will be able to gain access to or penetrate virtually any communications network used in the policy process. The hacking of the Democratic National Committee in 2016 and subsequent publication of campaign organization communications by WikiLeaks is only one illustration of this media dynamism combined with cyber war. At the same time, there will be almost simultaneous global access to events and decisions in the national security area. The entire national security policy process will be under tremendous pressure for immediate response and rapid implementation. In an age when video-driven national security and foreign policies are becoming almost commonplace (e.g., the emergence of al-Jazeera as a message center for various groups in the Arab and Islamic communities, and the use by the

Islamic State [ISIS] of social media for recruitment), the communications revolution carries serious implications for long-range national security planning as well as short-range policy and strategy. How all of this will affect the national security establishment in the next decade is not clear. Although policymakers are quick to blame the media for exciting the public with vivid images of atrocities, there is also a major impact of video on political leaders themselves. President George H. W. Bush's insistence on sending a US peace mission to Somalia in December 1992 was partly motivated by news footage of starving Somalis.[43]

US planning for future wars is once again emphasizing preparedness for conventional conflicts against the armed forces of other nation-states, including aspiring peer competitors. US military thinkers and planners must also consider the future requirements for fighting unconventional conflicts against nonstate adversaries, including insurgents, terrorists, and criminals linked to destabilizing political forces operating within or across state boundaries. Unfortunately there are never enough resources to cover all the "what ifs" that hostiles can imagine or carry out, so tradeoffs must be made. John Arquilla, an expert on information warfare at the US Naval Postgraduate School, believes that a better high concept is needed for US military planners, centered on fighting against networks instead of hierarchies. This adjustment in US thinking is necessary because "fighting networks" like al-Qaeda, the Islamic State, and insurgent groups can now accomplish large objectives that were previously available only to regular armed forces fighting in large formations. As he explains: "It grows more evident each day that warfare has been moving into a new realm where very small combat formations can achieve high levels of destruction and disruption. The only question is whether we will explore this new landscape with new kinds of forces and weapons, or march our twentieth-century military off into a twenty-first century wilderness, unchanged and unassisted."[44]

Conclusion

War retains its fundamental nature throughout history: as the use or threat of organized violence to accomplish a political aim. But the character of war changes across centuries for many reasons.[45] Factors such as geography, economic power, and the political viability of state actors, as well as those states' own definitions of allies and enemies, must be taken into account by US defense and foreign policy planners and political leaders. Culture, which includes religious and regional factors, also influences whether states will be friendly or hostile toward the United States and its allies. For example, when the Soviet regime collapsed and a nominally democratic Russia succeeded it, the political character of the regime changed, as did the political and social culture pertinent to Russian foreign policy. This change, in turn, had consequences for the United States and NATO as did the return of Russian authoritarianism under Vladimir Putin.

The information and communications revolutions illustrate the potential of technology to create a new context for military strategy. Persons tutored on so-called second-wave industrial military logic now find their ingrained habits stressed by "third-wave" or postindustrial technologies that leap across state borders at the speed of light. The new precision of the military art, made possible by the new technologies of smart warfare, has profound implications for the character and size of armies and to the vulnerability of legacy military systems.[46] Smart and highly specialized soldiers, fighting in smaller units and with a greater degree of self-direction, and capable of centrally directed or bottom-up "swarming" tactics, will pose challenges to enemies as well as to their own superiors higher up the chain of command.[47] This change to smaller, smarter armed forces in turn creates organizational, sociological, and political issues within US and other armed forces. With fewer regular soldiers and greater numbers of special operations forces, the US military may be better prepared for missions such as peace operations and low-intensity conflicts, but have greater problems against a peer competitor. Yet a smarter and more autonomous military can pose other problems in civil-military relations, including a possible incongruity with the values of the larger society.[48]

Notes

1. For an illustration of an inclusive approach to defining national security, see Barry R. Posen, *Restraint: A New Foundation for U.S. Grand Strategy* (Ithaca, N.Y.: Cornell University Press, 2014).

2. Colin S. Gray, *The Strategy Bridge: Theory for Practice* (Oxford: Oxford University Press, 2010), esp. pp. 154–155 and passim. See also, Sam C. Sarkesian and Robert E. Connor Jr., *The US Military Profession into the Twenty-First Century: War, Peace, and Politics* (London: Cass, 1999), esp. pp. 67–69; see also the 2nd edition, 2006.

3. Military force includes both kinetic such as combat and nonkinetic actions such as information warfare.

4. Colin S. Gray, *Theory of Strategy* (Oxford: Oxford University Press, 2018), p. 52.

5. Rupert Smith, *The Utility of Force: The Art of War in the Modern World* (London: Penguin/Allen Lane, 2005), p. 3.

6. See, for example, Defense Intelligence Agency, *Russia: Military Power—Building a Military to Support Great Power Aspirations* (Washington, DC, 2017), https://search.usa.gov/search?query=%22Russia:%20Military%20Power%E2%80%94Building

%20a%20Military%20to%20Support%20Great%20Power%20Aspirations%20%22&affiliate=defenseintelligenceagency&utf8=%26%23x2713%3B; Christopher S. Chivvis, Andrew Radin, Dara Massicot, and Clint Reach, *Strengthening Strategic Stability with Russia* (Santa Monica, Calif.: RAND, 2017), https://www.rand.org/pubs/perspectives/PE234.html; Alexander Golts and Michael Kofman, *Russia's Military: Assessment, Strategy, and Threat,* https://www.russiamatters.org/sites/default/files/media/files/GoltsKofman_CGI_Jun2016_RussiasMilitary_PDF.pdf; and Lester W. Grau and Charles K. Bartles, *The Russian Way of War: Force Structure, Tactics, and Modernization of the Russian Ground Forces* (Ft. Leavenworth, Kans.: Foreign Military Studies Office, 2016). https://www.armyupress.army.mil/portals/7/hot%20spots/documents/russia/2017-07-the-russian-way-of-war-grau-bartles.pdf.

7. Smith, *The Utility of Force,* p. 3.

8. Numerous examples appear in Max Boot, *Invisible Armies: An Epic History of Guerrilla Warfare from Ancient Times to the Present* (New York: Liveright, 2013); and in John Arquilla, *Insurgents, Raiders, and Bandits: How Masters of Irregular Warfare Have Shaped Our World* (Chicago: Ivan R. Dee, 2011). US intelligence challenges in dealing with unconventional conflicts during and after the Cold War are reviewed in Tim Weiner, *Legacy of Ashes: The History of the CIA* (New York: Doubleday, 2007).

9. Martin van Creveld, *The Transformation of War* (New York: Free Press, 1991).

10. Joseph R. Biden Jr., *Renewing America's Advantages: Interim National Security Strategic Guidance* (Washington, DC: White House, March, 2021).

11. Anthony H. Cordesman with Grace Hwang, *U.S. Competition with China and Russia: The Crisis-Driven Need to Change U.S. Strategy,* working draft (Washington, DC: Center for Strategic and International Studies, August 11, 2020), p. 6.

12. Samuel P. Huntington, "The Clash of Civilizations?" *Foreign Affairs* 72, no. 3 (Summer 1993), p. 40.

13. Dmitri Volkogonov, *Autopsy for an Empire: The Seven Leaders Who Built the Soviet Regime,* edited and translated by Harold Shukman (New York: Free Press, 1998).

14. Mikhail S. Gorbachev, "Document: The Revolution and Perestroika," *Foreign Affairs* 66, no. 2 (Winter 1987–1988), p. 425 (excerpts from his speech; English translation distributed by the Soviet press agency TASS).

15. See Angela E. Stent, *Putin's World: Russia Against the West and With the Rest* (New York: Twelve-Hachette, 2019) and Richard Lourie, *Putin: His Downfall and Russia's Coming Crash* (New York: St. Martin's, 2017) for perspective and appraisals.

16. James Dobbins, Howard J. Shatz, and Ali Wyne, *Russia Is a Rogue, Not a Peer; China Is a Peer, Not a Rogue* (Santa Monica, Calif.: RAND, October 2018).

17. See Lt. Gen. Robert P. Ashley Jr., Director, Defense Intelligence Agency, "Russian and Chinese Nuclear Modernization Trends," remarks as prepared for delivery, Hudson Institute, May 29, 2019, https://www.dia.mil/News/Speeches-and-Testimonies/Article-View/Article/1859890/russian-and-chinese-nuclear-modernization-trends; Hans M. Kristensen, "New Missile Silo and DF-41 Launchers Seen in Chinese Nuclear Missile Training Area," September 3, 2019, https://fas.org/blogs/security/2019/09/china-silo-df41; Lyle J. Goldstein, "China Is Learning from Russian Military Interactions with the United States," *The National Interest,* May 9, 2019; Mark Schneider, "Nuclear Weapons in Chinese Military Strategy," May 3, 2019, https://nipp.org/information_series/schneider-mark-nuclear-weapons-in-chinese-military-strategy-information-series-no-441; Graham Allison, "What Xi Jinping Wants," *The Atlantic,* May, 2017, https://www.theatlantic.com/international/archive/2017/05/what-china-wants/528561; and Stephen J. Cimbala, "Chinese Military Modernization: Implications for Strategic Nuclear Arms Control," *Strategic Studies Quarterly,* no. 5 (USSTRATCOM 2016), pp. 110–117.

18. On Chinese capabilities for, and interest in, information warfare, see David E. Sanger, *The Perfect Weapon: War, Sabotage, and Fear in the Cyber Age* (New York: Crown, 2018), pp. 100–123; Fred Kaplan, *Dark Territory: The Secret History of Cyber War* (New York: Simon and Schuster, 2016), pp. 221–235; Timothy L. Thomas, *Three Faces of the Cyber Dragon: Cyber Activist, Spook, Attacker* (Ft. Leavenworth, Kans.: Foreign Military Studies Office, 2012); Timothy L. Thomas, *The Dragon's Quantum Leap: Transforming from a Mechanized to an Informatized Force* (Fort Leavenworth, Kans.: Foreign Military Studies Office, 2009); and Timothy L. Thomas, *Dragon Bytes: Chinese Information-War Theory and Practice from 1995–2003* (Fort Leavenworth, Kans.: Foreign Military Studies Office, 2004). See also Kevin Pollpeter, "Towards an Integrative C4ISR System: Informationization and Joint Operations in the People's Liberation Army," chap. 5 in Roy Kamphausen, David Lai, and Andrew Scobell, eds., *The PLA at Home and Abroad: Assessing the Operational Capabilities of China's Military* (Carlisle, Pa.: Strategic Studies Institute, US Army War College, June 2010), pp. 193–235.

19. J. Berkshire Miller and Benoit Hardy-Chartrand, "Russia and China's Strategic Marriage of Convenience," *The National Interest,* August 27, 2019.

20. Lawrence Korb and Katherine Blakely, "This Deal Puts the Nuclear Genie Back in the Bottle," *Bulletin of the Atomic Scientists,* July 15, 2015, http://the bulletin .org/experts-assess-iran-agreement-20158507.

21. Eric Schmitt, Farnaz Fassihi, and David D. Kirkpatrick, "Focus on Iran in White House After Oil Strike," *New York Times,* September 16, 2019, pp. A1, A11.

22. US Census Bureau, "Quick Facts," https://www.census.gov/quickfacts/fact /table/US/PST045218.

23. These latter are usefully and collectively referred to by Eliot A. Cohen as "ungoverned spaces." See Cohen, *The Big Stick: The Limits of Soft Power and the Necessity of Military Force* (New York: Basic, 2016), chap. 7.

24. Colin S. Gray, *Modern Strategy* (Oxford: Oxford University Press, 1999), pp. 165–166.

25. David E. Sanger, "In Confronting Iran, Trump May Find the World Is Wary," *New York Times,* September 18, 2019, pp. A1, A10.

26. For a discussion of the various geopolitical theories, see Norman J. Padelford and George A. Lincoln, *The Dynamics of International Politics,* 2nd ed. (New York: Macmillan, 1967), pp. 106–112. See also Sir Harold J. Mackinder, *Democratic Ideals and Reality* (New York: Holt, 1919); and Nicholas J. Spykman, *The Geography of Peace,* edited by Helen R. Nickel (New York: Books, 1969).

27. Hans J. Morgenthau, *Politics Among Nations: The Struggle for Power and Peace,* 6th ed. rev. by Kenneth W. Thompson (New York: Knopf, 1985), p. 634.

28. Ibid.

29. Quincy Wright, *The Study of International Relations* (New York: Appleton-Century-Crofts, 1955), p. 348.

30. For the potential of cruise missiles in WMD attacks, see Lt. Col. Rex R. Kiziah, USAF, *Assessment of the Emerging Biocruise Threat,* Counterproliferation Papers, Future Warfare Series no. 6 (Maxwell Air Force Base, Ala.: USAF Counterproliferation Center, August 2000).

31. Expert commentaries on these topics appear in Russell D. Howard and James J. F. Forest, eds., *Weapons of Mass Destruction and Terrorism* (New York: McGraw-Hill, 2008). On the likelihood and risks of nuclear terrorism, see Brian Michael Jenkins, *Will Terrorists Go Nuclear?* (New York: Prometheus, 2008); and Graham Allison, *Nuclear Terrorism: The Ultimate Preventable Catastrophe* (New York: Holt/Times, 2004).

32. See, for example, Zac Rogers, "In the Cognitive War, the Weapon is You!" July 1, 2019, https://madsciblog.tradoc.army.mil/158-in-the-cognitive-war-the-weapon

-is-you; Chris C. Demchak, "China: Determined to Dominate Cyberspace and AI," *Bulletin of the Atomic Scientists,* no. 3 (2019), pp. 99–104, https://doi.org/10.1080/00963402.2019.1604857; Nina Kollars and Jacquelyn Schneider, "Defending Forward: The 2018 Cyber Strategy Is Here," September 20, 2018, https://warontherocks.com/2018/09/defending-forward-the-2018-cyber-strategy-is-here; and Martin C. Libicki, "The Convergence of Information Warfare," *Strategic Studies Quarterly,* no. 1 (Spring 2017), pp. 49–65.

33. Adam Lowther and Curtis McGiffin, "America Needs a 'Dead Hand,'" August 16, 2019, https://warontherocks.com/019/08/america-needs-a-dead-hand. See also Matt Field, "Strangelove Redux: US Experts Propose Having AI Control Nuclear Weapons," *Bulletin of the Atomic Scientists,* August 30, 2019, https://thebulletin.org/2019/08/strangelove-redux-us-experts-propose-having-ai-control-nuclear-weapons.

34. US Defense Intelligence Agency, *Challenges to Security in Space* (Washington, DC, January 2019), https://www.dia.mil/Portals/110/Images/News/Military_Powers_Publications/Space_Threat_V14_020119_sm.pdf.

35. Robert Gilpin, *War and Change in World Politics* (Cambridge, Mass.: Cambridge University Press, 1983), p. 177. See also Paul Kennedy, *The Rise and Fall of the Great Powers: Economic Change and Military Conflict from 1500 to 2000* (New York: Random, 1988).

36. Gilpin, *War and Change in World Politics,* p. 173.

37. See, for example, P. W. Singer, *Wired for War: The Robotics Revolution and Conflict in the Twenty-First Century* (New York: Penguin, 2010). Historian Max Boot traces the relationship between technology and warfare from early modern times to the present in his *War Made New: Technology, Warfare, and the Course of History, 1500 to Today* (New York: Gotham, 2006), including the information revolution discussed on pp. 307–436.

38. Gilpin, *War and Change in World Politics,* p. 185.

39. Wright, *The Study of International Relations,* p. 385.

40. M. Thomas Davis and Nathaniel C. Fick, "America's Endangered Arsenal of Democracy," *Joint Force Quarterly* 62 (3rd Quarter 2011), pp. 89–95. Important historical perspective is also provided in James Ledbetter, *Unwarranted Influence: Dwight D. Eisenhower and the Military-Industrial Complex* (New Haven, Conn.: Yale University Press, 2011), esp. pp. 15–44, 118–163.

41. Bruce D. Berkowitz and Allan E. Goodman, *Best Truth: Intelligence in the Information Age* (New Haven: Yale University Press, 2000), esp. pp. 58–98.

42. Thomas A. Keaney and Eliot A. Cohen, *Revolution in Warfare? Air Power in the Persian Gulf* (Annapolis, Md.: Naval Institute, 1995), pp. 188–226.

43. For perspective on this issue, see Christopher Jon Lamb, "The Impact of Information Age Technologies on Operations Other Than War," in Robert L. Pfaltzgraff and Richard H. Shultz Jr., eds., *War in the Information Age* (Washington, DC: Brassey's, 1997), pp. 247–278.

44. John Arquilla, *Worst Enemy: The Reluctant Transformation of the American Military* (Chicago: Ivan R. Dee, 2008), p. 213.

45. Colin S. Gray, *Another Bloody Century: Future Warfare* (London: Weidenfeld and Nicolson, 2005), pp. 29–37.

46. T. X. Hammes, "Cheap Technology Will Challenge U.S. Tactical Dominance," *Joint Force Quarterly* 81 (2nd Quarter 2016), pp. 76–85.

47. John Arquilla and David Ronfeldt, *Swarming and the Future of Conflict* (Santa Monica, Calif.: Rand, 2000).

48. Sarkesian and Connor, *The US Military Profession,* esp. pp. 50–62. See also the 2nd edition, 2006.

15

Improving the System

Americans are a restless, dynamic, forward-looking, and results-oriented people.[1] Because national security deals with the highest matters of state policy, including issues of life and death, Americans' expectations for the performance of their government with respect to security are high. Popular disillusion and disappointment are inevitable when the expectations for government performance on national security clash with the reality of policymaking in the US government. Given the variety of agencies and individuals involved in US national security, it is no wonder it is such a confusing system with a muddled process. Designing coherent policy and strategy in such a system is like trying to complete a complex jigsaw puzzle: to be assembled correctly, it takes time, and sometimes the actors merely "muddle through." This is a disturbing overall view. To complicate matters, national security issues lead to serious disagreements within Congress, and the US public may be confused as to how the system actually works. Furthermore, bureaucratic turf battles, weak or erratic leadership, and power struggles can lead to internecine warfare, causing serious problems in defining national interests and designing national security policy and strategy.[2] As the coronavirus epidemic of 2020 illustrates, some "medical" or "epidemiological" issues also have national security implications that affect US national defense and security.

The terrorist attacks of September 11 seemed to change all this, at least in the short term. The US war on terror was coupled with a policy of transformation in Middle Eastern politics, emphasizing the promotion of democracy in that troubled region. The George W. Bush administration admitted that the struggle to defeat transnational terrorism and promote

Middle Eastern democracy would be a protracted conflict for many years. Against this appearance of a singular focus in Bush policy and strategy, critics charged that the war against Iraq in 2003, and the insurgency in that country that roiled its postwar reconstruction, diverted US policy and strategy into wrong directions.[3] The Bush administration felt that US efforts against global terrorism, and against state sponsors of terror or states that might give terrorists weapons of mass destruction, were all important pieces in solving the same puzzle. Regardless of how these arguments played out in US domestic politics, there was no question that social change and political upheavals in the Arab and Islamic worlds would involve the United States, its allies, and its adversaries for years to come. For example, despite the reassurances of President Trump and his advisors, the US military engagement in Afghanistan, although whittled down in size and in numbers of American combat troops deployed there, remained active in 2020. President Biden withdrew all US troops from Afghanistan in August 2021 and announced that the combat role of the some 2,500 troops in Iraq would end by the end of the year and their mission would change to advisory.

Despite outward appearances, US national security policymaking takes place within a well-established pattern and institutional system. The problem is that competing interests and divergent political perceptions and mind-sets often disorganize the policy process. This was especially the case in the post–Cold War era and into the beginning of the twenty-first century. The passing of the Cold War swept away familiar frameworks of analysis and categories of thought about international politics. Scholars and policymakers alike have slipped their previous intellectual moorings, and certainties from the past have given way to ambiguities in the present.[4]

Coherence and effective functioning within the national security system are not self-executing—they require hard work. The sense of directed purpose and implementation is established only as a result of effective leadership putting the major pieces together. Without such leadership, the system becomes a series of political-bureaucratic fiefdoms, with each trying to dominate the other. Regardless of how well that is done, there are always disagreements and bureaucratic infighting.

The national security establishment and those in the policy process should not be viewed simply in theoretical terms or as intellectual curiosities. People are the foundation of the national security system, whether bureaucrats, staff members, political appointees, or members of Congress. They all come with imperfections, previous social and political relationships, divergent goals, ambitions, and human frailties. To speak of "the establishment" and "the process" in clinical terms, isolated from the real world and human dynamics, is to neglect the most important dimension of national security: the human factor. A realistic sense of the national security establishment and the effectiveness of the policy process must include how the human factor shapes and influences US national security. This dimension is too often neglected by those studying US national security.[5]

We have examined the scope and complexity of national security, its institutional and political dimensions, and the variety of interests and forces involved in the policy process. That process has become even more complicated and broader in scope owing to the US determination to fight the war on terrorism and the return of peer and near peer great power competitors. Thus there are many challenges to US security interests that defy simple solutions. A good example of this is the ongoing global pandemic beginning in 2019 and its consequences for US national health and security policies. Also, US national security policy has a great deal of continuity. Important elements shaping national security are part of the US political culture and are embedded in US history. What can be made of all of this? How can these be properly weighed and shaped to design the most effective US national security policy?

The search for answers can begin by applying what has been discussed in this book, recognizing that there are no perfect or immutable solutions to US national security. On the one hand, there are persistent US national interests; on the other, those interests must come to grips with a changing international security environment. Future presidents, defense secretaries and professional soldiers will have to be equipped with knowledge of American history and military art, but also be able to improvise and adapt to unforeseen circumstances, unpredictable threats for enemies or events abroad, and previously unknown domestic conditions at home. The last point is perhaps the least often appreciated. Many take for granted the consensus on American values that led to victory in World War II and a favorable outcome in the Cold War. But as the attack on the US Capitol on January 6, 2021, and the almost inconceivable politically driven responses to it showed, the existence of an enduring American value consensus cannot be taken for granted.

The President and National Security: Retrospections

The scope of presidential power expanded over the course of the twentieth century, and the prerogatives of the presidency have been closely guarded by the executive, regardless of party affiliation and political orientation. Presidential power in national security, resting on such concepts as executive privilege and the interests of national security, was rarely challenged. In the period from the end of World War II until the Vietnam War, roles and responsibilities as commander in chief, chief executive, and head of state gave the president control over policy and intelligence-gathering instruments, and he was seen by most US citizens as the focal point for foreign and national security policies. President Trump's disregard for precedent and distain for professional advice have caused some to wonder whether the checks and balances established by the Founders are sufficiently robust in the face of increasing presidential power—especially if the Congress is unable politically to assert its institutional prerogatives. The answer is in two parts: Congress must effectively exercise its powers of legislation, investigation, and oversight with respect to the executive branch. The President, in turn, must provide the policy objectives and benchmarks that set his or her administration apart from others and that define America's role in the world, thereby setting the predicates for an effective national security strategy (or lack thereof).

The International Arena in the Twenty-First Century

Compounding the problems of presidential leadership is the international security environment of the twenty-first century. It has become more difficult to define US national interests, national security policy, and strategy in the domestic and international arenas. After September 11, combating transnational terror networks such as al-Qaeda and their state sponsors became a consensus priority among US policymakers. But in the US system of government, the "how" is always controversial, even after broad agreement on the "what." Past controversies such as Vietnam and Watergate, as well as arguments over the domestic surveillance policies of George W. Bush and Barack Obama, have increased concerns about the extent of executive power. In the aftermath of Vietnam and Watergate, for example, Congress reasserted itself in foreign and national security affairs—the period of the "imperial Congress." Although the White House regained some initiative during the Ronald Reagan presidency, the assertive role of Congress remains an important factor in limiting presidential power. George W. Bush's claims for broad presidential power were urged by Vice President Dick Cheney, who served as chief of staff for President Gerald Ford when presidential power suffered under the hammering of an aroused post-Vietnam and post-Watergate Congress. Perceived failures by that administration, controversies over Obama administration policies in the Middle East, and numerous Trump administration policies and his administrative style have strengthened the case for Congressional oversight. Of course, Congressional "oversight" can be dysfunctional if it becomes especially partisan or verges into policy micromanagement.

The Vietnam legacy underpinned much of the congressional opposition and public resistance to extending US commitments beyond well-established treaty obligations. That legacy remains as an important conditioner of US attitudes, even though the victory in the 1991 Gulf War helped to erode the so-called Vietnam syndrome.[6] But the failed US efforts in Somalia in 1993, with the humiliating deaths of US soldiers in Mogadishu, as well as US involvement in Bosnia and Kosovo, seemed to have rekindled the Vietnam syndrome, at least until September 11. Initially favorable military campaigns in Afghanistan and Iraq positioned the United States for a more influential profile in Middle Eastern politics, but they did not necessarily make the case for a fundamental change in US grand strategy. US grand strategy would have to reflect postconflict staying power and the willingness to follow success in battle with endurance in the murky political waters of the greater Middle East, from West Africa to Southwest Asia. That proved to be very difficult.[7] In addition, grand strategic thinking in the Obama, Trump, and Biden administrations also had to acknowledge the renewed challenges from great power competition by Russia and China and the resulting need, in the case of China, to shift the weight of military-strategic planning and commitments to the Indo-Pacific theater. China remains a serious geopolitical rival, but the US shift to this theater will likely be slowed by the 2022 Russian invasion of Ukraine.

Many academic, policy community, and public attitudes also reflect a growing conviction that there are limits to US ability to influence events around the world. Indeed, some even argue for a revived isolationism. But the United States still remains the lone global military superpower, at least for the moment, with responsibilities that extend beyond the US homeland, and, arguably, the United States must assume the role of international "sheriff" for want of better candidates.[8] Nevertheless, the United States cannot succeed alone. The preceding point is underpinned by efforts to extend US values into other parts of the world and to establish democratic systems. Combined with lessened fears of major wars and the increasingly accepted idea that democracies do not fight one another, there is the notion that the United States should attempt to shape events around the world. But old demons have been replaced by new ones: "It was May 1989. . . . 'I'm running out of demons,' complained Army General Colin Powell. However, on 1 September 1993 when he unveiled [the Defense Department's] Bottom-Up review, he amended his original statement: 'Fortunately, history and central casting have supplied me with new ones along the way.'"[9] That trend continues, reinforcing the need for wise and prudent leadership.

The new demons range from religious, ethnic, and nationalistic conflicts to international terrorism, drug cartels, and the rise of nations that can challenge US military power at least regionally, and in some respects, globally. Add to this the proliferation of nuclear weapons and increasing concerns about chemical and biological terrorism and information warfare, and the number of potential threats is daunting.[10] From the relatively well-structured (if dangerous) threat system of the Cold War, we have evolved into a less well-structured and more chaotic system of "distributed" threats. War can now take place in any of five dimensions or domains: land, sea, air, space, and cyberspace.

The National Security System

Actors in the policy process have their own political constituencies and perspectives. They jealously guard their prerogatives, jockeying for power among themselves. The national security establishment, although directly under the control of the president, is still subject to congressional initiatives and oversight. Congressional initiatives, for example, have focused on intelligence activities and the defense budget process. Part of this effort reflects congressional reluctance to allow the establishment to operate outside congressional control. The legislation passed in 1996 requiring the secretary of defense to complete a Quadrennial Defense Review (now the National Defense Strategy) every four years is but one example. Other examples include the post–September 11 reorganizations of US intelligence and national security to create a director of national intelligence and a cabinet-level Department of Homeland Security.

The US political system divides power and responsibility in ways that impact on national security. Bureaucratic concerns about agency power and programs, interagency power struggles, budget constraints, congressional criticism, and domestic political considerations have solidified efforts to constrain the national security establishment. In the current period, presidential power in national security is being redefined by executive assertions, congressional initiatives, and court decisions, all happening simultaneously. Furthermore, various components of national security policy, such as budget considerations, weapons acquisitions, and strategy, are not solely determined by military and national security issues. The defense budget, for example, reflects a series of political compromises, many shaped by domestic policy issues rather than political-military policy and strategy. Nonetheless, the president has been given a broad mandate to fight terrorists, even preemptively. President Barack Obama's authorized raid in May 2011 by US Navy SEAL Team Six to kill Osama bin Laden in his Pakistani redoubt, without prior notice to Pakistan's political or military leadership, reminded the world that the United States has unique kinds of striking power and a willingness to use them when the stakes are high enough.

The complexity and potential volatility of national security issues make the problem of presidential control and guidance even more difficult. The technical aspects of strategic weapons, information-age technology, the complexities of logistics, and the evolution of the electronic battlefield, for example, require that the president rely heavily on military and civilian experts for interpretations of weapons requirements, as well as advice on their strategic implications. One consequence is that there is a built-in propensity for struggle among agencies to gain access to the president and to convince him or her of the wisdom of their views.[11] The internal struggles are compounded by gatekeepers who can be primarily concerned about preserving existing power structures. All of this makes the president vulnerable to agency biases, in some instances making him or her a captive of technicians, strategists, and midlevel bureaucrats. President Donald Trump dismissed John Bolton as his national security advisor in September 2019 after seventeen months for a number of reasons, including a desire by Trump to act more assertively as his own "decider" on national security issues without feeling constrained by advisors and other principals.[12]

In summary, the power and posture of agencies within the national security establishment, the interplay of personalities within the executive branch, the power of Congress in the national security policy process, the tendency to politicize national security issues, and the demands of domestic politics and policy—all in a changing international security environment—have created a far more complicated situation compared to previous eras. The apparent simplicity of Cold War definitions of allies, enemies, and "others" has given way to a confused landscape of possibilities, including a resurgence of Cold War concerns about conventional and even nuclear war.

In the search for answers and a systematic approach to national security, we need to focus on two dimensions: the presidential power base and constituency, and the national security environment.

The Presidential Power Base and Constituency

The presidential power base is shaped by four elements: the legal and political dimensions of the presidency in national security; democratic ideology and open systems; domestic political actors and the domestic political environment; and presidential character and accountability.

First, since the end of World War II, presidents have assumed office with similar legal and political power to shape and implement national security policy. The president is the commander in chief, head of state, and chief executive. The legal and political powers derived from those roles have provided presidents the opportunity to engage in a range of initiatives in the conduct of national security. How such powers are used is a function of the president's personality and character, leadership style, and worldview. Thus each president placed his own stamp on national security policy and the way that the national security establishment functions.

Second, the extent of presidential power in national security policy is limited and conditioned by the values and expectations of democratic ideology. The moral and ethical content of democracy and the public expectations that US international behavior should reflect values such as individual freedom, justice, and dignity limit the options in national security strategy, temper the means used, and establish an overarching presence in the way the president conducts national security policy. Proof of this self-conscious ethical auditing can be found in the extensive debates within and outside the US government about the George W. Bush administration use of "enhanced interrogation," extraordinary rendition, and indefinite detention without trial of enemy combatants as part of the US war on terror. All this creates an inherent moral and ethical dichotomy between the ends-means relationships in US national security policy and strategy, that is, moral and ethical means to reach moral and ethical ends—a gap that presidents often find difficult to bridge. A commitment to democratic ideology and its moral and ethical imperatives should shape a presidential mind-set and determine the boundaries within which he or she functions. In short, there are must be self-imposed restraint and limitations on those reaching the Oval Office.

Third, there are several domestic political actors with power to affect national security policy, chief among them the Congress—and congressional assertiveness is reinforced by the power of congressional incumbency—one that often goes beyond presidential terms in office.

Vietnam, Watergate, Iran-Contra, and Iraq, among other events, caused Congress to become more assertive in the national security policy process. This was shown by criticism of the Bill Clinton administration's efforts in Somalia, Haiti, Bosnia, and Kosovo, among others. Before and after September 11, the George W. Bush administration received its own criticism on several issues, including the perception that it tended to ignore international opinion and to avoid commitments on issues ranging from global warming to the control of antipersonnel mines.

The Barack Obama administration similarly encountered congressional opposition on national security and defense matters. The Donald J. Trump administration found itself under criticism and ultimately, impeachment, by the US House of Representatives over the president's behavior on national security issues, including alleged arm twisting by President Trump of Ukrainian government officials in order to improve the president's chances for reelection. President Trump would be impeached a second time for his alleged role in instigating the January 6, 2000, attack on the US Capitol by a crowd of his followers and failing to take action to stop it once it began. President Biden found himself widely criticized in congress and elsewhere for his sudden and chaotic withdrawal of US forces from Afghanistan.[13]

In addition, the mass and "niche" media have a significant influence in the policy process. Investigative reporting, brought to public notice in the Watergate affair, has become an institutionalized method of analyzing and reporting the news as well as for setting the agenda. Combined with the traditional adversarial relationship between the media and the national security establishment, this makes it difficult for the president to remain unchallenged in his policy pronouncements and in shaping that establishment. Equally important, representatives of the media have their own informal networks of information that penetrate the core of the national security establishment, making it difficult to develop policy out of public view when necessary.

Modern communications technology creates enormous pressures for transparency and immediacy in the flow of information, but also invites confusion and emotional enlargement. Political predispositions of the media elite—whether perceived as left- or right-leaning—add to the problem, with certain national security policy and strategy issues presented in highly critical terms.[14] For example, some media critics suggested that reporters subdued their criticism of the Clinton administration because of Clinton's domestic social agenda, which conformed to the media's view.[15] The allegedly liberal bias of reporters and editors must be viewed within a broader context, however. Owners of flagship media tend to be wealthy entrepreneurs, not populists.

In addition, the bias in news gathering and reporting is toward the negative and sensational, not necessarily against either political party. All presidents have felt the sting of negative reporting that sometimes crossed the line between professional criticism and personal abuse. Indeed, outrageous and even irresponsible press behavior is a US tradition: it was even worse in the early years of mass-market newspapers when the rambunctious publisher William Randolph Hearst once boasted that he had started a war (the Spanish-American War)—an overstatement with a grain of truth. How far the Internet and social media will take us and in what direction is unknown—perhaps, toward a new era of "yellow journalism" or something even more unmanageable, yet powerful. President Barack Obama was not in office for very long before he had experienced both Obama-mania from his supporters and Obama-phobia from his detractors, and the same was true for Presidents Trump and Biden.

Several groups and institutions in the public domain, including special interest groups and single-issue groups, seek to influence national security policy and strategy. Their involvement has increasingly politicized and publicized nuclear, environmental, and human rights issues, linking them to national security. Special interest or single issue groups even made their views apparent relative to the new war on terrorism. This creates a greater awareness about national security issues and defense policy, and the public becomes more sensitive to presidential postures on such issues. Furthermore, power clusters within the administration can also act as rallying points for advocating one or the other direction in national security policy and strategy, leading to struggles between the president's inner circle and the rest of the administration.

Fourth, even though the president has a freer hand in national security policy compared to some other issues, the nature of the office makes him accountable to the public. This is not limited to elections every four years but includes a variety of measures and instruments that counterbalance presidential actions and require explanation. There is a constant political and intellectual interplay that tests presidential accountability, including opinion polls, interest group activity, party cohesion, congressional-executive relationships, and the president's own staff and bureaucracy. The power clusters and the policy and strategic options they represent force the president to develop some degree of consensus and political coherence in policy and strategy, regardless of his leadership style and decisionmaking preferences.

Although several conclusions can be drawn from these observations, one stands out: the importance of presidential leadership. The president's leadership style and how he or she organizes and directs the national security establishment are primary determinants of the effectiveness of US national security policy and strategy. This shapes national will, political resolve, and staying power. The myriad national security challenges present or on the horizon will surely test the resolve of the US public and the president's leadership.

Presidential Leadership and the National Security Establishment

The boundaries within which the president must operate and the latitude of his or her legal and political power to deal with the national security system and environment are a consequence of his power base and constituency. The president's leadership, personality, and character are critical in determining his effectiveness. In this respect, *system* is defined to include the variety of political actors, the national security establishment, the policy process, and all of the informal procedures that are important parts of national security policymaking. The *environment* refers to the characteristics of the international security dimension of world politics, the capabilities and policies of external political actors, and the dynamics created by their interactions.

Although model-building and theoretical frameworks for analyzing the presidency are a useful and necessary undertaking, they have limits. Presidential performance is not bound by any single model; neither does the president necessarily engage in a conscious effort to adopt a particular model and shape his or her performance accordingly. Furthermore, the rigid application of models and theories can fail to account for a president's ability to change his or her approach and adopt new perspectives to address problems and challenges, as President Ronald Reagan did during his second term with respect to the Soviet Union in expediting a peaceful end to the Cold War. Finally, choosing a single model as the sole basis for examining the presidency tends to favor a managerial perspective, highlighting the mechanics rather than the political and social components of presidential performance.

Given the dominance of the political component in virtually all national security considerations, the political-psychological framework can provide a useful and pointed focus on the president and his role in shaping the national security environment. This approach is based on the premise that presidential leadership—fashioned by the president's personality and character—and keen political instincts and intuitiveness are key elements in presidential performance.[16] These qualities should flow from a mind-set that projects a national security posture that is coherent, purposeful, and in accord with the norms of democracy.

This is not to suggest that national security policy needs to be an uncontentious process. It does mean that a reasonably effective national security policy requires wise and effective presidential leadership in the policy process. Leadership, especially broad vision and goal setting, is more important than mastery of organizational procedures, managerial techniques, and knowledge of the details and technical aspects of national security issues. On the other hand, the President and his or her staff cannot be oblivious to the "nut and bolts" of decisionmaking in a highly charged political milieu and bureaucratic mine field. Broad vision upward and relentless monitoring downward are both necessary to ensure that the President's national security objectives are realized in actual policy outcomes.

Presidents who expect to be successful should begin with a deep understanding of the organizational dynamics within the national security establishment. This must go hand in hand with political acumen in dealing with Congress and the public. Put simply, the president must be a creative political leader. It also helps for a President to have had at least some prior experience in government, preferably at a national level. Jumping directly from a profession other than politics into the highest level of national political involvement carries great risks for leaders, including presidents, as Ulysses S. Grant and Donald J. Trump both learned more than a century apart.

To ensure that the presidential presence permeates the national security establishment and that presidential views of national interests, policy, and strategy guide the system, presidents must begin by placing their mark on three areas: the national security triad, the national security establishment, and the national security system. It must be remembered that none of this can be accomplished unless presidents are first able to deal with their power base and constituency to create a domestic environment that is receptive and conducive to their leadership.

The president must appoint the policy triad—secretary of state, secretary of defense, and national security advisor—making sure that they are capable and that their worldviews are in general accord with his own. In the 2001 Bush administration, Secretary of State Colin Powell, Secretary of Defense Donald Rumsfeld, and National Security Advisor Condoleezza Rice fit the Bush worldview. President Obama's choices for cabinet officers resulted in part from Obama borrowing Abraham Lincoln's concept that a "team of rivals" equally committed to the goals of the administration, but with different experiences and perspectives, would provide the president with the best possible advice. Initial appointees by President Trump who did not share his worldview were later replaced by others whom he found to be more supportive of his policies and who seemed to be more loyal to him personally. President Biden's national security team appears to be both capable and harmonious. There is nothing necessarily dysfunctional about a presidential need for staff who share his or her overall perspective and worldview, provided the president is self-confident and strong enough to avoid converting his staff into "yes men" (or women).

Regardless of presidential style, each member of the national security policy triad must possess the personality, character, and leadership style that allow for effective engagement in the difficult process of designing national security policy and strategy. Each must be able to deal with the others in the triad in a firm but prudent way, with commitment to the most effective policy overriding any agency or department loyalty. Each member of the triad must combine political skills with national security expertise to ensure that the best posture will emerge as a result of the dynamics within the triad. Finally, members of the triad must not only be competent advisors to the president but also skilled managers and leaders to deal effectively with their own bureaucracies.

Although not part of the triad, the director of national intelligence and the chairman of the Joint Chiefs of Staff are also key players in shaping the national security system. In each case, the president must appoint individuals whose philosophical range and worldviews are compatible with his own. As is the case with the triad, the director of national intelligence (DNI) and the chairman of the Joints Chiefs of Staff (JCS) must be skilled advisors, politically wise, and capable of effective control and direction of their own structures. This also applies to the secretary of the Department of Homeland Security (DHS). The DNI and DHS are both post–September 11 creations and thus, from the standpoint of their relationships to other agencies and to the White House and Congress, works in progress.

The personal strength and power of the members of the triad, the DNI, and the JCS chairman do not guarantee smooth dynamics or relationships within the national security establishment and system. But the strength of character, political acumen, and leadership skills of these individuals, combined with their commitment to the president and effective national security, are essential ingredients for coherent policy, realistic strategy, and effective implementation. The Harvard JFK School makes an important distinction in their curriculum that is relevant here: between policy formulation and policy implementation. Policy formulation is the process of agreeing upon a preferred definition of political and/or military objectives: at the highest levels, this encompasses what some call "grand strategy." Policy implementation, on the other hand, is the process of getting it done: turning broadly stated concepts and goals into actual legislation or action programs. The skill sets that are required for policy formulation are not necessarily those that are necessary for policy implementation: and vice versa. In addition, the "friction" that Carl von Clausewitz wrote about with respect to military operations also characterizes the fate of many policy proposals as they work their way through bureaucratic channels or congressional committees.[17]

The president must also establish cooperative links with key members of the congressional establishment—members who have important roles in intelligence oversight, budget policies, and armed services matters. Cooperative links are established and reinforced by the president's sincere efforts to provide timely and relevant information on national security matters, articulate national interests and policy goals, and nurture and expand the president's power base and constituency support. The triad, DNI, and JCS chairman are important persons in these cooperative links; they are an extension of the president but are not presidential clones.

Finally, all of these matters must coalesce into workable ways to deal with the external security environment, in which sovereign states—many of which have differing ideologies and political systems contrary to US values and norms—look to their own self-interests. Their policies and conceptions of national security can contradict the goals of US security policy. The difficulties are magnified by the emergence of regional and eventually global powers with their own policy agendas and relative immunity to pressure from external powers. With globalization complicating all of these developments, conflict in some form is inevitable.

The complexities of the external security environment are compounded by the shifts in the locations of power (e.g., the power center emerging in Asia and the regional issues in relations among China, Japan, Russia, India, and other Asian powers). Europe has already established itself as a major power system and one averse to the use of armed force as a means of settling disputes, even though it faces bickering over the sovereignty of individual members within the framework of the European Union and a resurgent military threat from Russia. Russia's effort to regain its position of power within the international community, after a shattering post–Cold War decade in the 1990s, and the rapid rise of China are other complicating factors in world politics.[18]

Most worrisome for the United States would be an enduring strategic partnership between Russia and China. The security cooperation between these two aspiring peer competitors to the United States would create dangers to US and allied vital interests, especially in Asia. A dramatic example of growing Sino-Russian security cooperation occurred in July 2019 when Russian and Chinese long-range, nuclear capable bombers conducted a first ever joint patrol and entered the air defense identification zones of Japan and South Korea. South Korea scrambled fighter jets and reported firing several hundred warning shots against the intruders, which provoked criticism from Japan because the South Korean maneuvers took place over islands that Japan considers its sovereign territory. Whether Russia and China deliberately staged this event for the purpose of exacerbating already existing Tokyo-Seoul frictions, or whether that friction was incidental, the result illustrated the complexity of a growing Sino-Russian entente against the backdrop of a more complicated security canvas in multipolar Asia.[19]

It is conceivable that new power relationships will emerge alongside legacy partnerships. These might include, for example, United States–Europe (excluding Russia); United States–Canada–Mexico; China-Japan; Russia and the states of Central Asia, the Caspian basin, and the Caucasus; and perhaps India and other Southeast Asian states. India's support on foreign policy issues is sought after by Moscow and Washington. China's Belt and Road Initiative aims to increase its influence across Eurasia and beyond by supporting infrastructure projects, including transportation and manufacturing, that tie states and business to Chinese-dominated networks of influence. In the Middle East, competition for regional hegemony between Iran and Saudi Arabia creates cross-currents of involvement by outside powers, including the United States and Russia, each with its own regional objectives. For example, the United States must balance its support for Israel against its desire to maintain good relations with Saudi Arabia, Egypt, and Turkey, all consumers of American military assistance and at least partially convergent with US political aims in the region. Regional political turmoil is also encouraged by the global spread of identity politics.[20] This may lead to Samuel Huntington's proposed "clash of civilizations."[21] Combined with existing competing power clusters, the security environment is undergoing major changes in the twenty-first century, requiring rethinking of the US security position and its global strategy.

The interplay of innumerable factors makes the national security arena difficult to understand and respond to. US politicians tempt the public to adopt simplistic perspectives and policy postures. This reflects frustration over the limited ability to achieve moralistic goals emerging from US ideology, a response to the complexity of such issues as nuclear strategy and weapons proliferation, and the complexities of information-age technology, as well as the public's lack of understanding and knowledge regarding the nature and character of modern warfare—especially unconventional conflicts and operations other than war.

To respond to the complex international security environment and its challenges, the United States must have an effective national security policymaking process. This requires a national security establishment staffed by skilled people who are properly organized and directed to respond and act effectively. But a perfectly skilled and bureaucratically efficient establishment and a Congress perfectly supportive of the president would be for naught if presidential leadership were found to be lacking. As Presidents George W. Bush, Barack Obama, Donald Trump, and Joseph Biden discovered in formulating national security policy, presidents can exploit the power of initiative to great effect before countervailing political forces assert themselves. Absent presidential leadership, the policy process recycles its own inertia. President Biden assumed office with sufficient prior experience to realize this.

The Establishment and the Policy Process

Regardless of how well policy is articulated and world conditions are understood, the process by which policy is determined and strategies designed has an important impact on the end result. Furthermore, how well the national security establishment and the policymaking machinery function determines the relevance of policy and the effectiveness of strategy. If it is true that a flawed policy process is likely to lead to bad policy and strategy, how effective are the national security establishment and the policymaking process in coming to grips with the policy and strategic issues outlined here? How well can the establishment incorporate the necessary understanding and knowledge, balanced by keen intuitive insights, into the policymaking process? Is the current structure of the national security establishment relevant to the strategic landscape of the twenty-first century? These are the critical questions to which we now turn.

From the analyses of the national security policy process and the way the establishment functions, we offer three conclusions: (1) the national security process is cumbersome; (2) the diffusion of power makes it difficult to adopt innovative policies and strategies, except during crises; and (3) the national security establishment as it existed in the Cold War is outmoded.

With respect to the way the US political system operates, Roger Hilsman concluded some time ago: "Reflecting on the complexity and difficulty of these problems and the untidy, frequently stalemated American political system, one wonders how the system can cope."[22] He pointed out the fact that the policy process involves many power centers within the system. With respect to foreign and national security policy, "a long-run increase in the number of power centers, finally, seems inevitably to work to lessen the power of presidents in foreign affairs."[23] The same is true in national security policy. Hilsman went on to say, "So many centers of power make building a consensus for positive action a formidable task. . . . What is discouraging is how difficult it will be to get such a disparate myriad of power centers to agree on policies for meeting these complex problems."[24] This comment remains valid today. The exception is when there is a clear and serious threat to US national interests, which increases national will, political resolve, and staying power.

On the other hand, Hilsman and others who comment upon the incoherence of the policymaking process and the diffusion of power in the American system must be reminded of the mind-sets of the framers of the US Constitution. They were willing to trade some efficiency for the sake of liberty: liberty is the cornerstone of the American historical experiment. No other country has grown so large from a foundation based on an idea: a free people with inherent rights not granted by government but innate in their persons, making government a limited delegation of authority for specifically defined purposes only. As the United States has grown into a global economic and military superpower, it is well to remember the national origins that got us there: not militarism, but military prudence rooted in the American national character and based on consensual notions of America's world role. As Henry Kissinger has noted, leaders in foreign policy must balance between two aspects of order—power and legitimacy:

> Calculations of power without a moral dimension will turn every disagreement into a test of strength; ambition will know no resting place; countries will be propelled into unsustainable tours de force of elusive calculations regarding the shifting configuration of power. Moral prescriptions without concern for equilibrium, on the other hand, tend toward either crusades or an impotent policy tempting challenges; either extreme risks endangering the coherence of the international order itself.[25]

The emergence of Congress as a more powerful component in the national security policy process has shifted power away from the executive and into the hands of bureaucratic gatekeepers, midlevel managers, and other decisionmakers. Gatekeepers filter information and policy recommendations and determine what can flow upward, usually for the purpose of protecting bureaucratic power bases, maintaining the status quo, and gaining more power. Furthermore, a coalition of a few members of Congress can frustrate the design of national security policy, because the policy process is extremely susceptible to a veto by small groups. Their power is strengthened by the inevitable iron triangle (key members of Congress, small sections of the bureaucracy, and special interest groups). Congressional power to conduct investigations of the executive branch can be used not only to gather important facts for oversight, but also to derail the entire agenda of an administration.

The irony, according to Hilsman, is that on the one hand there must be a certain concentration of power to come to grips with problems of policy; on the other hand, there must also be a balance of power to limit and restrain adventurous policy and strategy in foreign and national security affairs. "But power diffused can lead to evil as surely as power concentrated. Here is the irony."[26] In short, there are structural problems as well as political and intellectual ones not only in the national security establishment but also in the national security policy process. With respect to the virtual stalemate between tendencies toward presidential assertiveness and auto-limitation, the Congress and media reflect the shifting popular moods and changing tides of Washington, DC. Members of Congress call for presidential leadership but resent loss of initiative to the White House. Media and political pundits complained about the imperial presidency under Richard Nixon, about the accidental presidency of Gerald Ford, and about the lack of Washington insider savvy among the White House staff of Jimmy Carter, and the ad hoc, improvisational leadership style of Donald Trump.

Given the nature of the international security environment and the need to clarify national interests, policy choices, and strategic alternatives, what needs to be done to develop an effective structure for dealing with national security policy? The current establishment and policy process have evolved over six-plus decades, reflecting the lessons learned from the issues and crises during the Cold War and post–Cold War periods, the politics of congressional-executive struggles, and entrenched habits.

Periodically, there have been calls to change the national security establishment and the way that policy is formulated. Many such calls have been initiated by scholars, think tanks, and government sources. For example, during the Reagan presidency, recommendations came from the president's Special Review Board (the Tower Commission) in a 1987 report on the Iran-contra affair and from the 1986 report "National Security Planning and Budgeting: A Report to the President by the President's Blue Ribbon Commission on Defense Management" (the Packard Commission), whose suggestions focus on the national security establishment and its role in the national security policy process.[27] The Packard Commission recommended structural and procedural changes to the National Security Council, the secretary of defense, the JCS, Congress, and the defense budget process.[28] This was followed in 1988 by the Commission on Integrated Long-Term Strategy, which focused on national security strategy in the context of a changing international security environment.[29]

Written years ago, these reports tried to create a starting point for shaping the national security establishment into a more efficient structure and for providing a more effective and expeditious national security policy process. Little was done to implement their recommendations.

Other studies called for structural and organizational changes in the national security establishment and the policy process.[30] For example,

Carnes Lord suggested two changes: "The establishment of a separate staff component charged with planning, and the breakdown of compartmentalization throughout the staff."[31] Put simply, the establishment must be more streamlined and responsive to the initiatives of the executive as well as more adept at anticipating potential problems and creating innovative national security policy and strategy formulations. These include new procedures to develop a more integrated and long-range defense budget that is realistic as to resources, requirements, and capabilities. Congress must streamline the committee system and the budget process, and new structures should be developed to provide closer executive-legislative coordination and cooperation. Furthermore, these reports and others have noted that national interests and national security objectives need to be spelled out more clearly and that a US strategy that can respond to conflicts across the spectrum on a long-term basis should be designed. These reports and studies are especially relevant in the new century. But there is little to indicate that these changes and procedures will be implemented.

One of the more interesting periodic reports related to national security policy was the former congressionally mandated Quadrennial Defense Review (QDR). The QDR was the Department of Defense's statement of its priorities in strategy and policy planning as well as its assessment of how well it was doing to meet prior goals. In the QDR, changes in terminology could signal important shifts in political and military thinking by the top brass. For example, in the 2006 version of the QDR, the Department of Defense referred to the "long war" instead of the "war on terror" as a description of its major security challenge. The "long war" language suggested that the Department of Defense now acknowledged the need for a protracted conflict against terrorists and their state sponsors very much like the Cold War. Instead of a clear military solution, the "long war" of the twenty-first century might, like the Cold War, have to be endured for generations until it burned itself out. The likelihood of eliminating or controlling all terrorism and terrorist threats is small, even assuming a best case for US-allied cooperation. The words of the first QDR remain relevant, with a caveat about the resurgence of state-based security challenges:

> Since 2001 the U.S. military has been continuously at war, but fighting a conflict that is markedly different from wars of the past. The enemies we face are not nation-states but rather dispersed non-state networks. In many cases, actions must occur on many continents in countries with which the United States is not at war. Unlike the image many have of war, this struggle cannot be won by military force alone, or even principally. And it is a struggle that may last for some years to come.[32]

For better or worse, the QDR, as noted earlier, has been replaced by the National Defense Strategy (NDS) as of 2018, which, unlike the QDR, is classified.[33] More recent policy guidelines recognize the need to prepare for possible state-on-state conflict with a peer competitor.

What seems most instructive in all of these analyses and recommendations is that the role of the president is central to the policy process. He or she is charged with taking the initiative and developing the mechanisms to make the national security policy process work effectively. To be sure, Congress has a major role, but a body of 500-plus members cannot establish the necessary consensus and direction to lead the nation in national security policy. The fact remains that national security policymaking and execution rest primarily with the president and are dependent on presidential wisdom and effectiveness for their success.

A president empowered by seemingly unprecedented and uniquely dangerous international threats or crises gets cooperation across party lines in Congress, cooperation that is missing in more normal times. Both the president and Congress have been charged with being "too weak" or "too strong" from various perspectives and at various times, depending on the issues at hand and the biases of the observer.

And thus we come back to the issue raised by Hilsman—the irony of the diffusion of power and the concentration of power. Where is the balance between enough power for effective national security and not enough to override the will of the people?

Conclusion

Reform of structures and procedures is not enough for successful national security policy. A compelling political vision that provides coherent definitions of national interests, clarity of national security policy, and clear directions to the design of strategy is also required. The president must provide this vision leading to clear policy and strategy. Equally important, there must be a new formulation and rethinking of the meaning of national interests and national security, including whether national security issues should include such items as the environment and climate change, refugee assistance, economics, transnational criminal networks, peacekeeping, digital ransom and other acts of cyber war by state or nonstate actors, social media disinformation, breakthrough technologies including artificial intelligence and hypersonic missile technology, pandemics, and space deterrence.

The many power centers and forces involved in the political-psychological dimensions of national security make it difficult to give the necessary coherency to national security policy as it responds to the changing international security environment. But by exercising effective leadership and by providing a sense of purpose and vision in articulating national interests, the president can shape the boundaries, determine the directions, and establish the critical points to map out an effective US national security policy and strategy.

Notes

1. For example, see Sean Wilentz, *The Rise of American Democracy: Jefferson to Lincoln* (New York: Norton, 2005) and Thomas E. Patterson, *We The People: An Introduction to American Government* (New York: McGraw-Hill, 2022).

2. David C. Unger, *The Emergency State: America's Pursuit of Absolute Security at All Costs* (New York: Penguin, 2013).

3. Various assessments of George W. Bush defense policy and strategy are offered in Stephen J. Cimbala, ed., *The George W. Bush Defense Program: Policy, Strategy and War* (Washington, DC: Potomac, 2010).

4. Stephen M. Walt, *The Hell of Good Intentions: America's Foreign Policy Elite and the Decline of U.S. Primacy* (New York: Farrar, Straus, and Giroux, 2018), esp. pp. 255–291. See also Gregory F. Treverton, *Intelligence for an Age of Terror* (Cambridge, Mass.: Cambridge University Press, 2009), which compares the environment for US intelligence during the Cold War and after September 11. For additional pertinent analysis and discussion, see the essays in Robert J. Art and Robert Jervis, eds., *International Politics: Enduring Concepts and Contemporary Issues* (Boston: Longman, 2011).

5. An important exception, and an indispensable Cold War memoir, is Robert M. Gates, *From the Shadows: The Ultimate Insider's Story of Five Presidents and How They Won the Cold War* (New York: Simon and Schuster, 1996).

6. Stanley Karnow, *Vietnam: A History* (New York: Penguin, 1997). See also Marvin Kalb and Deborah Kalb, *Haunting Legacy: Vietnam and the American Presidency from Ford to Obama* (Washington, DC: Brookings Institution, 2011). For an appraisal by a historian specializing in US intelligence policy, see John Prados, *Vietnam: The History of an Unwinnable War, 1945–1975* (Lawrence: University Press of Kansas, 2009).

7. A judicious assessment of US challenges appears in Eliot A. Cohen, *The Big Stick: The Limits of Soft Power and the Necessity of Military Force* (New York: Basic, 2016), pp. 31–61.

8. Colin S. Gray, *The Sheriff: America's Defense of the New World Order* (Lexington: University Press of Kentucky, 2004).

9. David Silverberg, "Old Demons, New Demons," *Armed Forces Journal International* (November 1993), p. 14.

10. For example, see Steve Lambakis, "Thinking About Space Deterrence and China," Information Series no. 443 (Fairfax, VA: National Institute for Public Policy, July 9, 2019); and Kathleen Hicks, "Russia in the Gray Zone" (Aspen, Colorado: Aspen Institute, July 19, 2019) in Johnson's Russia List 2019, #115, July 21, 2019, davidjohnson@starpower.net.

11. For example, see Bob Woodward, *Obama's Wars* (New York: Simon and Schuster, 2010), chaps. 20–23.

12. Michael Crowley and Lara Jakes, "President Ousts Bolton Amid Rifts on Foreign Policy," *New York Times,* September 11, 2019, pp. A1, A17.

13. Although the withdrawal was incomplete and many US allies in Afghanistan were left behind, the logistical performance of the airlift was outstanding. Without that the results would have been far worse.

14. S. Robert Lichter, Stanley Rothman, and Linda S. Lichter, *The Media Elite* (Bethesda, Md.: Adler and Adler, 1986), pp. 20–21, 23, 294.

15. See ibid.

16. An example of this approach appears in Nigel Hamilton, *American Caesars: Lives of the Presidents from Franklin D. Roosevelt to George W. Bush* (New Haven, Conn.: Yale University Press, 2010).

17. Friction, according to Clausewitz, was the concept that "more or less corresponds to the factors that distinguish real war from war on paper." Carl von Clausewitz, *On War,* edited and translated by Michael Howard and Peter Paret (Princeton, N.J.: Princeton University Press, 1976), p. 119.

18. See James Dobbins, Howard J. Shatz, and Ali Wyne, "Russia Is a Rogue, Not a Peer: China Is a Peer, Not a Rogue" (Santa Monica, Calif.: RAND, October 2018), https://www.rand.org/pubs/perspectives/PE310.html.

19. J. Berkshire Miller and Benoit Hardy-Chartrand, "Russia and China's Strategic Marriage of Convenience," *The National Interest,* August 27, 2019.

20. Relevant projections by the US intelligence community appear in Daniel R. Coats, Director of National Intelligence, "Worldwide Threat Assessment of the US Intelligence Community," statement for the record, Senate Select Committee on Intelligence, January 29, 2019, https://www.odni.gov/index.php/newsroom/congressional-testimonies/item/1947-statement-for-the-record-worldwide-threat-assessment-of-the-us-intelligence-community. Major challenges to US security are expertly summarized in Eliot A. Cohen, *The Big Stick: The Limits of Soft Power and the Necessity of Military Force* (New York: Basic, 2016).

21. Samuel P. Huntington, "The Clash of Civilizations?" *Foreign Affairs* 72, no. 3 (Summer 1993), pp. 22–49.

22. Roger Hilsman, *The Politics of Policymaking in Defense and Foreign Affairs: Conceptual Models and Bureaucratic Politics* (Englewood Cliffs, N.J.: Prentice-Hall, 1987), p. 313.

23. Ibid., p. 316.

24. Ibid., p. 317.

25. Henry Kissinger, *World Order* (New York: Penguin, 2014), p. 367. See also the discussion of positive versus negative liberty in John Lewis Gaddis, *On Grand Strategy* (New York: Penguin, 2018), esp. pp. 310–311.

26. Ibid., p. 318.

27. "Report of the President's Special Review Board" (Washington, DC: US Government Printing Office, February 26, 1987); and "National Security Planning and Budgeting: A Report to the President by the President's Blue Ribbon Commission on Defense Management" (Washington, DC: US Government Printing Office, June 1986).

28. *National Security Planning and Budgeting,* p. 1.

29. *Discriminate Deterrence: Report of the Commission on Integrated Long-Term Strategy* (Washington, DC: US Government Printing Office, January 1988).

30. See, for example, James C. Gaston, ed., *Grand Strategy and the Decision-Making Process* (Washington, DC: National Defense University Press, 1992).

31. Carnes Lord, "Strategy and Organization at the National Level," in Gaston, *Grand Strategy,* p. 156.

32. US Department of Defense, *Quadrennial Defense Review Report* (Washington, DC, February 6, 2006), p. 9

33. An unclassified summary of the National Defense Strategy can be found at https://dod.defense.gov/Portals/1/Documents/pubs/2018-National-Defense-Strategy-Summary.pdf.

Further Reading

Readers may find the suggestions below helpful in exploring topics of interest in greater depth and, in the case of the online recommendations, keeping up with rapidly changing developments in national security issues. No book remains current indefinitely, but we hope that this one will provide a solid foundation for further study and analysis of the evolving strategic landscape and the US response to it.

Arquilla, John, and David Ronfeldt. *In Athena's Camp: Preparing for Conflict in the Information Age.* Santa Monica, Calif.: RAND, 1997. https://www.rand.org/pubs/monograph_reports/MR880.html.

———. *Swarming and the Future of Conflict.* Santa Monica, Calif.: RAND, 2000. https://www.rand.org/pubs/documented_briefings/DB311.html.

Associated Press, "NATO Chief Urges China to Join Nuclear Arms Control Talks." September 6, 2021. https://apnews.com/article/europe-china-9374925e86e57697c264ed0a74587b88.

Astorino-Courtois, Allison, Robert Elder, and Belinda Bragg. *Contested Space Operations, Space Defense, Deterrence, and Warfighting: Summary Findings and Integration Report.* Arlington, Va.: Strategic Multilayer Assessment (SMA), July 2018.

Barton, Jacob. "China's PLA Modernization Through the DOTMLPF-P Lens." *Mad Scientist Blog.* US Army Training and Doctrine Command, May 24, 2021. https://madsciblog.tradoc.army.mil/330-chinas-pla-modernization-through-the-dotmlpf-p-lens.

Barzashka, Ivanka. "Biden Should Guide Missile Defense His Own Way." *Bulletin of the Atomic Scientists,* September 9, 2021. https://thebulletin.org/2021/09/biden-should-guide-missile-defense-his-own-way.

Belfer Center for Science and International Affairs, Harvard JFK School. https://www.belfercenter.org.

Bensahel, Nora. "Transforming the US Army for the Twenty-First Century." *Parameters* no. 1 (2021). https://press.armywarcollege.edu/parameters/vol51/iss1/6.

Blank, Stephen. "Arms Control and Russia's Global Strategy After the INF Treaty." June 19, 2019. https://www.realcleardefense.com/articles/2019/06/19/arms_control_and_russias_global_strategy_after_the_inf_treaty_114513.html.

Boot, Max. "Twenty Years of Afghanistan Mistakes, but This Preventable Disaster Is on Biden." *Washington Post,* August 15, 2021. https://www.washingtonpost.com/opinions/2021/08/15/twenty-years-afghanistan-mistakes-this-preventable-disaster-is-biden.

Boulanin, Vincent. "Regulating Military AI Will Be Difficult; Here's a Way Forward." *Bulletin of the Atomic Scientists,* March 3, 2021. https://thebulletin.org/2021/03/regulating-military-ai-will-be-difficult-heres-a-way-forward.

Bracken, Paul. *Deterrence in a Second Nuclear Age.* Testimony before US Senate Committee on Armed Services, Subcommittee on Strategic Forces. Washington, DC, April 28, 2021.

———. *The Second Nuclear Age: Strategy, Danger, and the New World Politics.* New York: Holt/Times, 2012.

Broad, William J., and David E. Sanger. "A 2nd New Nuclear Missile Base for China, and Many Questions About Strategy." *New York Times,* July 6, 2021. https://www.nytimes.com/2021/07/26/us/politics/china-nuclear-weapons.html.

Burns, Nicholas, and Kathryn Cluver Ashbrook. *A Decade of Diplomacy: The Future of Diplomacy Project at 10.* Cambridge, Mass.: Harvard Kennedy School, Belfer Center for Science and International Affairs, Autumn 2021. https://www.belfercenter.org/publication/decade-diplomacy-future-diplomacy-project-10.

Che, Chang. "All the Drone Companies in China: A Guide to the 22 Top Players in the Chinese UAV Industry." June 18, 2020, https://supchina.com/2021/06/18/all-the-drone-companies-in-china-a-guide-to-the-22-top-players-in-the-chinese-uav-industry.

Cheng, Dean. "China's Nuclear Forces Swell: A Tri-Polar World?" August 3, 2021. https://breakingdefense.com/2021/08/chinas-nuclear-forces-swell-a-tri-polar-world.

Chow, Brian G., and Brandon W. Kelley. "Peace in the Era of Weaponized Space." *Space News,* July 28, 2021. https://spacenews.com/op-ed-peace-in-the-era-of-weaponized-space.

Cimbala, Stephen J. *The United States, Russia, and Nuclear Peace.* New York: Palgrave Macmillan, 2020.

Clausewitz, Carl von. *On War.* Edited and translated by Michael Howard and Peter Paret. Princeton, N.J.: Princeton University Press, 1976.

Cohen, Eliot A. *The Big Stick: The Limits of Soft Power and the Necessity of Military Force.* New York: Basic, 2016.

Cooper, David A. *Arms Control for the Third Nuclear Age: Between Disarmament and Armageddon.* Washington, DC: Georgetown University Press, 2021.

Cooper, Helene. "China Could Have 1,000 Nuclear Weapons by 2030, Pentagon Says." *New York Times,* November 3, 2021. https://www.nytimes.com/2021/11/03/us/politics/china-military-nuclear.html.

Cooper, Julian. "Russia's Updated National Security Strategy." July 19, 2021. NATO Defense College, Russian Study Series no. 2/21. https://www.ndc.nato.int/research/research.php?icode=704.

Cordesman, Anthony H. *The Reasons for the Collapse of Afghan Forces.* Washington, DC: Center for Strategic and International Studies, August 17, 2021. https://www.csis.org/analysis/reasons-collapse-afghan-forces.

Cordesman, Anthony, and Grace Hwang. *Chinese Strategy and Military Power in 2021.* Washington, DC: Center for Strategic and International Studies, June 87, 2021. https://www.csis.org/analysis/updated-report-chinese-strategy-and-military-forces-2021.

Defense Intelligence Agency. *China: Military Power—Modernizing a Force to Fight and Win.* Washington, DC, 2019. https://www.dia.mil/Portals/110/Images/News/Military_Powers_Publications/China_Military_Power_FINAL_5MB_20190103.pdf.

Fravel, M. Tayor. *Active Defense: Chinese Military Strategy Since 1949.* Princeton, N.J.: Princeton University Press, 2019.

Futter, Andrew. *Ballistic Missile Defence and US National Security Policy.* London: Routledge, 2013.

Gaddis, John Lewis. *The Cold War: A New History.* New York: Penguin, 2005.
Galdorisi, George, and Sam Tancredi. "Algorithms of Armageddon: What Happens When We Insert AI into Our Military Weapons Systems." Lexington Park, MD: The Patuxent Partnership, July 14, 2021. https://paxpartnership.org/wp-content/uploads/2021/07/Algorithms-of-Armageddon-what-happens-when-we-insert-AI-into-our-military-weapons-system_white-paper.pdf.
Garretson, Peter. *What War in Space Might Look Like Circa 2030–2040?* Washington, DC: Nonproliferation Policy Education Center, August 28, 2020. https://www.afpc.org/publications/articles/what-war-in-space-might-look-like-circa-2030-2040.
Gates, Robert M. *Duty: Memoirs of a Secretary at War.* New York: Knopf, 2014.
Geist, Edward. "The U.S. Doesn't Need More Nuclear Weapons to Counter China's New Missile Silos." *Washington Post,* October 18, 2021. https://www.washingtonpost.com/outlook/2021/10/18/china-silos-missiles-nuclear.
Gibbons-Neff, Thomas. "Documents Reveal U.S. Officials Misled Public on War in Afghanistan." *New York Times,* December 9, 2019. https://www.nytimes.com/2019/12/09/world/asia/afghanistan-war-documents.html.
Gordon, Michael R. "U.S., Russian Officials to Hold Talks on Future Arms Control Agenda." *Wall Street Journal,* July 28, 2021. https://www.wsj.com/articles/u-s-russian-negotiators-to-hold-talks-on-arms-control-in-future-11627423200.
Goswami, Namrata. "China's Steady Space Progress Takes Another Leap." *The Diplomat,* June 18, 2021. https://thediplomat.com/2021/06/chinas-steady-space-progress-takes-another-leap.
Goure, Daniel. "Can the US Army Transform Itself Without a New Approach to Warfare?" September 16, 2021. https://breakingdefense.com/2021/09/can-the-us-army-transform-without-a-new-approach-to-warfare.
Grau, Lester W., and Charles K. Bartles. *The Russian Way of War: Force Structure, Modernization, and Tactics of the Russian Ground Forces.* Ft. Leavenworth, Kan.: Foreign Military Studies Office, 2016. https://community.apan.org/wg/tradoc-g2/fmso/m/fmso-books/199251.
Gray, Colin S. *The Future of Strategy.* Cambridge, UK: Polity, 2015.
———. *Theory of Strategy.* Oxford: Oxford University Press, 2018.
Greentree, Todd. "What Went Wrong in Afghanistan." *Parameters* 51, no. 4 (Winter 2021–2022), pp. 7–22. https://press.armywarcollege.edu/parameters/vol51/iss4.
Herun Sheng and Cheng Shuangping. "Instant Aggregation of Superiority: Key to Achieving Victory in Modern Warfare" (in Chinese). *Jiefangjun Bao* (official newspaper of the People's Republic of China's Central Military Commission), February 9, 2021. http://www.81.cn/yw/2021-02/09/content_9983700.htm.
Hicks, Kathleen. "Russia in the Gray Zone." Aspen, CO: Aspen Institute, July 19, 2019. https://www.aspeninstitute.org/blog-posts/russia-in-the-gray-zone.
Hitchens, Teresa. "Army, Navy Units to Transfer to Space Force Soon." September 21, 2021. https://breakingdefense.com/2021/09/army-navy-units-to-transfer-to-space-force-soon.
———. "New SPACECOM Strategy Will Define 'Space Combat Power.'" August 24, 2021. https://breakingdefense.com/2021/08/new-spacecom-strategy-will-define-space-combat-power.
———. "US Must Build Space 'Superhighway' Before China Stakes Claims: Senior Space Force Officer." October 20, 2021. https://breakingdefense.com/2021/10/us-must-build-space-superhighway-before-china-stakes-claims-senior-space-force-officer.
Huntington, Samuel P. *The Soldier and the State: The Theory and Politics of Civil-Military Relations.* New York: Vintage, 1964.

Imperato, Anthony, Peter Garritson, and Richard M. Harrison. *U.S. Space Budget Report.* Defense Technology Program Brief no. 23. Washington, DC: American Foreign Policy Council, May 2021. https://www.afpc.org/publications/policy-papers/u.s-space-budget-report.

Insinna, Valerie. "US Air Force Gears Up for First Flight Test of Golden Horde Munition Swarms." *Defense News,* July 13, 2020. http://www.defensenews.com/air.

Janowitz, Morris. *The Professional Soldier: A Social and Political Portrait.* New York: Free Press, 1971.

Joshi, Sameer. "Drone Swarms Are Coming, and They Are the Future of Wars in the Air." *The Point,* February 2, 2021. https://theprint.in/defence/drone-swarms-are-coming-and-they-are-the-future-of-wars-in-the-air/596842.

Judson, Jen. "US Army to Fund Extended-Range Precision Strike Missile Starting in FY22." *Defense News,* June 14, 2021. https:///www.defensenews.com/land/2021/06/14/us-army-to-fund-extended-range-precision-strike-missile-starting-in-fy22.

Kalb, Marvin, and Deborah Kalb. *Haunting Legacy: Vietnam and the American Presidency from Ford to Obama.* Washington, DC: Brookings Institution Press, 2011.

Kania, Elsa. "'Intelligentization' and a Chinese Vision of Future War." December 19, 2019. https://madsciblog.tradoc.army.mil/199-intelligentization-and-a-chinese-vision-of-future-war.

———. "Quantum Surprise on the Battlefield." September 20, 2018. https://madsciblog.tradoc.army.mil/84-quantum-surprise-on-the-battlefield.

Kaplan, Fred. *The Bomb: Presidents, Generals, and the Secret History of Nuclear War.* New York: Simon and Schuster, 2020.

Kilcullen, David. *Counterinsurgency.* New York: Oxford University Press, 2010.

———. *The Dragons and the Snakes: How the Rest Learned to Fight the West.* London: Hurst, 2021.

Kofman, Michael. "Drivers of Russian Grand Strategy." *Russia Matters,* April 23, 2019. https://www.russiamatters.org/analysis/drivers-russian-grand-strategy.

Kollars, Nina, and Jacquelyn Schneider. "Defending Forward: The 2018 Cyber Strategy Is Here." September 20, 2018. https://warontherocks.com/2018/09/defending-forward-the-2018-cyber-strategy-is-here.

Korb, Lawrence J. "The Focus of US Military Efforts in Outer Space Should Be . . . Arms Control." *Bulletin of the Atomic Scientists* no. 4 (2019), pp. 148–150.

———. "A Path Toward Renewing Arms Control." *Bulletin of the Atomic Scientists,* July 18, 2019. https://thebulletin.org/2019/07/a-path-toward-renewing-arms-control.

Kristensen, Hans M., and Matt Korda. "China's Nuclear Missile Silo Expansion: From Minimum Deterrence to Medium Deterrence." *Bulletin of the Atomic Scientists,* September 1, 2021. https://thebulletin.org/2021/09/chinas-nuclear-missile-silo-expansion-from-minimum-deterrence-to-medium-deterrence.

———. "Russian Nuclear Weapons, 2021." *Bulletin of the Atomic Scientists* no. 2 (2021), pp. 90–108.

———. "United States Nuclear Weapons, 2021." *Bulletin of the Atomic Scientists* no. 1 (2021), pp. 43–63.

Lambakis, Steven. *Space as a Warfighting Domain: Reshaping Defense Space Policy.* Information Series no. 499. Fairfax, VA: National Institute for Public Policy, August 12, 2021. https://nipp.org/information_series/steven-lambakis-space-as-a-warfighting-domain-reshaping-defense-space-policy-no-499-august-11-2021/.

Langeland, Krista, and Derek Grossman. *Tailoring Deterrence for China in Space.* Santa Monica, Calif.: RAND, 2021. https://www.rand.org/pubs/research_reports/RRA943-1.html.

Lehman, John F., with Steven Wills. *Where Are the Carriers? U.S. National Strategy and the Choices Ahead.* Philadelphia, Pa.: Foreign Policy Research Institute, 2021.

Lin, Herbert S. *Cyber Threats and Nuclear Weapons.* Stanford, Calif.: Stanford University Press, 2021.

Lopez, C. Todd. "Aboard Commercial Rocket, Space Defense Agency Sends Up Satellites for First Time." *DOD News,* June 23, 2021. https://www.defense.gov/Explore/News/Article/Article/2668483/aboard-commercial-rocket-space-defense-agency-sends-up-satellites-for-first-time.

———. "Nuclear Posture Review, National Defense Strategy Will Be Thoroughly Integrated." *DOD News,* June 25, 2021. https://www.defense.gov/Explore/News/Article/Article/2671471/nuclear-posture-review-national-defense-strategy-will-be-thoroughly-integrated.

Lowther, Adam B., ed. *Guide to Nuclear Deterrence in the Age of Great Power Competition.* Bossier City: Louisiana Tech Research Institute, September 2020. https://atloa.org/wp-content/uploads/2020/12/Guide-to-Nuclear-Deterrence-in-the-Age-of-Great-Power-Competition-Lowther.pdf.

Lye, Harry. "Space Attacks Could Trigger Article 5." June 17, 2021. https://www.army-technology.com/features/space-attacks-could-trigger-article-5-nato.

Mad Scientist Laboratory. "Disrupting the 'Chinese Dream': Eight Insights on How to Win the Competition with China." February 25, 2021. https://madsciblog.tradoc.army.mil.

Marchand, Thibaut, and Sylvie Lanteaume. "With Cutting-Edge Hypersonics, Russia Leads in New Arms Race." *Agence France-Presse,* July 22, 2021. https://www.themoscowtimes.com/2021/07/22/with-cutting-edge-hypersonics-russia-leads-in-new-arms-race-a74588.

Martin, Marissa, Kaila Pfrang, and Brian Weeden. *Chinese Military and Intelligence Rendezvous and Proximity Operations.* Washington, DC: Secure World Foundation, April 2020. https://swfound.org/media/207179/swf_chinese_rpo_fact_sheet_apr2021.pdf.

Mattis, Peter. "China's 'Three Warfares' in Perspective." January 30, 2018. https://warontherocks.com/2018/01/chinas-three-warfares-perspective.

McClintock, Bruce, Jeffrey W. Hornung, and Catherine Costello. *Russia's Global Interests and Actions: Growing Reach to Match Rejuvenated Capabilities.* Santa Monica, Calif.: RAND, June 2021. https://www.rand.org/pubs/perspectives/PE327.html.

McMaster, H. R. *Battlegrounds: The Fight to Defend the Free World.* New York: HarperCollins, 2020.

Miller, Frank. "Talking About 'Strategic Stability.'" June 8, 2021. https://www.realcleardefense.com/2021/07/08/talking_about_strategic_stability_784615.html.

Moskos, Charles C., John Allen Williams, and David R. Segal, eds. *The Postmodern Military: Armed Forces After the Cold War.* New York: Oxford University Press, 2000.

Noonan, Michael P. *Irregular Soldiers and Rebellious States.* Lanham, Md.: Rowman and Littlefield, 2021.

North Atlantic Treaty Organization. "NATO 2030: Fact Sheet." June 2021. https://www.nato.int/nato_static_fl2014/assets/pdf/2021/6/pdf/2106-factsheet-nato2030-en.pdf.

Office of the Secretary of Defense. *Military and Security Developments Involving the People's Republic of China, 2020.* Annual Report to Congress. Washington, DC: US Department of Defense, 2020. https://media.defense.gov/2020/Sep/01/2002488689/-1/-1/1/2020-Dod-China-Military-Power-Report-Final.pdf.

O'Rourke, Ronald. *Renewed Great Power Competition: Implications for Defense—Issues for Congress.* Washington, DC: Congressional Research Service, February 22, 2022. https://crsreports.congress.gov/product/pdf/R/R43838.

Panda, Ankit. "The Dangerous Fallout of Russia's Anti-Satellite Missile Test." November 17, 2021. https://carnegieendowment.org/2021/11/17/dangerous-fallout-of-russia-s-anti-satellite-missile-test-pub-85804.

Pavel, Barry, and Christian Trotti. "New Tech Will Erode Nuclear Deterrence. The US Must Adapt." November 4, 2021. https://www.defenseone.com/ideas/2021/11/new-tech-will-erode-nuclear-deterrence-us-must-adapt/186634.

Payne, Keith B. "Realism, Idealism, Deterrence, and Disarmament." *Strategic Studies Quarterly* no. 3 (Fall 2019), pp. 7–37.

———. *Shadows on the Wall: Deterrence and Disarmament.* Fairfax, Va.: National Institute Press, 2020.

Payne, Keith B., and Michaela Dodge. *Stable Deterrence and Arms Control in a New Era.* Occasional Paper vol. 1, no. 9. Fairfax, Va.: National Institute Press, September 2021.

Posen, Barry R. *Restraint: A New Foundation for U.S. Grand Strategy.* Ithaca, N.Y.: Cornell University Press, 2014.

Ricks, Thomas E. *The Generals: American Military Command from World War II to Today.* New York: Penguin, 2013.

Sanger, David E., Julian E. Barnes, and Nicole Perlroth. "Preparing for Retaliation Against Russia, U.S. Confronts Hacking by China." *New York Times,* March 7, 2021. https://www.nytimes.com/2021/03/07/us/politics/microsoft-solarwinds-hack-russia-china.html.

Sarkesian, Sam C. *America's Forgotten Wars: The Counterrevolutionary Past and Lessons for the Future.* Westport, Conn.: Greenwood, 1984.

Sarkesian, Sam C., and Robert Connor. *The US Military Profession into the Twenty-First Century.* London: Routledge, 2006.

Schelling, Thomas C. *Arms and Influence.* New Haven, Conn.: Yale University Press, 1966.

Singer, P. W., and Emerson T. Brooking. *LikeWar: The Weaponization of Social Media.* New York: Houghton Mifflin Harcourt, 2019.

Smith, James M., and Paul J. Bolt, eds. *China's Strategic Arsenal: Worldview, Doctrine, and Systems.* Washington, DC: Georgetown University Press, 2021.

Sokolski, Henry D., ed. *China Waging War in Space: An After-Action Report.* Arlington, Va.: Nonproliferation Policy Education Center, August 2021. http://npolicy.org/article_file/2104_China_Space_Wargame_Report.pdf.

Starling, Clementine G., Mark L. Massa, Christopher P. Mulder, and Julia T, Siegel. *The Future of Security in Space: A Thirty-Year U.S. Strategy.* Washington, DC: Atlantic Council, Scowcroft Center for Strategy and Security, 2021. https://www.atlanticcouncil.org/content-series/atlantic-council-strategy-paper-series/the-future-of-security-in-space/.

Stent, Angela A. *Putin's World: Russia Against the West and With the Rest.* New York: Twelve-Hachette, 2019.

Sun Tzu. *The Art of War.* Edited and translated by Samuel B. Griffith. Oxford: Oxford University Press, 1963.

Thayer, Bradley A. *The PRC's New Strategic Narrative as Political Warfare: Causes and Implications for the United States.* Fairfax, Va.: National Institute Press, 2021. https://nipp.org/papers/the-prcs-new-strategic-narrative-as-political-warfare-causes-implications-for-the-united-states/.

Trachtenberg, David J. *Clarifying the Issue of Nuclear Weapons Release Authority.* Information Series no. 503. Fairfax, Va.: National Institute Press, September 22, 2021.

US Air Force Space Command. *The Future of Space 2060 and Implications for U.S. Strategy: Report on the Space Futures Workshop.* September 5, 2019. http://www.spaceref.com/news/viewsr.html?pid=52822.

US Army, Training and Doctrine Command. *The Changing Character of Warfare: The Urban Operational Environment.* April 9, 2020. https://adminpubs.tradoc.army.mil/pamphlets/TP525-92-1.pdf.

———. *The Operational Environment and the Changing Character of Warfare.* October 7, 2019. https://adminpubs.tradoc.army.mil/pamphlets/TP525-92.pdf.

US Congressional Budget Office. *The U.S. Military's Force Structure: A Primer, 2021 Update.* May 2021. http://www.cbo.gov/publication/57088#data.

US Department of Defense. *U.S. Space Defense Strategy, Summary.* June 2020. https://media.defense.gov/2020/Jun/17/2002317391/-1/-1/1/2020_Defense_Space_Strategy_Summary.pdf.

US Department of Defense, Office of the Under Secretary of Defense (Controller) Chief Financial Officer. *Fiscal Year 2022 Budget Request.* May 2021. https://comptroller.defense.gov/Budget-Materials/Budget2002.

van Loon, Margot, Larry Wortzel, and Mark B. Schneider. *Hypersonic Weapons.* American Foreign Policy Council, Defense Technology Program Brief no. 8. 2019. https://www.afpc.org/publications/policy-papers/hypersonic-weapons.

Vergun, David. "Spacecom Attains Initial Operational Capability, Commander Says." *DOD News,* August 24, 2021. https://www.defense.gov/Explore/News/Article/Article/2744166/spacecom-attains-initial-operational-capability-commander-says.

Walt, Stephen M. *The Hell of Good Intentions: America's Foreign Policy Elite and the Decline of U.S. Primacy.* New York: Farrar, Straus, and Giroux, 2018.

Williams, John Allen. "The Military and Society Beyond the Postmodern Era." *Orbis: A Journal of World Affairs* 52, no. 2 (Spring 2008), pp. 199–216.

Woolf, Amy F. *U.S. Strategic Nuclear Forces: Background, Developments, and Issues.* Washington, DC: Congressional Research Service, updated December 14, 2021. https://sgp.fas.org/crs/nuke/RL33640.pdf.

Internet Resources

Updated Reading Lists

Several organizations interested in national security, including the US military, issue updated reading lists periodically. These are some we find particularly interesting.*

Australian Army Reading List: http://www.researchcentre.army.gov.au/sites/default/files/2019-12/army_reading_list_190328_combined_lo_res.pdf

British Army Reading List: http://www.militaryreadinglists.com/reading_list/61-centre-for-historical-analysis-and-conflict-research.html

Canadian Army Reading List: http://www.publications.gc.ca/site/archivee-archived.html?url=https://publications.gc.ca/collections/collection_2010/forces/D2–249–2009-eng.pdf

Chairman of the Joint Chiefs of Staff: http://www.amedd.libguides.com/c.php?g=566155&p=3905794

US Air Force Chief of Staff: http://www.static.dma.mil/usaf/csafreadinglist

US Army Chief of Staff: http://www.history.army.mil/CSA-reading-list/index.htmly

US Army Training and Doctrine Command (TRADOC), Mad Scientist Laboratory: https://madsciblog.tradoc.army.mil

US Coast Guard Office of Leadership: http://www.dcms.uscg.mil/Our-Organization/Assistant-Commandant-for-Human-Resources-CG-1/Civilian-Human-Resources

*We are indebted to Michael P. Noonan for his suggestions for this section.

-Diversity-and-Leadership-Directorate-CG-12/Office-of-Leadership-CG-128/Reading-List
US Marine Corps Commandant: http://www.grc-usmcu.libguides.com/usmc-reading-list-2020
US Navy Chief of Naval Operations: http://www.navy.mil/CNO-Professional-Reading-Program
US Special Operations Command, USSOCOM Commander's Reading List: https://jsou.libguides.com/readinglists

Websites and Web Magazines

The strategic environment is constantly evolving. These sources are helpful for keeping up with strategic issues and developments.

Atlantic Council: https://www.atlanticcouncil.org
Belfer Center for Science and International Affairs, Harvard JFK School: https://www.belfercenter.org
Breaking Defense: https://breakingdefense.com
Brookings Institution: https://www.brookings.edu
Bulletin of the Atomic Scientists: https://thebulletin.org
Canadian Forces College, SOMNIA (Spotlight on Military News and International Affairs): https://www.cfc.forces.gc.ca/254-eng.html
Carnegie Endowment for International Peace: https://carnegieendowment.org
Center for American Progress: https://americanprogress.org
Center for a New American Security: http://www.cnas.org
Center for Strategic and International Studies: http://www.csis.org
Center for Strategic and Budgetary Assessments: https://csbaonline.org
Congressional Budget Office: https://www.cbo.gov
Congressional Research Service: https://crsreports.congress.gov
Defense News Newsletters: https://defensenews.com/newsletters
Foreign Military Studies Office: https://community.apan.org/wg/tradoc-g2/fmso
Foreign Policy Research Institute: http://www.fpri.org
Government Accountability Office: https://www.gao.gov
Hoover Institution: https://www.hoover.org
Modern War Institute at West Point: http://www.mwi.usma.edu
National Institute for Public Policy: https://nipp.org
Power Institutions in Post-Soviet Societies: https://journals.openedition.org/pipss
Pritzker Military Museum and Library: http://www.pritzkermilitary.org
RAND Corporation: https://www.rand.org
Strategic Studies Quarterly, US Air War College: https://www.airuniversity.af.edu/SSQ
The Strategy Bridge: http://www.thestrategybridge.org
Union of Concerned Scientists: https://www.ucsusa.org
US Army Training and Doctrine Command (TRADOC), Mad Scientist Laboratory: https://madsciblog.tradoc.army.mil
US Army War College Strategic Studies Institute: http://www.ssi.armywarcollege.edu
US Naval War College Review: https://digital-commons.usnwc.edu/nwc-review
US Institute of Peace: http://www.usip.org
War on the Rocks: http://www.warontherocks.com

Index

About the Book

The main focus of US national security policy has shifted dramatically since the years of the Obama administration, moving away from nation building and counterinsurgency efforts and toward preparing for traditional state-on-state conflict with powerful peers. The sixth edition of *US National Security* reflects that change. It also addresses such current issues as the impact of an increasingly partisan political process, sharp divisions in public opinion, the ongoing challenges of homeland security, and developments in cyberspace and other possible domains of future warfare.

Retaining the successful structure and approach of the previous editions, the book clearly introduces and explores the full range of actors, processes, and politics involved in maintaining US national security.

John Allen Williams is professor emeritus of political science at Loyola University Chicago. **Stephen J. Cimbala** is Distinguished Professor of Political Science at Penn State Brandywine. The late **Sam C. Sarkesian** was most recently professor emeritus of political science at Loyola University Chicago.